THE RECOVERY OF NATU
ENVIRONMENTS IN ARCHITECTURE

The argument for the recovery of natural environments in architecture challenges the modern practice of sealing up and mechanically cooling public scaled buildings in whichever climate and environment they are located. This book unravels the extremely complex history of understanding and perception of air, bad air, miasmas, airborne pathogens, beneficial thermal conditions, ideal climates and climate determinism. It uncovers inventive and entirely viable attempts to design large buildings, hospitals, theatres and academic buildings through the 19th and early 20th centuries, which use the configuration of the building itself and a shrewd understanding of the natural physics of airflow and fluid dynamics to make good, comfortable interior spaces. In exhuming these ideas and reinforcing them with contemporary scientific insight, the book proposes a recovery of the lost art and science of making naturally conditioned buildings.

C. Alan Short was educated at Trinity College Cambridge and Harvard University's Graduate School of Design. He has been The Professor of Architecture at the University of Cambridge since 2001 and his work focuses on the design of sustainable buildings. He also designs buildings, one of very few architects in the UK deeply involved in higher education and research, maintaining Short & Associates Architects as a vehicle for putting research concepts into practice.

BRI Book Series

New interdisciplinary and transdisciplinary approaches need a forum for information and discussion.

This book series shares similar aims and scope to the journal, *Building Research & Information*, but allows for a deeper discussion, together with more practical material.

SCOPE: This book series explores the linkages between the built, natural, social and economic environments, with an emphasis on the interactions between theory, policy and practice. Emphasis is on the performance, impacts, assessment, contributions, improvement and value of buildings, building stocks and related systems: i.e. ecologies, resources (water, energy, air, materials, building stocks, etc.), sustainable development (social, economic, environmental and natural capitals) and climate change (mitigation and adaptation).

If you wish to contribute to the series then contact the series editor Richard Lorch at richard@rlorch.net with a short note about your ideas.

The Rebound Effect in Home Heating
A Guide for Policymakers and Practitioners
Ray Galvin

The Recovery of Natural Environments in Architecture
Air, Comfort and Climate
C. Alan Short

Forthcoming:

Professionalism for the Built Environment
Simon Foxell

Sustainable Retrofit
Building Professional Capabilities
Sarah Sayce

THE RECOVERY OF NATURAL ENVIRONMENTS IN ARCHITECTURE

Air, Comfort and Climate

C. Alan Short

LONDON AND NEW YORK

First published 2017
by Routledge
2 Park Square, Milton Park, Abingdon, Oxon OX14 4RN

and by Routledge
711 Third Avenue, New York, NY 10017

Routledge is an imprint of the Taylor & Francis Group, an informa business

British Library Cataloguing-in-Publication Data
A catalogue record for this book is available from the British Library

Library of Congress Cataloging-in-Publication Data
Names: Short, C. Alan (Charles Alan), 1955- author.
Title: The recovery of natural environments in architecture : air, comfort and climate / C. Alan Short.
Description: New York : Routledge, 2017. | Includes bibliographical references and index.
Identifiers: LCCN 2016029513| ISBN 9780415824408 (hardback : alk. paper) | ISBN 9781138651463 (pbk. : alk. paper) | ISBN 9781315765853 (ebook)
Subjects: LCSH: Natural ventilation. | Buildings–Environmental engineering.
Classification: LCC TH7674 .S56 2017 | DDC 697.9/2--dc23
LC record available at https://lccn.loc.gov/2016029513

ISBN: 978-0-415-82440-8 (hbk)
ISBN: 978-1-138-65146-3 (pbk)
ISBN: 978-1-315-76585-3 (ebk)

Typeset in Univers
by Saxon Graphics Ltd, Derby

CONTENTS

Tables

ACKNOWLEDGEMENTS

Grateful thanks are due to Cambridge University for granting an extended sabbatical in 2012–13 which enabled some traction to develop on the outline for the book as conceived in the lobby of the Fairmont Hotel in Toronto at a major 'green' conference in 2011. It draws on three decades of thinking and invention. William Curtis, my former tutor at Harvard in 1980, Slade Professor at Cambridge in 2003, released me from the rather historicist line taken in Britain in the mid 1970s, our adherence to which was relieved by periodic guerilla publications from Professors Watkin and Scruton. Christopher Alexander, Lionel March and Phil Steadman were available to sharpen concentration and add some quantification to our thoughts. Ken Frampton supervised me for a while, as I recall, briefly, as did Anthony Blunt, Ron Lewcock, even Eric Fernie. Trinity would fund supervisions with anyone, anywhere for its scholars. Gary Tinterow and Professor Jean Sutherland Boggs also broadened our conception of the Modern Era in the Fogg Institute. More recently, membership of the Cambridge History and Philosophy of Science Department's 'Coffee with Scientists' seminar group, granted by the Raussing Professor, Hasok Chang, has been revelatory. Dean Hawkes introduced us to environmental things as Cambridge undergraduates in architecture. However, the work of Otto Koenigsberger and Steven Sozokoly at the Architectural Association had particularly interested me, both of whom I met, sent as Sir Basil Spence's office boy to learn about desert climates and then brief him.

As architects in Edward Cullinan's cooperative practice we were regularly exposed to Max Fordham's extraordinary grip on the Natural Sciences, his manual calculations faster than any computer, and to Dr Randall Thomas, later to be the Royal Academy of Engineering Visiting Professor at Cambridge, both to become highly pro-active and courageous co-authors of many of our early buildings, and a youthful Brian Ford, fresh from the Royal College of Art and the Centre for Land Use and Built Form Studies at Cambridge, co-author of all our projects until the mid 1990s. Through Brian Ford, I met Drs Nick Baker and Simos Yannas, to become the ultimate arbitrators of success in our work for us. Professors Kevin Lomas and Malcolm Cook at Loughborough, very long term and much valued research and practice collaborators, were first encountered at what became the Institute of Energy and Sustainable Development under the late Professor Neil Bowman at De Montfort University in Leicester. Building Research Establishment monitoring of the Queens Building design process introduced us to Dr Frank Mills. The complexity of fire and smoke control in naturally ventilated buildings brought us to kneel before

Geoff Whittle, the fire engineer and pioneer smoke modeller from Arup. Professor Paul Linden in the dank basement of Applied Maths and Theoretical Physics on Silver Street introduced us to the powerful water-bath technique before becoming Blasker Professor of Environmental Science and Engineering at the University of California San Diego, happily returning to Cambridge as Taylor Professor of Fluid Mechanics and a renewed collaboration in his new Grand Challenges project.

The Director of Cambridge University's BP Institute, Professor Andrew Woods has also become a long term and very productive collaborator, unraveller of the potential perils of passive downdraught cooling and more recently the flows within the operating theatre. His colleague Dr Shaun Fitzgerald identified the useful phenomenon of mixing ventilation, in part through monitoring our theatre buildings. The Professor of Sustainable Engineering, Peter Guthrie, has made available his apparently unending supply of first-class researchers in the Centre for Sustainable Development, Eleni Soulti, Stephi Hirmer, Kirsten Macaskill, Roberta Mutschler, Jill Solanki, Dr Maria-Christina Georgiadou and Dr Daniel Godoy-Shimizu, amongst others, to help with work for the Department of Health, and of course his own very considerable wisdom and insight. More recently I have enjoyed a truly fascinating collaboration with the Cambridge Institute of Atmospheric Science on the nature of climate change and the complexities of its prediction with my college colleague Professor Hans Graf and our research assistant Stephen Xiajong Xiu. Our China connections have blossomed through Professors Runming Yao at Reading University and Baizhan Li, Dean at the University of Chongqing, leading to the recent award of the major Anglo-Chinese research councils' Low Carbon Cities project, 'Low carbon environmentally responsive heating and cooling of cities', LoHCool. I have been the happy beneficiary of a string of excellent research students and post-docs who have graduated into important emerging academics, including Dr Alistair Fair, Dr Henrik Schoenefeldt, Dr Nicola Mingotti and more recently Dr Andrew Acred.

The author is grateful to the Society of Architectural Historians for the award of the George Collins Conference Fellowship 2014, which enabled oral presentation of the work on the Johns Hopkins Hospital. Two Trinity College Cambridge Newton Trust Awards in 2011 and 2013 funded the re-drawing of Billings' and Folsom's proposals and their translation into the digital model reconstructions included in this book. Grateful thanks also to Professor Sonia Horn, medical historian, who, as a Visiting Scholar from Vienna at 'HPS' in Cambridge provided invaluable guidance on the history of antisepsis and the history of the operating suite, the 'bloody rooms'. The computational analyses of the later 19th century hospital schemes were conducted at the University of Leeds' Pathogen Control Research Centre by Dr Catherine Noakes and Dr Carl Gilkeson, who have become regular and much valued collaborators. Important commentary came from Professor Malcolm Cook and Dr Faisal Durani at the Department of Civil Engineering at Loughborough University.

Helen Wells, Tutor at Oundle, very kindly devoted much time to identifying all too elusive literary accounts of performance in interior spaces to illuminate Chapter 5. Many archivists and librarians have found themselves unpacking rarely, if ever consulted boxes of papers. The author is indebted to the staff of the Rare Books Room at the Cambridge University Library who pride themselves on retrieving anything in 20 minutes, an extraordinary facility to have at one's fingertips. I must thank staff at the Cornell University Library Division of Rare and Manuscript Collections for locating Willis Havilland Carrier's Deceased Alumnus records, most particularly Hilary Dorsch Wong, Reference Coordinator, and her colleague Marcie Farwell, whose discoveries illuminate Chapter 2. Staff at The Brooke Russell Astor Reading Room in the New York Public Library Manuscripts and Archives Division have been hugely helpful, with particular thanks to Weatherly Stephan, Manuscripts Specialist, whose hunch paid off.

The book exploits the findings of a series of major UK Research Council funded projects from a pleasing variety of Councils: Design and Delivery of Dynamic Environments for the Performing Arts, DEDEPA, was funded by the Arts and Humanities Research Council, work which informs Chapter 5, in collaboration with Professor Peter Barrett, Pro-Vice-Chancellor, at the University of Salford. Lady Rachel Cooper, formerly at Salford and now at Lancaster University, has been a very valued long-term mentor since my time as a Dean of Art and Design. The UK Engineering and Physical Sciences Research Council funded the project, 'Design and Delivery of Robust Hospital Environments in a Changing Climate', jointly with Loughborough, Leeds and the Open University, delivered within the highly effective and impactful 'Adaptation and Resilience to a Changing Climate' programme, introducing us to the Adaptation Sub-Committee of the Climate Change Committee. The project, and the preceding National Institute of Health funded research into exemplar low energy hospital design, from which Chapters 6 and 9 derive much material, has also received funding from the Department of Health and the Newton Trust. Davis Langdon AECOM provided very detailed elemental costs for this and a number of UK based research projects.

Robert Smith and subsequently Peter Sellars as Directors of Estates and Facilities at the Department of Health; the DH Chief Architect Christopher Farrah and Cambridge University Hospitals NHS Foundation Trust; Bradford Teaching Hospitals NHS Foundation Trust, where the energetic Ian Hinitt was Deputy Director of Estates; West Hertfordshire NHS Trust led by Jan Filokowski, Chief Executive and former NHS Fellow at Wolfson College Cambridge and University Hospitals of Leicester NHS Trust; the access they permitted to clinical areas for two years were an absolutely essential part of this work. Subsequently 49 successful NHS Trusts reported on the progress of their grants to our NHS Energy Efficiency Fund team in the Centre for Sustainable Development adding considerably further to our still limited understanding of this immense and infinitely fascinating estate.

The book reports on built buildings, their progress from first concepts to measured and analysed performance and their reception. Extraordinarily committed and loyal clients have been instrumental: Louis Farrugia and the Board of the Simonds Farsons Cisk Brewery in Malta as it was known, who ignored their own project managers' advice to stick to 'business as usual'; David Cefai, Chief Brewer and niece to the Marquis Scicluna, Romina Scicluna Testaferrata Moroni Marshall; Professors Kenneth Barker and David Chiddick as Vice-Chancellor and Pro-Vice-Chancellor at De Montfort University and the then-Professor of Architecture George Henderson, steadfast commissioners of the Queens Building, negotiating the Polytechnic and Colleges Fund with immense skill to seize much of that year's budget; Vikki Heywood, to become

Director of the Royal Shakespeare Company and Chair of the Royal Society of Arts and her colleague Brigid Larmour, who with Professor Roger Stonehouse and Director of Estates Richard Furter selected the only architects out of 30 competing teams with no relevant experience to design their Contact Theatre, Sir Martin Harris becoming instrumental in enabling its construction, later to become President of my college; Dr Pat Noon, Chief Librarian, William Woolhead Director of Estates and their colleagues at Coventry University; at UCL, Professor George Kolankiewicz, Director of the School of Slavonic and East European Studies, Dr Robin Aizlewood Deputy Director, Professor Julian Graffy, Lesley Pitman, Librarian and Maria Widdowson the School Secretary discovered what commitment to an innovative building really involves in the context of the contemporary construction industry, supported throughout by Professor Malcolm Grant, Provost, and former colleague as Head of Land Economy at Cambridge; at Judson University in Chicago very grateful thanks are certainly due to the President, Jerry Cain, Professor Keelan Kaiser the Chair of Architecture, John Cinelli, indefagitable and scrupulous project architect for Burnidge Cassell Associates, Larry Pithan and Wade Ross of the environmental engineers KJWW and William Fawcett, Chadwick Fellow in Architecture at Pembroke College, Cambridge who applied his understanding of net present value to the project; Mrs Li at Future House in Beijing and the Ministry of Housing and Urban Development who designated the building a Ministry Case Study Project; the Headmistress, Bursar and Governors of Berkhamsted School and Professor Aldwyn Cooper, Vice-Chancellor of Regents University, London.

The buildings are innovative and very demanding to design and build in every detail, so that very great thanks go to the many architects who helped realize them from 1987, many of whom now practise successfully in their own right. The practice policy has always been to credit every member of the team in all architectural publications about each project, one cannot understand why others do not, but in the context of this book, particular acknowledgement must be given to Professor Brian Ford, Anthony Peake, Anne Goldrick, Peter Sharratt, Philip Meadowcroft, Iona Foster, Louise Pritchard, Tim Hewitt, Jennifer Jeffries, Elaine Toogood, Quinton Pop, Shaila Amin, Sura al-Maiyah, Jamileh Manoochehri and with emphasis to Michael Ritchie and Adam Whiteley, the late Georgina Livingston, and our excellent landscape architect, plantswoman and botanist Slaine Campbell. It has always been exhilarating working with the structural engineer Stephen Morley, engineer to much of our earlier work. However, the principle thanks must be reserved for my extraordinarily conscientious and tenacious editor Richard Lorch, who has read the emerging manuscript critically many, many times, scrupulous in his criticism, firm but always fair in his responses.

And so, one can only hope, as so disarmingly expressed in 1733 by the 'aerologist' Dr Arbuthnot (p. xi), 'I may venture so far as to affirm, that he who reads the whole over with due Attention will find it not quite an useless Speculation.'

Introduction

The Author ... in none of the Works hitherto executed under his directions, has he ever had the opportunity, either in Buildings or in Ships, of introducing his plans, with all the advantages of which he considers them susceptible, were they incorporated in the original design, instead of being merely appended to designs or works already executed.

(D.B. Reid, 1844, Illustration of the Theory and Practice of Ventilation with remarks on warming, exclusive lighting and the communication of sound)

Dr Reid's complaint is common enough as an excuse for failure in the contemporary world of 'green' or 'sustainable' design but the present author has no such complaint: the projects described here are fully realized with the full support of their sponsors. The more surprising, perhaps, because the attempts to make natural environments in buildings as described in the following chapters can legitimately be described as experiments. The buildings are in themselves full size pieces of 'apparatus'. It may seem strange to confess to experimentation on real clients who are commissioning new buildings and major rebuildings in good faith, perhaps confirming every client's worst suspicions about architects. All built works are somewhat experimental as the outcomes cannot be fully predicted. Shapin and Schaffer's retelling of Robert Boyle's attempts to evacuate a glass vessel strikes a certain resonance with the author's own experiments, the improvisation of a 'pneumatic engine' at the very limits of contemporary materials and, more particularly of expertize, met by the hostile responses of the 'anti-experimentalists' (Boyle 1660).[1] Shapin and Schaffer (2011, p. 29) describe Boyle's use of experimental means to investigate the possibility of a complete vacuum:

> Among the chief difficulties (as experienced by Robert Boyle in 1660) was the problem of leakage. Great care had to be taken to ensure that external air did not insinuate itself back into the (vacuum) pump or receiver through a number of possible avenues. This is not at all a trivial and merely technical point. The capacity of this machine to produce 'matters of fact' crucially depended on its physical integrity, or, more precisely, upon collective agreement that it was air-tight for all practical purposes.

Christiaan Huygens' diagram of his parallel and competing experimental apparatus is shown in Figure 0.2. In this book (Huygens, 1932), many such

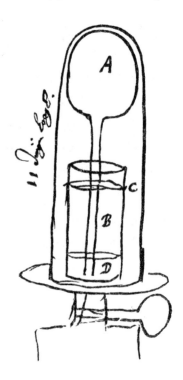

Figure 0.2
Huygens' diagram of his new pump configured for the void-in-the-void experiment in December 1661. A: flask full of water; D: water in outer vessel B; C: water level in both A and B after exsuction of air from receiver. Huygens Oeuvres, Volume xvii, p. 317 Figure 39. With grateful thanks to Edinburgh University Library for permission to reproduce.

simple diagrams for buildings from throughout history will be explored. These buildings attempt to make and contain specific 'atmospheres' in closely controlled chambers to regulate their internal environments using the natural physics Boyle and his colleagues were revealing and manipulating.

The book commences by reviewing the evolving beliefs and understandings of what constitutes 'air'. There is an urgent need to know something of this because ancient beliefs about 'air' persist. These theories and beliefs are explored in relation to their effects on human beings: the effects of climate, the history of weather, vapour in air, 'tainted' air, etc. An understanding of the progression of these theories and beliefs is vital for interpreting the counter-intuitive practices in making indoor climates.

The rediscovery of the deep cultural associations and beliefs about what constitutes 'bad air' and 'good air' allows an examination and questioning of many current practices in building design. Some of these aspirations, beliefs and understandings will be shown to be erroneous or misguided. This is closely aligned to beliefs and practices about the internal atmosphere in buildings, particularly the ventilation and internal temperature. For example, an underlying implicit environmental determinism underpins many current practices to maintain an artificial indoor climate at one fixed temperature, when this is unnecessary, expensive and unhealthy.

This book presents an alternative to artificial indoor weather. It exhumes lost ideas, interrogates them with the benefit of contemporary scientific insight and proposes a recovery of the lost art and science of making naturally-conditioned buildings.

An urgency exists as civil societies and governments recognize the need to reduce the amount of energy used in buildings. The UK Climate Change Committee exhorts the UK Government to both maintain investment in 'low carbon heat', the winter heating challenge that all policy has focused upon, and also to 'promote passive cooling in existing buildings' to confront the burgeoning cooling problem.[2] Mechanical cooling is highly energy intensive. It is clear that buildings in the future will need to be fundamentally reconfigured from current types to be very, very much more naturally resilient to a changing climate, whilst using significantly less energy in the process. The problem is not 'merely' an engineering problem, it requires assault across many fronts.

Shackled to a highly complex 19th and 20th century provenance, which can only be superficially outlined here, contemporary designers persist in making less and less resilient buildings, by demanding and creating increasingly energy-intensive artificial environments. The design professions (and also those involved in valuing, marketing and selling buildings) have little appreciation and no real 'will' to explore the opportunities to make resilient buildings that can condition their own environments. The potential outcomes, largely unknown, are too frightening and threaten to burst the genteel boundaries around the later 20th and early 21st century professions' custom and practice.

For the last five or six decades, the majority of buildings, however audacious their form, actually conform to a dull recurring constructional type: they have steel or concrete structural frames, highly glazed façades, substantial service voids between the floors, lightweight materials internally for lining, and mechanical plant which pipes conditioned air and refrigerant through themselves at an enormous cost in energy and carbon dioxide (CO_2) emissions. Can the conundrum be fixed by bolting new energy technologies onto these 'business as usual' building types? The European Union and national governments have invested hugely in this belief, individually and severally, fuelled by the promise of new high-tech industries generating the right kind of employment, skilled and clean, but as the former Chief Scientist to the Department of Energy and Climate change the late Professor David MacKay (2008) reported, to no great effect. Few policy-makers and their advisers understand that in the built environment gadgets are supplementary to rigorous low carbon design in their capability to deliver the huge reductions in carbon emissions required. Order-of-magnitude savings lie in design and redesign. The analysis of the UK NHS Estate in Chapter 9 shows just how effective the redesign of one of the world's largest estates of building stock could be in reducing emissions, quite viable politically and economically, if the will is there. A number of exemplars will be used to demonstrate the futility of bolting on gadgetry to 'business as usual' buildings. The nature and scale of the problem means that a fundamental rethink is needed. Counter-intuitively, there is evidence of astonishingly prescient naturally-conditioned public buildings from the early modern era, 'before air conditioning' (BaC) that show what can be achieved without gadgetry, as well as an oeuvre of recent buildings that prove viability, resilience and cost-effectiveness. None of the buildings described hereon rely on active renewable devices for their performance. Much has been learnt, but there is a great deal more to research and understand, and a whole architecture to invent/ reinvent.

The prevailing forms for our principle building types have evolved since the mid 20th century out of a myriad unrelated colliding interests. The types now need to be fundamentally reinvented. Viable prototypes for this reinvention were developing in very interesting ways until the introduction of 'artificial weather' in the late 1920s. The aggressive marketing of 'artificial weather' released the design of buildings from any need at all to be responsive to climate. Climate determinists from the late 19th century into the 1930s argued that 'civilization' depended on the provision of specific climatic conditions. Did this desire to make a preferred climate help to make palatable a dark environmental determinism?

There was, and continues to be, an increasingly global commercial endeavour to suppress all the undeveloped and developing world's climates, a 'moral climatology' by implication.

The book is illustrated with a sample of passive and hybrid non-domestic buildings designed, built and monitored by the author's research-based practice and its many distinguished academic collaborators since the late 1980s. The range of these buildings embraces libraries, laboratories and research institutes in temperate climates and in the less temperate urban heat island of Central London; densely occupied theatres and acute hospitals; buildings for industrial processes in the Mediterranean; academic and educational buildings in the Continental climates (Beijing and Chicago), with an additional wrinkle for the hot dry summers of Gujurat in India. A glimpse is offered into the rich adaptive design opportunities afforded by the existing building stock, an unglamorous and therefore tragically under-researched field in architecture.

It will be argued that a wider acceptance is needed of the concept of 'social practice', but this raises the stakes. Buildings are altogether more culturally complex than a simple consumer commodity. The behaviours and attitudes of the supply side are atrophied in a persistent social practice, a way of building whose historical intent and roots are long forgotten but stuck in the psyche of an indifferent commissioning class. This is maintained by networks of risk-shedding consultants, bound by intensely defensive contractual agreements and served by an aggressive industry driven by an immense vested interest in maintaining the *status quo*.

The accounts are candid, describing the actual recorded performance of each building, 'warts and all'. But, the author is an architect and Professor David Watkin's note of caution may well apply. Paraphrasing E.H. Carr's observation that the historian is also a product of history and of society as well as the events he describes, Watkin (1980, p. ix) writes:

> … so much of it (architectural history) has been written by practising architects: a parallel would be if most history books were written by professional politicians,' all 'with an axe to grind'.[3]

In an attempt to counter this tendency, reception histories are included for each building account, some quite robust, making for some uncomfortable reminiscences.

The resonance with Shapin and Schaffer's account of Boyle's predicament in recording a 'convincing' experiment is in the importance of limiting the engine's leakage of air. The integrity of the knowledge it produced was vitally dependent on the physical integrity of the machine. The reaction of the critics, gleefully exploiting its lack of physical integrity in order to challenge its inventor's findings, accords with the experiences recorded here. Projects were dogged by the extreme difficulty of achieving airtightness. The many conduits and simple opening and closing elements within them, coupled to controls of different degrees of sophistication have, in some cases, badly compromised outcomes. Even temporary failures during disappointingly extended commissioning periods have immunized clients against embarking on similar exercises again (ibid. Ch. II, p. 30). Poor outcomes, even if recorded during incomplete commissioning, have been broadcast to cast doubt on the integrity of the enterprise and its underlying theoretical approach, just as in the 17th century.

In Boyle's time, the philosopher Hobbes raised many objections to the actual principle of experimentation: the problem of credible witnessing; the suspiciously closed shop of the Royal Society; the difficulty in achieving replication of results and the 'failure' of certain experiments and the integrity of the apparatus. A nasty debate ensued. But now there is a quite different problem with gathering and broadcasting experimentally-derived data from real buildings in a risk-averse society: the data itself generates risk. It becomes evidence of contractual non-compliance, of negligence. In a recent public lecture, Dr Bill Bordass (the pre-eminent building performance analyst) pointed out that there is virtually no verified data on contemporary building performance in the public realm (Bordass, 2013). Chapter 4 reports on an exception, a peer-reviewed account of construction and commissioning failure that gives an indication of the capacity of the construction industry to deliver our 'recovery' in the widest sense of the term. It is poor and probably diminishing.

This book visits 'natural' buildings across the world's principal climate zones. It will develop a sense of how to evolve an authentic 'environmental design strategy' specific to each place and environment. This is very much an evolving art and science still in its infancy. Despite a plethora of building performance modelling in recent years, one senses that most of what one hopes to know and design is yet to be realized. This presents a huge opportunity for research and invention that few architects have pursued. Ironically, architects have become potentially important instruments of global destiny, just as Le Corbusier had hoped, but with a very different and unwelcome imperative: adapting to climate change rather than the exciting urgency of post-war reconstruction.

The invention that is promoted in this book is based on understanding the complex behaviour of air as a fluid and the potential for architectural form to interact and influence these dynamics for human comfort. In doing so, a path of discovery based on successful lessons from older buildings is used to reveal knowledge of how architects effectively harnessed air through architectural form.

Shapin and Schaffer deconstruct the established 'triumphalist', Whiggish accounts of the unfolding of events, accounts which they refer to as the civilian equivalent of the 'General Staff Histories' produced by the winning side (Butterfield, 1931).[4] Building services engineers may not realize that they have pursued the classic Historicist line for half a century. This is the template, incidentally, for much architectural history as delivered to students through the mid and later 20th century, the progression of the great generals towards High Modernity, easily identifiable by their recurring formal attributes, as catalogued, stripped of meaningful intent, by Henry-Russell Hitchcock and Philip Johnson (1932) in *The International Style*, confounding as Scott observes, the true relation of taste to ideas (Scott 1914). Less obvious perhaps is the implicit aspiration toward a universal and constant environmental artificiality. Was this a conscious neutralizing of the popular theories of environmental determinism and their bad portrayal of hot climates, turning sour as the 20th century progressed? This possibility is explored in Chapter 2. 'Naturalness' is simply depicted as primitive in the Whiggish histories of architecture, its promotion heretical, unethical even in depriving aspirant developing economies of progress and so 'un-modern'. 'Difference' became ideologically unacceptable. Perhaps the thinking about the making of built environments became ensnared in this. For example, in his discussion of 'glazed apertures', Frampton (1985, p. 26) points to climate-differentiated nuances, railing against what he terms the 'universal system':

> The way in which such openings provide for appropriate ventilation also constitutes an unsentimental element reflecting the nature of local culture. Here, clearly, the main antagonist of rooted culture is the ubiquitous air-conditioner, applied at all times and all places, irrespective of the local climatic conditions ... wherever they occur, the fixed window and the remote-controlled air-conditioning system are mutually indicative of domination by universal technique.

Rather sadly, Frampton did not pursue this moment of extreme clarity in connecting climate, place and architecture, opting instead for *Die Tektonik* as the 'primary principle' in the recovery of a place-conscious poetic. This late-modern 'structural fallacy', in effect Scott's 'Mechanical Fallacy', is less and less persuasive. Clients are surely more likely to be interested in what it is actually going to be like in and around their forthcoming buildings.

My perspective is that of an architect who, in the course of the work, has attempted to acquire some understanding of the many diverse fields of knowledge invoked in building. Hopefully, this is just sufficient to broker all the established and emerging knowledge from these specialisms into a fully imagined design, resolved enough for it to be complete as a form when built, to be prospered within and then suffer the 'gaze', be observed by a critical audience.

Boyle is reported (Shapin and Schaffer, 1985) to have been unconcerned by experimental failure. He deemed it to be useful. The contractual implications of failure in the contemporary construction industry are considered so terrible that genuine innovation, i.e. experimentation, is now rare. This was found to be particularly the case in hospital projects being delivered under the UK Private Finance Initiative. The author's work falls into a complex and fitful stream of thought and experimentation, identifying a catholic continuity that Whiggish notions of Modernity would render quite impossible. The modern 'era of suspicion' and its practice of 'denunciation' is unhelpful to our work.

Bruno Latour (1993, p. 44) suggests: 'To unmask: that was our sacred task, the task of us moderns.'

Society can learn a great deal from some of its 'losers'. Perhaps as Latour suggests, 'we have never been modern' (ibid. p. 47). I am certainly not 'anti-modern', although my architectural work has been interpreted as such. How can it be? It depends on absolutely current understanding. As Latour (1993) observes, 'the anti-moderns even accept the chief oddity of the moderns, the idea of a time that passes irreversibly and annuls the entire past in its wake'. Sadly, the profession of architecture habitually discards its research.

The evidence for continuity is explored in Chapters 1 and 2 and then episodically throughout the book. William Curtis's teaching at Harvard in the late 1970s lies very much behind this position. But my architectural work is not 'post-modern' although also often described as such, sometimes in well-meaning laudatory terms. My aim is to avoid what Latour describes as 'the hint of the ludicrous that always accompanies post-modern thinkers; they claim to come after a time that has not even started' (1993, p. 47).

Rather, I like to think we are 'non-modern'.

Notes

1. Boyle, R. (1660) 'New Experiments Physico-Mechanicall, Touching the Spring of the Air and its Effects (Made, for the most part, in a New PNEUMATICAL ENGINE) Written by way of Letter To the Right Honorable Charles Lord Viscount of Dungarvan, Eldest Son to the Earl of CORKE', Oxford') available digitally at http://echo.mpiwgberlin. mpg.de/ECHOdocuView?url=/permanent/archimedes_repository/large/boyle_exper_ 013_en_1660/index.meta&viewMode=image&pn=5&characterNormalization=orig (accessed 5 November 2013).
2. Committee on Climate Change June 2015, 'Reducing emissions and preparing for climate change: 2015 Progress Report to Parliament. Summary and Recommendations.'

Presented to Parliament pursuant to sections 36 and 59 of the Climate Change Act 2008.

3. Professor Watkin quotes from E.H. Carr (1962) *What is History?*

4. The spirit of Whig politics cast itself as reforming, progressive and humane within the bounds of Protestant reasonableness. Reinforced by its old school aristocratic supporters, it wholly opposed notions of divine right absolutism, arguing for constitutional monarchy and steady parliamentary reform. Butterfield identified a pattern in then recent histories in which Whig beliefs and the strategies to achieve them were unquestioned, 'the fallacy of the Whig historian' as he cut through, or abridged, what he calls the complex process of mutation through which the past becomes the present (p. 24), the organizing of historical knowledge upon an unrecognized assumption, the Whig 'vision', from which inferences are drawn and then claimed as the authentic voice of history. Sir Karl Popper developed this theme more generally in *The Poverty of Historicism* first delivered in 1936, published in book form 1957, deeply critical of the portrayal of the present as being historically predestined by progressives tempered by those who opposed them. With notable exceptions, architects' accounts of architecture in the 'modern' era have been bedeviled by the propogandizing of 'progressive' fallacies. Geoffrey Scott (1884–1929) picked a number off in *The Architecture of Humanism: A Study in the History of Taste* (1914), a 'natural history of our opinions', from which the 'Mechanical Fallacy' might be of particular interest in our context. https://archive.org/stream/cu31924014760353#page/n183/mode/2up, provided by Cornell University Library (accessed 1 January 2016), a 'natural history of our opinions'.

1

Airs, fears, dangers

Figure 1.1
'An Allegory of Air', Studio of Johann Breughel the
Younger 1601–78. Oil on copper, 69 × 87cm.
Reproduced with grateful thanks to Sotheby's.

The building must be designed so as to separate putrid exhalations from currents of fresh air, in the same way that currents of fresh water had to be divided from used water. The idea was that the shape of the building itself would ensure satisfactory ventilation, thus rendering traditional methods redundant. Cupola and dome were transformed into machines to draw up miasmas: experts climbed onto rooftops to breathe the invisible, evil-smelling spirals they created. ... The degree of stench was a measure of the architect's efficiency.

(Alain Corbin, 1996, *The Foul and the Fragrant:*
Odour and the Social Imagination, p. 98)

Anxiety has long existed about the real and imagined condition of the 'air', the elastic fluid in which we bathe and which we actually take into our bodies. Interest in 'air' has developed into a broader fascination with atmospheric circulation, the weather, the 'history of weather', what we call 'climate', and its apparent effect on our physical and intellectual vigour. This constant immersion in air, the unavoidable business of 'being' in air, has preoccupied commentators for more than two millennia. Figure 1.2 reproduces John Griscom's particularly vivid image of human respiration from his 1850 manifesto *The Uses and Abuses of Air...*, accompanied with the equally vivid comment (1850, p. 6):

> By the exercise of a certain corporeal function, which is carried on without our notice or aid, it flows, surely and steadily, deep into the body ... it is no sooner inhaled then its work is begun.

It is surprising, then, that modern Western societies have contracted out the responsibility for delivering 'good' air in their interior spaces to a specialized component of the construction industry, the world of mechanical and electrical engineers and subcontractors with roots deep into the 19th century. As Griscom (1850, p. 6) asks: 'But when was a deficient supply of air ever known, except through the agency of man himself in his folly and ignorance?'

The industry that coalesced around the rise of mechanical and electrical engineers is very defensive of its custom and practice beliefs and technologies, with a small number of notable exceptions. It is devoted to the technology of making artificial environments and it justifies this position by thorough 'denunciation' of pre-modern practices as 'unsafe', 'uncomfortable' and 'primitive'. Architects have been complicit in this. As Chapter 6 will explain, some designers by the later 19th century were working closely with heating engineers and 'ventilation experts'.

Before this appropriation of indoor environments, commentators could imagine very direct architectural means to make the 'atmospheres' within buildings. Corbin (1996, p. 98) reminds us, 'The influence of aerist theories on Enlightenment architecture is well known.' This may be an overstatement but he quotes Beguin (1979, p. 40) referring to the 18th and early 19th centuries: 'Planners aimed "to use nothing but architectural resources to capture the air, cause it to circulate, and expel it"'. [Emphasis added by current author.]

This intent is exactly what this book attempts to explore. Could it have been achieved? Was it, and if it was, can it be effected now and, more importantly, in the future, in a changing climate?

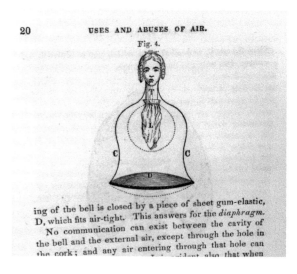

Figure 1.2
Griscom, J.H. (1850) *The Uses and Abuses of Air*, 2nd Edn. New York: J.S. Redfield, Clinton Hall, p. 20 Figure 4.

Classical architectural motifs were being adapted, subversively, to serve these purposes. The environmental phenomena being manipulated demanded large-scale architectural interventions even if the forms bent to perform the task concealed their actual function. The site plan of architect C.N. Ledoux's late 18th century model settlement at Chaux in France indicated that he pursued 'aerist trends', being very deliberately 'exploded' apart into discrete pavilions, with hygienic separation. Such anxieties crystallized in the problem of the hospital, the military hospital most particularly. Deodorization was the primary intent, achieved eventually through progress in ventilation, but not least in the management of excrement and of those expelling it through increasingly disciplinarian regimes which Foucault describes in some detail.[1] A significant part of this book's enquiry is devoted to the analysis of hospital environments (Chapters 6 and 9). Their social redemption as a building type in the 19th century is tracked through progress in disinfection alongside the consequences of Western society's olfactory 'advancement': a lessened tolerance to odours, and tendencies to disproportionate, obsessional preoccupations with cleanliness, with indications from contemporary advertisers that the condition characterizes displaced populations.

Vinikas (1992) reports on the 'awakening' of the American public in the early 20th century, in particular recent immigrants, to undetected personal odours. He writes:

Lever Brothers under President Francis Countway ran large ads with photographs in Milwaukee's Polish newspaper Kuryer Polski. One explained the plight of a young woman whom men avoided because of underarm odour after only a few dances.

The message was reinforced by relentless campaigning by the newly established 'Cleanliness Institute' from mid 1930 to late 1931, which 'exploited women's worst fears of giving offense' (Hoy, 1995). Aspirant working men and women came to believe they could also join the new 'sweatless, odorless and successful business class'. For example, the spectacular increase in sales of Listerine (a mouthwash) by the Lambert Company in the early 1900s was engineered through its manufacturer's announcement of a concocted disease, halitosis, presenting itself through socially crippling bad breath. The company, through the entrepreneurial younger Jordan Wheat Lambert, flattered Joseph Lister into giving his personal endorsement for the fluid, hence the brand name Listerine. Without even changing the design of the bottle, the campaign ignited what became the huge industry dedicated to the deodorization of the self (Vinikas, 1992, pp. 145–146). Did the manufacturers of building environments also embrace this new sales opportunity to respond to what Vinikas (1992, p. 95) describes as the population's 'fearful bashful preoccupation with their own natural odours … pervade[ing] the common consciousness'?

Were there other consequences for Western architecture? Did a 'clean aesthetic' wrapping a 'clean atmosphere' suppress other promising emerging architectural propositions?[2]

Bad air

By the early 17th century a commentator on the human condition, Robert Burton (1621, p. 75), could declare in exasperation: 'We scorn all that is cheap ... This air we breathe is so common, we care not for it; nothing pleaseth but what is dear.'

Burton speculates on what forces might determine the weather in the essay *A Digression of the Nature of Spirits, Bad Angels, or Devils, and how they cause Melancholy*, recording various scholars' collective conclusion that there were nine kinds of bad spirit, variously inducing dishonesty, mischief, vengefulness, war-mongering, falsely accusatory, 'rejoicing at suicide', but (1621, p. 30):

> The sixth are those aerial devils that corrupt the air and cause plagues, thunders, fires, etc.; spoken of in the Apocalypse, and Paul to the Ephesians names them the princes of the air ... they cause whirlwinds on a sudden, and tempestuous storms; which though our meteorologists generally refer to natural causes, yet I am of Bodine's mind ... they are more often caused by these aerial devils.

Perhaps this is one of the earliest recorded instances of scepticism about meteorologists' explanations of the cause of extreme weather phenomena (ibid. p. 30). In this theory, the 'aerial devils' bind themselves to the victim so that the individual is the epicentre of personalized weather systems, a current research interest at the University of California – Berkeley's Center for the Built Environment.

A century later, Leoni published his translation of Alberti's rueful commentary of the 1440s on the ill effects of poorly judged environmental design in 1726 (Book IX, p. 91):

> What are the causes which make the air unhealthy, we have already shewn sufficiently at length in the proper place. We may only observe here in general, that for the most part those causes are either the too great power of the Sun, or too much shade: some infectious winds from neighbouring parts, or pestilent vapours from the soil itself: or else something in the very Climate itself that is noxious.[3]

The noxious vapours refer to what became known variously as fixed air, phlogiston, air saturated in fomites, gasified carbonic acid or generic 'miasma'. John Arbuthnot MD FRS (1733, p. vi), writing a little later,[4] was astonished by what he saw as the lack of enquiry into the physiology of the effects of air by his colleague physicians:

> The Reason of which Neglect may be, that air is one of those *Ingesta*, or things taken inwardly, which neither can be forborn nor measured in Doses; But the use of air being unavoidable, is no Reason against inquiring into its Effects ... and there are many more Useless Inquiries than this, about the Effects of Air, which are daily the Subject of Human Curiosity.[5]

For the avoidance of any doubt about the importance of the topic in his readers' minds he adds, a little patronizingly: 'Abstinence from Air is not, the Sort of Air which they use, is [sic] in the Power of a great many People' (ibid.).

The above quote stresses the importance of understanding as much as possible about, 'the Effects of a Substance that we take inwardly every moment'. He concluded his and other physicians' patients were so little interested because they were obsessed with the minutiae of 'every drug rarely taken'. He refers warmly to the 'first Founder of our Art' the great Hippocrates, paraphrasing: 'Air is what he means by the Powers of the Universe which he says Human Nature cannot overcome' (ibid.).

But, as we shall see, human ingenuity has indeed attempted to change and redistribute the air intended 'to be taken inwardly' by people for centuries. Arbuthnot offered precise calculations of the volumes of air required to sustain life, reporting Hales' experiments: one gallon of air for a minute of human life (Hales, 1727).

This 18th century dictum equates rather poorly with an already outdated later 20th century standard of 8 litres of air/per second/per person of supply air, equating to over 105 gallons of air/minute, 6335 gallons/hour. The current recommendation by the UK Chartered Institute of Building Services Engineers (CIBSE) for public buildings is higher: 10 litres per second.

In self-experimentation, Hales subjected himself to near asphyxiation, determining 74 cubic inches of air barely sustained him for half a minute. He concluded 1 hogshead of air (i.e. 63 gallons) would just enable a person to survive an hour, but 500 people locked up with 500 hogsheads of air 'would be dead or in Convulsions' after only 20 minutes. Hales investigated the gases produced by decomposing vegetal matter. His interest derived from a finding made by chemist and physicist Robert Boyle (1627–91) a century earlier:

> … a good quantity of Air was producible from Vegetables, by putting Grapes, Plums, Gooseberries, Cherries, Pease, and several other sorts of fruits and grains into exhausted and unexhausted receivers, where they continued for several days emitting great quantities of Air.

It was assumed this sustaining gas was the product of combustion, 'fixed air', in effect carbon dioxide (CO_2). Joseph Black (1728–99) collected CO_2 but deduced that plants needed another constituent of air to survive.[6] This odourless, colourless, indetectable product of combustion became known as phlogisticated air (Conant, 1950). It is possible that 'modern' practice still feels the dead hand of phlogiston theory.

The theory, attributed to Becher (1667), christened by Stahl (1703), deriving the term from the Greek word for flame, *phlox*, proposes an invisible element that enables combustion. Experiments in combusting material in a closed vessel appeared to confirm that the air itself had a limited capacity to absorb phlogiston and support combustion. As the material extinguished itself the air was described as being fully phlogisticated and in this state could not support life. A room of phlogisticated air was a dangerous place to be. Air with no phlogiston, dephlogisticated air, supported vigorous combustion, later to be identified as being oxygen-rich. Subsequent work revealed, of course, that far from elements expelling a substance of some kind when burning, or oxidizing, they actually acquired oxygen, gaining mass, not losing it.

Thomas Henry's finely judged Preface to his 1776 translation of Lavoisier's *Essays Physical and Chemical*, made for the Fellows of the Royal Society, proposed that his own countrymen Stephen Hales, Joseph Black, David Macbride, Henry Cavendish and Joseph Priestley had been setting the pace in the analysis of this fluid (air), '... pursued much further than could possibly have been expected'.[7] This was very necessary because Lavoisier and his well-connected, influential and, from the perspective of Henry's countrymen, poisonous but exceptionally able chemist collaborator and wife Marie-Anne Paulze, were aggressive denouncers of contemporary British scientific endeavour. The orchestrated destruction of phlogiston theory was brutal. However, revenge was served cold when she foolishly married the bizarre Benjamin Thompson, Count Rumford, of the patent fireplace design, in 1804 after Lavoisier's death at the hands of the Revolutionaries. Lavoisier wrote (as translated by Henry, 1776, p. x):

> Hence it was, that Dr. Hales did not always draw just conclusions from his experiments, and this was the source of errors, of which he was by no means apprised, and which will make it necessary to repeat his trials, some time, with particular precautions.

Thomas Henry (1776, p. 407) adds an Appendix, the account of Lavoisier's report to the Royal Academy on 26 April with a description of the November 1774 experiments concerning the production of a gaseous substance during the calcification of metals and this is:

> ... the air itself, undivided, without alteration, without decomposition, to such a degree, that if after having been engaged this combination, it be set at liberty, it is separated more pure, more respirable, if I may be permitted to use the expression, than the air of the atmosphere, and it is more proper to support flame and the combustion of bodies.

Henry (1776, p. 427) concludes by claiming that Lavoisier has 'discovered a method of compounding common respirable air' by adding the correct quantity of phlogiston to dephlogisticated air. Henry shrewdly delays mention of phlogiston until page 256. His contemporary Richard Kirwan FRS (1733–1812) kept faith with the theory of phlogiston longer than was professionally sensible. His *Essay on Phlogiston and the Constitution of Acids* was published in 1787 but, notwithstanding his unfashionable views, he won the Royal Society's Copley Medal in 1782. He believed phlogiston was hydrogen. Lavoisier's wife translated Kirwan's *Essay*, to which Lavoisier and others added withering, sarcastic criticism of its contents. Kirwan ultimately confessed his mistaken loyalty to the theory in 1791, a victim of a fashionable 'modern denunciation' as the modern French philosopher and anthropologist Bruno Latour might describe it. In fact there was more to the theory than meets the eye. Professor Chang (2009, p. 249) makes the argument that phlogiston can be equated with negative electricity, an energy potential:

> ... a truly Whiggish historian ought to lament the premature death of the phlogiston theory, rather than celebrate the highly engineered and disciplined takeover of chemistry by Lavoisier's band of brilliant and self-important chemist-publicists.

An anti-Lavoisierian consensus formed on the composition of hydrochloric acid, due in large part to the work of Humphry Davy in the early 19th century, 'Lavoisier's oxygen theory of acidity was clearly dead, never to be revived again.' Worse still, Lavoisier is suspected of simply inverting various phlogiston theories and Chang reminds us that it was Priestley who discovered that dephlogisticated air was 'eminently respirable'. Extraordinarily, Chang (2009) suggests that the phlogistonists, in asserting that all metals were rich in phlogiston, anticipated the revelation that all metals have a sea of free electrons. He records that phlogiston was almost renamed 'electron' in 1780.

The question is for how long beyond Lavoisier's systematic assault on the phlogistonists did a 'healthy building' constituency retain the idea that bad air was continuously invading interior spaces and therefore required efficient flushing away? One could read rather more with regard to attitudes to forming 'successful' building environments into these reported experiments. The early 18th century anxieties about the freshness of air and its supply may have prefigured a whole discipline around what is now known as 'air quality'. Griscom (1850, p. 74) was moved to devote his Chapter VIII to:

> Effects of Vitiated Air upon the Human system ... Air ... in its vitiated state, pregnant with the hidden causes of physical, moral and social degeneracy and decay.

The 18th century physician Arbuthnot (1733) recommended engaging a good carpenter to ensure all windows and doors closed as tightly as possible to exclude the 'miasmas'. Nonetheless it was clearly recognized by all the experimentalists and their public that a good reliable supply of 'air', whatever it comprised, was essential to life, whilst 'bad air' was dangerous, be it phlogiston, 'free air' or a miasma, and should be evacuated as expeditiously as possible. Soon the design and configuration of buildings would become thoroughly implicated, most particularly hospitals. This will be explored in more detail in Chapters 6 and 9.

In considering 'Contagion and Effluvia', Arbuthnot explains exactly why his patients should develop a close interest in the air they are planning to breathe. He speculates on whether air is capable of introducing the plague without 'Infection' in places with no history of the disease. He thinks it quite likely, firstly because of some unknown peculiar quality of the air, that 'Enormities, Combinations and Alternations of the Common Qualities of the Air' can induce the 'utmost Degree of Putrefaction'. Arbuthnot's influence is significant for historically reconstructing a line of reasoning about 'bad air' This obsession and fear foreshadows a line of reasoning that would eventually deliver the vision of a completely artificial environment: '... still more extraordinary Effects may be produced by some Contagion of the Air, by uncommon Effluvia from Bodies near the Surface of the Earth'.

Arbuthnot (ibid. p. 180) refers with deference to earlier work by 'the Honourable Mr. [Robert] Boyle, a philosopher most learned in the Physiology of the Air' who:

> ... gives Instances of Steams of a particular Nature, being emitted from the Earth at particular times ... then those Effluvia may be carried and mix'd with the Air of that Place by Winds.

The concept of 'miasma' is fully realized. This is a terrible prospect and Arbuthnot gives a graphic account of the 1346 Plague in Marseilles arriving from Cathay in a vapour: 'most horribly foetid, that breaking out of the Earth, like a kind of subterranean Fire … infected the Air after a wonderful manner'.

Writing in 1733, after the immediate 1732–33 winter measles epidemic in England, Arbuthnot reports from recent memory as having been very foggy, strangely warm with little precipitation brought by a southerly air stream spreading across to Saxony in an atmosphere resulting in 'stinking Fogs'. These are the typical indicators of an unusually prolonged inversion, in which warm air becomes wedged beneath cold.

Corbin reports a very useful and succinct definition of miasma as envisaged given by French physician and pathologist François Boissier de Sauvages (1706–67) in the 'Dissertation' (1754):

> A vapour rises from the whole surface of the earth as a result of subterranean heat −10° Reamur. It is to a greater or lesser degree abundant, denser than air, and it spreads out when nothing prevents it, and falls again in the evening.[8]

Chang *et al.* (2007) explain that later speculation envisaged suspended particles of putrified or fermented organic matter of plant or animal origin including the effluvia of lungs, giving the miasma its foul smell. An unidentified agent was suspected of being latent in living bodies tending to putrefaction. This was countered by a good circulation of the blood and, ironically in terms of later social mores, by vigorous perspiration. Chang *et al.* (2007, p. 189) write:

> These beliefs about contagion implied that ventilation and fumigation were required to remove effluvia from the rooms of the sick. The concern expressed by Johnstone was typical of medics at the time: 'The necessity of changing the air in the sick room … arises … from the atmosphere being filled with the excrementious streams which fly off from the patient's body continually, and which putrefy in a stagnant un-renewable air, and render it truly poisonous, a pabulum morbid rather than of life'.[9]

Fainting provided an immediately identifiable symptom of the presence of miasma. Dr Anthony Meyler alarmed an already weary audience at the Dublin Society in 1818 by observing:

> … in the frequent faintings, which take place in close and crowded rooms, and in the pallid and languid faces of those who have passed a few hours in those places, we witness the deadly influence of the unwelcome atmospheres within which they are filled.[10]

The helpful irony in this history is that the actions one might rationally take to dispel a 'miasma' also dilute and vent away airborne bacteria and their ciphers, Fomites, tiny particles of decaying material, the conceptual precursors to the germ.[11] John Snow's deductions based on evidence from the pattern of cholera infections around the shared Broad Street water pump in Soho, London in 1854 turned sanitarians' attention, reluctantly, away from the miasma theory about the spread of disease to the possibility of waterborne contagion (Halliday, 2001).[12] Chapter 6 will show that the miasma and germ theories co-existed in hospital building circles in the United States at least into the late 1880s, satisfying trustees, customers and medical personnel simultaneously to no bad effect.

The defence: organized ventilation in buildings

Progress in configuring buildings for effective ventilation was already far advanced, but the focus of effort was on the design of glass structures for the propagation of rare and expensive botanical specimens, not for human beings (Hix, 1996). The stakes were high. It is reported that in the 18th century growing a pineapple in Europe under glass would have cost the modern equivalent of £5000.[13] Historian Lewis Mumford (1938, p. 414ff) saw precisely the importance of this horticultural imperative:

> The first evolved pattern of the new architecture (the cubists' attempts to incorporate the machine into architecture) appeared in the glasshouse: the Crystal Palace in London was but the monumental embodiment of this mutation. ... Thus the new methods of construction, the new materials, and the new means of regulating the air of a building in order to adapt it more perfectly to the needs of living occupants came directly from the biotechnics of gardening.

There were ambitious collaborations between botanists, engineers, iron and glass manufacturers and, in a subservient role, architects, which led to the design of the important glasshouses of the era, in particular the Palm House at the Royal Botanic Gardens at Kew (Schoenefeldt, 2007). By 1817 Loudon could observe that the accepted rules of architecture were inapplicable in housing horticulture, the plants came first. Horticulturalists were keen users of thermometers, barometers and hygrometers and were obsessive data gatherers. Light transmission was clearly critical to success in propagation and, most importantly, the control of solar radiation. By 1817 Loudon was publishing schemes for glasshouses entirely made of sash windows achieving up to 50% free area.

Head gardener and architectural designer Joseph Paxton's (1803–65) early 'ridge and furrow' glasshouses at Chatsworth House date from the mid 1830s. The relatively vast Palm House at Kew delivered a yet greater level of transparency with yet more slender glazing bars and with a ventilation system scaled up in an attempt to prevent solar heat build up in summer. For the Palm House, the architect Decimus Burton participated in an exemplary collaborative manner with the other disciplines and trades. Sketches sent by Turner the iron founder to the Director of Kew, W. J. Hooker, in March 1844 already show a generous capacity for through-ventilation (Schoenefeldt, 2007). Plants had been heavily implicated in the 'Chemical Revolution', it being concluded that they emitted either 'good' or 'bad' air in terms of human consumption. Research into plant physiology and the effects of the environment upon it intensified. Schoenefeldt reports that Paxton's own *Magazine of Botany* published from 1839–42 carried a series of articles on environmental influences on plant physiology for the guidance of horticulturalists. Commerce was an important driver. Kew was funded through the Treasury to stimulate growth in trade in Imperial palm products through the identification of the most productive types and environments. By the mid 1840s diagrams of hothouses show confidently placed arrows predicting the movement of air, at least as precise and detailed as Dr Reid's diagrams of the Parliamentary Chambers for human beings.[14]

Schoenefeldt's (2007) reconstruction of Burton's May 1847 scheme for the Kew Palm House shows an air supply plenum formed below the working floor level, perimeter heating and centre-pivot ventilators placed mid-height at the gallery level and above at the apex of the glazed vault above. The government contract required indoor temperature to be maintained at 80° F (26.6° C) even as external temperatures fell to 20°F (–6.6°C), a ferocious contractual requirement. The built glasshouse did not achieve this. The plants did not survive the third winter. The problem resided in the inadequacy of the boiler and the flue system to match the colossal heat losses through the single-glazed envelope. There was a £500 annual bill for coke over the 1848–49 winter (and no palms to show for it).

By the early 1840s a Scottish medical doctor, Dr David Boswell Reid (1805–63), felt confident in compiling a comprehensive text on the ventilation of buildings, explaining (Reid, 1844, p. 172, para. 398):

> … the skin evolves [sic] carbonic acid gas, and moisture charged with animal matter, in minute quantity: the precise manner in which the carbonic acid so evolved [sic] is produced, has not hitherto been determined.

Reid is known in part through his unfortunate and probably misinterpreted scheme for 'improving' conditions in the holds of slave ships. He was nothing if not naively pragmatic. He produced detailed proposals to ventilate the new UK House of Commons and the temporary Upper House.[15] His design strategies are considered in Chapter 5 on 'theatres'. His schemes became derailed. The architect of the new Houses of Parliament, Charles Barry, could not have been less interested. Unlike Kew, this is a story of deep interdisciplinary dysfunctionality and sabotage, summarized in graphic detail by Dr Elisha Harris, a loyal Reid evangelist, over a decade later in 1858, in his introduction to Reid's rather more interesting American publication, *Ventilation of American Dwellings...* (Reid, 1858). Harris wrote gleefully:

> Dr. Reid's experience at the new Houses of Parliament affords a memorable example of the difficulties attending the introduction of improvements, even when their practicability and success have been sufficiently demonstrated, when an inventor has to co-operate with another, who has the power of construction in his hands, and who has also the means of opposition, and of sheltering such opposition under the undefinable limits of taste and decoration. [Emphasis added by current author.]

This complaint has a strangely contemporary ring reminiscent of the grievances expressed by environmental engineers who are dropped into architects' teams through parallel procurements, and the clue to Reid's (1844) rueful preface to his compendium:

But Dr. Reid had so deep a sense of the injustice done to him in the House of Peers, by the Whig first Marquis of Clanricarde, just returned as Ambassador to Russia and then as Postmaster-General, and of the impossibility of his plan having any fair play unless the Architect was controlled as the House of Commons had specified, that he never acted again in the Houses of Parliament except under protest.

The architect being referred to is Charles Barry (1795–1860). Reid eventually wrought vengeance in a tribunal hearing under Lord Derby's new administration, interested in embarrassing its predecessor. He was able to cross-examine Barry for seven days in a 30-day hearing which ended extremely profitably for Reid: £3250 compensation, actually thought to be miserly by one of the two arbiters, Sir John Forbes, in addition to £4400 paid in salary arrears. His nemesis was indeed the Marquis of Clanricarde who blocked Reid's proposals for the Upper House, perhaps as Barry's 'apparachik' (ibid., Harris's Introduction; Sturrock and Lawson-Smith, 2006).

It is not entirely clear how Reid was proposing to disseminate and monetize his ideas, if at all. Perhaps the prospects of enhanced social standing and public respect, all dashed, were adequate enough rewards in his mind. We shall see that later with ventilation entrepreneurs such as Robert Boyle Jr (1850–1930) formed global companies, his had multiple offices in London, Glasgow, Paris, Berlin and New York. Boyle had a clear business model, the manufacture and sale of ventilation-enhancing devices, principally terracotta ductwork and joints but also an intriguing ridge ventilator, supported by the extraordinarily dull and repetitive publication, *Natural and Artificial Methods of Ventilation* of 1899, with liberal reference to scientists, one Dr Parkes in particular.

The Reid vision surfaced from time to time (1844):

> It is no exaggeration to say, that along with those means of natural defence and seclusion which they naturally present, the great and primary object of architecture is to afford the power of sustaining an artificial environment …
> it is in reality, to every building what the breath of life is to the human frame – the Vivifying principle.

No doubt this confirmed Barry's worst suspicions of this gauche 'enthusiast'. Reid's case (1844, p. 76, para. 131) was supported by graphic accounts of the environmental conditions in which British citizens in the 1840s were immersed:

> At present however the intolerable oppression produced by excessive gas lighting (hence the full title) arises at times not only from the escape of unconsumed gas, but also from the return of the products' combustion to the zone of respiration.

In contrast 'good ventilation renders coal and gas smokeless'. City dwellers appear to have lived in dense atmospheres of particulate matter. Reid's para. 132 describes 'the anxiety to exclude the large amount of soot that at present penetrates all varieties of dwellings in large and densely populated cities'.

Dr Reid anchors his work firmly in 'the power and grandeur of Chemistry', more specifically on:

... the discoveries of Priestley, Scheele, Lavoisier, and Black ... the great lineaments which it presents might then have been unfolded, but the Chemistry of the numerous gases which have since been made known was a blank in the page of science. They too often surrounded or entered the habitations of men without being perceived.

Reid (1844, p. 85, para. 156) then describes what contemporary engineers call 'displacement ventilation' in a discussion of 'Ascending and descending atmosphere':

Here it may be observed that, in this country, air, vitiated by respiration, tends invariably upwards ... vitiated air collects above in any apartment ... and that an ascending movement should be given accordingly to the air which enters ... and never to return again to the zone of respiration ... but be continually succeeded by fresh accession of pure air. [Emphasis added by current author.]

Chapter 6 will show that particulate matter, even microscopic pathogens, will not necessarily join the happy upward motion of the medium in which it is suspended. Reid's perceived need for geographical qualification is interesting. Perhaps he was anxious that the principles may fail in other climate types which, indeed, they might. Two paragraphs further on in his text Reid makes a prescient aside: 'In the case of forced ventilation, where the ingress and egress of air is subject to the action of a power that may be regulated at pleasure ... It may be expedient ... to resort to a descending movement.'

This reversal of displacement systems continues to be a significant concern and will be considered with regard to present-day hospital wards, downdraught cooling and more sophisticated mixing ventilation ideas in succeeding chapters. Reid apologizes for the thin and diagrammatic nature of his figures, but it is the simple quality of the many line diagrams of ventilation strategies which is so immediately engaging.

Reid, injured by his experiences with the English Establishment, but with £7650 in his pocket, emigrated to the USA into the welcoming arms of more reliable champions, speaking at the Smithsonian Institution, advising on the ventilation for the US Capitol building, a continuing unresolved issue, some years before the May 1866 Wetherell Report as reported by Billings (1884, p. 120), being elected to a Chair at the University of Wisconsin, and publishing *Ventilation of American Dwellings* in 1858.[16] Here we find Dr Elisha Harris's touching and revealing introduction.[17] Harris refers to the unbroken chain of interest in 'air' from Hippocrates to Hales and Beddoes, it being down to Dr Reid, after two millennia: 'to determine by practical experiments made on living men, by means of apparatus, and in rooms specially constructed for this purpose, the precise amount of air for health and comfort.'

He places Reid's contribution to knowledge alongside that of Sir Humphrey Davy. He adds (Reid, 1858, p. xvii), perhaps with Reid at his elbow and Barry on the distant horizon:

Architecture must be studied and practiced as an art upon which the principles of vital chemistry, and human physiology, have claims, no less important than the laws of mechanics, and the strength of materials, and in which the laws of health and the claims of personal comfort should ever be paramount to architectural embellishment and artistic effect. [Emphasis added by current author.]

Reid gave clear and digestible form to remarkably prescient ideas but he was not a lone voice. For the entrepreneurial, the stakes appeared to be very high, both financially and perhaps even more importantly, socially, from achieving solutions to the problem of safely ventilating a grateful nation's building stock, all securely patented. Robert Ritchie's (1862) treatise on ventilation neatly rounds up the state of understanding in the early 1860s. His Preface refers to the importance of ventilation coming prominently to public notice in 1835 by the appointment of the Select Committee. He admires the late Thomas Tredgold CE, inventor of the siphon ventilator (the J-tube), Dr Reid, Dr Birkbeck, Professors Brande and Farraday, Mr Chadwick and Mr Gurney. In Part III, 'Example of Spontaneous Roof-ventilation', Ritchie reproduces a highly prescient diagram, a scheme for ventilating a church (Figure 1.3). These early principles will recur regularly through the coming chapters.

The idea that 'good' government would override human experience of the natural world had become unconvincing. Walter Bernan (1845), for example, wrote: 'The progress of ventilation, for instance, will show that methods have been practised long ago, and have become obsolete and been forgotten, which have recently been revived, and from their manifest advantage to the community are now rising in public estimation.'

Bernan attempts to catalogue all historical instances from ancient Egyptian homes to the directions given by the architect Andrea Palladio (1508–80). Dr

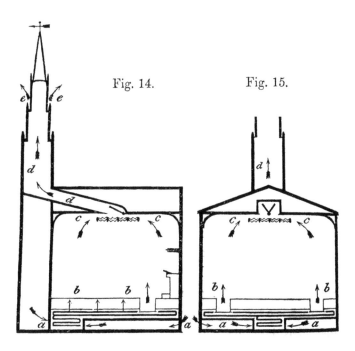

Figure 1.3
'Example of spontaneous roof-ventilation', in Ritchie, R. (1862) *A Treatise on Ventilation, Natural and Artificial*, p. 106. Anonymized proposal about which Ritchie writes, 'the author was much indebted to the architects for their co-operation'.

Harris had written off the interior environments of all ancient civilizations (Reid, 1858, p. v). Bernan writes bullishly in his Preface (1845, p. iv):

> Leaving Falconer and his disciples in possession of their dogma [Falconer's *On Influence of Climate*], that governments may stamp the manners, but it is the air they breathe which moulds the form, temper and genius of a people. ... The formation and regulation of artificial climate will then assume the character of an art for developing and expanding the mind and body, for prevailing health, and prolonging life; and the skillful practice of the art, as a means of saving fuel, will become essential not to the well-being only, but to the existence of many communities.

Here Bernan is traducing William Falconer's (1781) argument of 64 years earlier and the opinion of Hume, the arch-critic of environmental determinism, that good government will deliver healthy populations. But Bernan is promoting the idea, perhaps his original idea, of an inhabited environment entirely delivered and regulated artificially, the object being altruistic, the equitable 'redistribution' of access to a healthy environment, with a body of practitioners with new knowledge to deliver it. He describes himself in 1845 as a 'civil engineer' and his vision of a substantial heating, ventilation and air-conditioning industry eventually came to fruition.

By 1867 established heating and ventilation engineers were contributing to competition entries for important public buildings.[18] This is clearly evident in the competition entries for the Palace of Justice, the Law Courts on the Strand in London. There is no mention of the engineering co-submissions in any of the series of articles about the progress of the competition in the *Building News and Engineering Journal*.[19] Nor is there any content on heating and ventilation of any public buildings save for the 26 April 1867 issue which summarizes an article from the then-current issue of the *Chambers' Journal* reporting an experiment in one of the Philadelphia Almshouse wards in which the ventilation exhausts are placed at low level with apparently very beneficial effects. The two professions, engineers and architects, are already split apart.

Nor does the *Building News and Engineering Journal* carry any advertisements for the design services of ventilation and heating engineers or contractors or manufacturers of equipment except for the Hayward Brothers' domestic scaled iron hopper action vent. The worlds of 'architecture' and 'ventilation' in London were entirely separate. Was this just snobbishness on the part of an emerging profession over an emerging trade? The editors, writing anonymously, after savage reviews of some the entrants' designs and confusing shifts in opinion through the early months of 1867, are focused only on established architectural values despite the very complex functional brief. They bemoan the lack of a design which would, apart from dazzling with its: 'flash of expression, ... force of nature, development of sinew'. Its muscularity, would:

> ... at the same time mark the age to which it belongs, instead of being a mere *rechauffe* of world-gathering examples ... the reflection of a past century, not the genuine creation of the present.

On the whole, architects and their critics appeared to have no interest at all in the likely environments within their competing visions, nor in the 'tradesmen' who would be delivering them. The detachment of the profession from commerce was thought to be universal but Cook and Hinchcliffe (1995) record that Alfred Waterhouse, the architect appointed to design the Museum of Natural History in South Kensington, positively recommended to his client that a contract be entered into with W.W. Phipson based on Wilson Phipson's proposal for a low pressure steam heating system. He was cheaper than Haden and Son of Trowbridge who were nervous about building labour cost inflation.[20] Hadens of Trowbridge, founded in 1816, were the agents for Boulton and Watt's steam engines, subsequently applying their skills in forming pipework to heating and ventilation schemes.[21] Their posters imply they had invented a practical 'expansion joint', a critically important component (Billington and Roberts, 1982; Donaldson and Nagengast, 1994).[22]

Phipson worked for, or with, architects Matthew Digby Wyatt, G.G. Scott, Augustus Pugin, William Burges and clearly Alfred Waterhouse.[23] He heated and ventilated London's Royal Albert Hall in 1868 and Mount Stuart House, Cardiff Castle and Castell Coch for the Bute's, the wealthiest clients in Britain, the Royal Holloway College in 1882 and the Birmingham Town Hall in 1890 (Barber *et al.*, 2000). Architectural historian John Mordaunt Crook (2013) only mentions Phipson once, in a footnote, in his recently republished standard biography of architect William Burges. Phipson's drawings of 1880 for the 'Mansion for the Marquis of Bute at Rothesay' are amongst the few that survive.

Phipson was a protégé of the Belgian engineer Dr Van Hecke, who worked on hospital ventilation and he assisted with the system installed in Hospital Necker in Paris. Van Hecke came to the notice of Dr Pettenkofer in Berlin, a recognized authority on hygiene, whose name is carved into the façade of the London School of Hygiene and Tropical Medicine, but whose career ended disastrously as we shall find out in Chapter 6. On his return to London in 1859, Phipson rushed out *Remarks on Ventilation with Extracts from Official Reports on the Combination of Ventilation and Warming System Van Heck* (1859). The Van Hecke ingredient to otherwise straightforward principles was his design for a fan which would sit within the air channel, in line, monitored with an anemometer to stabilize the supply of air, set for example at 2200 cubic feet/hour/patient for a hospital.

Phipson's corresponding ventilation installation at the Royal Albert Hall, London was 28 000ft of 4-inch hot water pipe threaded and coiled through a tripartite plena below the arena fed from three boilers to give resilience to the system, warm air rising through interstices in the floors. Fresh air is forced into the system by two engine powered fans 5ft 9in in diameter. Phipson gave particular attention to the problem of opening doors to the exterior subverting the natural buoyancy of the warming air. He seems to have coupled the doors to the lantern, closing the exhaust when a massive additional free area opened at the base. This is still very much a problem, explored in Chapter 4 when considering the SSEES Building.

Ventilation provisions at an architectural scale were already in place in Waterhouse's design for the London Natural History Museum and they are prodigious (Cook and Hinchcliffe, 1995). Had Waterhouse learnt the principles from his engineer co-designer for the unsuccessful entry in the 1867 competition for the London Law Courts building? Two towers to the south, flanking the entrance of the Natural History Museum, contain exhaust stacks at each corner arriving at columnated terminations, as do the towers at the ends of the 'pavilions'. Two towers to the north are constructed as ventilation stacks encircling central smoke stacks, Cook and Hinchcliffe (1995) point to the innovation in their multi-sleeved design. One wonders if the difference in height between the various stack terminations induces more complex unintended airflows.

English Heritage reminds us that American mechanical engineers Nason & Briggs had already provided viable fans for the US Capitol building in Washington, DC in 1857, in which Reid had an involvement. However, the first multiblade fans were devised by Bennett Hotchkiss of New Haven, CT. Benjamin Franklin Sturtevant of Boston stole the emerging market, establishing a London presence in the 1890s. But Sturtevant (1906) considers the Sirocco fan invented by Samuel Cleland Davidson in Belfast in 1898 and subsequently sold as intellectual property to the American Blower Company (ABC) could reliably drive more than enough mass flow of air persuasively through large public buildings. Nicholas Pevsner (1970) wrote that already by 1876 the Philadelphia World's Exhibition demonstrated American supremacy in the field of mechanical ventilation, clearly evident from the 'old world's' reactions to the heating and ventilating exhibits, but offers no evidence.

Surgeon Major John S. Billings (1838–1913) does indeed illustrate the levels of sophistication being achieved in North America in his 1884 book, assembled from a series of papers, *Letters to a Young Architect on Ventilation and Heating*, written between 1880–82 in *The Sanitary Engineer*, the American version printed in New York (Billings, 1884). It is panoramic in its vision and supremely confident in the veracity of its content. He explains (ibid. p. 27) that it contains almost no mathematical equations or unfamiliar technical expressions to suit the intended audience of aesthetes: 'My architectural friend … said; "I do not care for scientific theorizing and speculations in this matter; what I want are practical rules."'

Billings makes the caveat that all that can be usefully achieved is a general impression of the relative importance of the subject but proceeds to provide rather more. He wrote that men eager to become doctors cannot short cut the need to develop understanding in anatomy and physiology, however impatient, and thus it is, Billings advises, in this field of building design.

The book reveals the state of progress in understanding building environments by the late 1880s. He refers reasonably warmly to Reid's work at St. George's Hall in Liverpool but is short on detail. He advises (ibid. p. 151) that the publication describing it is rare and not worth pursuing. Billings identifies cost as the principle barrier to delivering well-ventilated public buildings. There were very few such fit-for-purpose buildings and so there was a crisis of toxic public internal environments across the United States. Furthermore, he sees no generic solutions as each situation is different. We will return to this critical question. Can one safely disseminate 'design tools'? Can there be reliable exemplars? Chapter 6 reviews his involvement with Johns Hopkins' Trustees in Baltimore from 1875 in realizing their ambition to build the most advanced hospital in the world. That design was distilled from several rather promising finalists' schemes and findings from a very full European tour of hospitals and other building types, particularly theatres and auditoria, which will be considered in Chapter 5.

Billings was quite well connected; he knew the physician Joseph Lister and called on him in Edinburgh, and he certainly subscribed to germ theory. Billings (ibid. p. 20) was of the view that offensive and poisonous gases, carbonic acid or 'sulphureted hydrogen', are diluted to a harmless concentration by fresh air: 'Carbonic acid is usually found in very bad company', it is a good indicator of the presence of yet worse but largely indetectable gases.

However, he continues with the very latest findings (ibid. p.17):

The dangerous impurity in some air, such as that in a hospital ward for contagious diseases … is not a gas, and does not possess any very marked or unpleasant odour. It consists of very minute particles of organic matter, which are capable of producing disease when introduced in the living human body, and some of which are capable of growth and multiplication under certain circumstances … no amount of fresh air will dilute one of these particles … however … for the great majority of the disease germs, exposure to, and agitation in, large quantities of fresh air, especially if this contains ozone, will destroy or greatly impair their vitality and powers of doing harm.

He is scornful of the habit of removing carbonic acid by placing holes for exhaust at floor level. He explains that the level of carbonic acid is constant vertically to the outer limits of the atmosphere. He has a method of measurement, using Pettenkofer's portable apparatus and discovers that concentration was about 400 parts per million. Beware, he advises, of purveyors of tin ventilation enhancers and foul air exterminators, they tend to have no data or understanding on which to base their gadgets. However, Billings' figure seems quite high. Is it revealing of general atmospheric conditions within cities at the time? Extraordinarily, today the Intergovernmental Panel on Climate Change (IPCC) proposes that a rise in atmospheric CO_2 equivalent concentration to 400 parts per million would be a watershed in monitoring climate change, marking a potential global mean temperature rise of 2°C or more.

Billings provides guidance on effective air supply rates, 2120 ft^3/person/hour for hospitals, equating to 16.7 litres/person/second. This is about twice the 8 litres/person/second general UK standard in recent times. Billings recommends that this rate rises to 5300 ft^3/person/hour, nearly 42 litres/person/second, when the patient's circumstances demand it. In Chapter 6 we see in Billings' hospital design for the isolation ward, special rooms for the malodorous sick would achieve this rate. For (drama) theatres, Billings' recommended mean rate is 1585 ft^3/person/hour, about 12.8 litres/person/second, which seems eminently credible. The ventilation cooling of theatres is investigated in Chapter 5.

Billings discusses the correct location of air supply and exhaust in a space, he quotes the French engineer General Arthur Jules Morin's (1795–1880) opinion, specifically, promoting air introduced near the ceiling and discharged at low level (Morin, 1863). Morin's 1863 treatise is still a most impressive, even beautiful publication. His theoretical proposal for a generic Lyric Theatre, clearly unconstrained by budget, imagined for the Parisian climate of the time, indicates a startling level of sophistication for the early 1860s (Figure 1.4).

Billings observes that the desirable geometry of organized ventilation varies by purpose and season, discharge openings should also be at high level in large assembly halls, invoking contemporary ideas of mixing ventilation. He dismisses

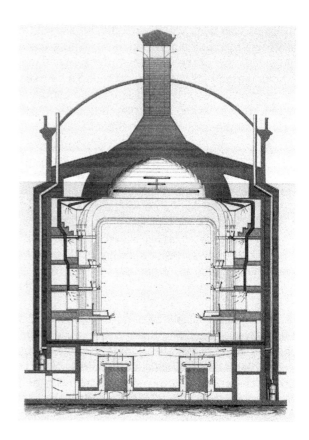

Figure 1.4
General Morin's scheme for a Theatre Lyrique (Morin, 1863).
Reproduced by courtesy of Cambridge University Library.

worries about humidity control. The human frame can survive very low humidities. Billings writes that the Boston City Hospital reported no ill effects from dry air in 1879. He suggested there is no need at all to invest in humidifying air to 70% relative humidity (ibid. p. 59 and p. 73).

His Figure 11 shows his modifications to the highly current idea of ground cooling supply air to barracks as devised by Dr Julius Jeffreys (1800–77) for the British army in India, summarized in Jeffrey's book of 1858.[24] Jeffreys devotes much of the book to army dress and in particular the design of helmets, but in the section on living accommodation he proposes that the upper 50 ft of the earth's crust should be regarded as an 'equalizing reservoir' with an annual cycle of natural pre-heating and pre-cooling. Billings' figure shows an array of dry wells, 500 ft^2 of exposed surface area per well, stopped up at the surface, all interconnected and charged with air by capturing the wind in revolving cowls, 'refrigerator well ventilation' (Figure 1.5). Jeffreys (ibid. p. 136) claims that if the wells are at 60° F at the commencement of the hot season, they will rise by only 6° F by the onset of cooler weather but will be capable of achieving a 30° F depression in air temperature at the point of intake, a covered verandah.

Jeffreys' argument is that in India wells cost from a mere 18 pence (1s/6d) to half a crown (2s/6d). He describes designs for ventilating machines. He reports that Desaguiller's centrifugal pump fan had been used successfully in Bengal with water sprays, comprising Dr Rankin's 'Thermantidote', a species of evaporative cooling (Figure 1.6).

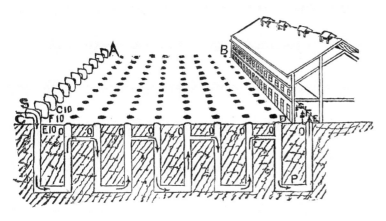

FIGURE II.

Figure 1.5
Billings' Figure 11, reproducing Dr Jeffreys' diagram of his 'refrigeration well ventilation' idea (Jeffreys' Figure 12 'The plan for cooling by subterranean action', p. 132). Billings' redrawing is faithful to the original.

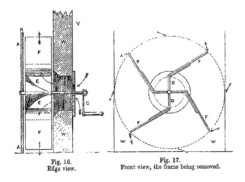

Fig. 16. Edge view.

Fig. 17. Front view, the frame being removed.

Figure 1.6
Jeffreys reproduces drawings (Figures 16 and 17) of a 'centrifugal ventilating machine'. His idea was to connect these to the subterranean refrigeration well to force a continuous airflow into the barracks connected to it or the simpler 'pendulum pump'.

Billings reviews other recent publications on ventilation and heating with little enthusiasm. He is particularly scathing about the Hon. Henry Rutton, 'Probably there are very few books which show a more complete and profound self-satisfaction on the part of the author than this'. Rutton was a failed entrepreneur with bogus patents for ventilation inlets and outlets turning his hand to theorizing as a new venture.

Billings is brief on the important question of what size air intakes and exhausts should be, which is surprising given the sheer scale of the ventilation infrastructure in the public buildings he reviews later in the book. Morin's diagrams give a reliable glimpse into the future. Billings provides a calculation for calculating the pressure difference in a stack: multiply the height from the intake opening to the exhaust by the temperature difference between inside and out and multiply the product by 1/491. The amount of air passing up the stack q is the sum of the cross-sectional area times the velocity v. The velocity is calculated as eight times the square root of the stack height multiplied by the difference between internal and external temperatures, all divided by 491. Billings explains that gas expands by 1/491 for every degree Fahrenheit rise in temperature. Comparison with a contemporary calculation suggests Billings' formula broadly doubles mass flow rates. Nonetheless, the parallel outcomes are of the same order of magnitude. A useful, applicable level of understanding is developing.

The consequence for the design of safe, comfortable buildings remains similar through all these theoretical scenarios: a building designed to rid itself of noxious vapours, or of miasmas emanating from damp ground or from living skin squamae, or carbonic acid, or clouds of minute particles, soot for example, must be configured to flush through every part of itself sufficient quantities of air to remove all that is thought to be noxious, malodorous, and ultimately fatal. Although medical and scientific understanding have immeasurably improved the understanding of spread of disease and maintenance of good health, access to 'good' air and ventilation remains important. And that is likely to require the dramatic reconfiguration of contemporary design 'recipes' for our familiar building types.

Notes

1. See also the literature review by Scott Drake (1977) The Architectural Antimephitic. *Architectural Theory Review* **2**(2), 17–28. (Published online 28 July 2009; accessed 27 June 2013).
2. Paul Overy investigates this possibility, offering striking evidence in the 1930 German Hygiene Museum, its popular exhibit on human physiology in transparent plastic, 'The Glass Man', revealing his internal organs and veins, all illuminated. Overy comments, 'The figure clearly embodied ideas about the implied relationship between transparency and hygiene and cleanliness.' Overy, P. (2007, p. 50) *Light, Air and Openness, Modern Architecture between the Wars.* London: Thames and Hudson Ltd.
3. Leoni (1726, Book IX, p. 91) *The Architecture of L.B. Alberti,* original edition in the Middleton Library of the Faculty of Architecture and History of Art, University of Cambridge.
4. Arbuthnot wrote with the full force of his Fellowships of the Royal Society and of the Royal College of Physicians of London and Edinburgh behind him.
5. The Norwich and Norfolk United Medical Book Society volume in the University Library, Cambridge.
6. See www.alchemyandchemistry.edwardworthlibrary.ie/Chymical-Experiments/Air-experiment (accessed 25 October 2013).
7. Thomas Henry (1776) translation of Lavoisier's *Opuscules Physiques et Chimiques* (*Essays Physical and Chemical*), for Joseph Johnson 1776, p. x., assisted by Mr Aikin of Warrington, who evidently had a good command of technical French.
8. Corbin (1996) quotes Boissier de Sauvage p. 22.
9. The authors Anna Lewcock, Fiona Scott-Kerr and Elinor Mathieson are referring to: Johnstone, J. (1758) *An Historical Dissertation Concerning the Malignant Epidemical Fever in 1756: With Some Account of the Malignant Diseases Prevailing in the Year of 1752, in Kidderminster.* London: W. Johnston.
10. 1818 Lectures to the Dublin Society by Dr Anthony Meyler, Observations on Ventilation and on the Dependence of Health on the Purity of the Air which we Respire, p. 45.
11. See succinct account within Science Museum web pages on the Sanitarian John Snow: www.sciencemuseum.org.uk/broughttolife/people/johnsnow.aspx (accessed 25 May 2015).
12. Revealing that Mr. Punch was ahead of the medical profession in understanding and depicting contagion.
13. http://news.bbc.co.uk/dna/place-lancashire/plain/A10697132 (accessed 9 November 2013).
14. Polmaise heating system for glasshouses, *Gardener's Chronicle*, 3 January 1846, p. 3.
15. Reid was equipped to speak with some authority, he had a full university medical education, performed as a research assistant in the Chemistry Department at Edinburgh, was elected a Fellow of The Royal College of Physicians, Edinburgh in 1831, and was well-informed as to latest progress in chemistry but not well enough connected to be elected to the Royal Society. Perhaps he was just too practical, or to survive the volatile interests of national politicians and peers. He built a small auditorium

alongside his laboratory and tested its ventilation efficiency on his students. The British Association visited the laboratory and Dr Reid was catapulted into public life by invitation to consult on the new Houses of Parliament.

16. The reviews were indeed enthusiastic. The publisher cites *Scientific American*: 'We can only say in conclusion, that the subject area is one of personal interest and national importance, and Dr. Reid has done a great service to the American people by the publication of this work', and the *New York Times*: 'Dr. Reid of England whose most excellent work entitled, 'ventilation in American Dwellings…'.

17. Harris was a distinguished former Physician in Chief to the New York Quarantine Hospital on Staten Island and a future Sanitary Superintendent of New York from 1868–70. The American Public Health Association minutes recording his demise in 1884: www.ncbi. nlm.nih.gov/pmc/articles/PMC2272621/ (accessed 7 November 2013).

18. www.helm.org.uk/guidance-library/heating-ventilation/heatingventilation.pdf (accessed 9 November 2013).

19. *Building News and Engineering Journal,* 22 February 1867. https://archive.org/stream/ buildingnewsengi14londuoft#page/198/mode/2up (accessed 8 November 2013).

20. The Chartered Institution of Building Services Engineers (CIBSE) Heritage Group records. www.hevac-heritage.org/victorian_engineers/phipson/phipson.htm (accessed 8 November 2013).

21. The Chartered Institution of Building Services Engineers (CIBSE) group archive contains material on Haden and Son.

22. English Heritage has gathered useful background, early and disastrous attempts at mechanically ventilating hospitals with bellows by Rev. Stephen Hales and others, notably York County Hospital of 1849, 'one of the most interesting examples of the complete failure of artificial ventilation in the country', H.C. Burdett (1893) *Hospitals and Asylums of the World*, London: J&A Churchill.

23. The CIBSE Heritage Group record Phipson as being 'unknown', until the recent gift of various papers.

24. The book is dedicated to a member of the Army Sanitary Commission. Florence Nightingale's attacks on the army's care of soldiers abroad were published in 1859 but delivered in 1857.

2

Climate and its annihilation

But suppose that there was an alteration in the state of the seasons, with the position in the sky currently occupied by the north wind and the winter being taken over by the south wind and the summer, and the position of the south wind being taken over by the north – and if that were to happen, then the sun, as it was driven out of the mid-heavens by the wintry blasts of the north wind, would surely pass over the skies above Europe, just as now it passes over those above Libya. Indeed I expect that the sun, with the whole of Europe to cover, would affect the Ister as it currently does the Nile.

(*Herodotus: The Histories,* Book 2, Para 26)

These (climate) models and calculations allow for little human agency, little recognition of evolving, adapting, and innovating societies, and little endeavor to consider the changing values, cultures, and practices of humanity. The contingencies of the future are whitewashed out of the future. Humans are depicted as 'dumb farmers,' passively awaiting their climate fate.

(M. Hulme, 2011, *Reducing the Future to Climate: A Story of Climate Determinism,* p. 256)

The history of thought about air, atmospheres and, in particular, 'diseased' air as reprised in Chapter 1 reveals centuries of speculation on what phenomena might give cause to a health-preserving environment, one that might promote 'civilization'. There were many speculations about how one might conjure up such a happy outcome, almost invariably using natural forces, being the primary means available. Climate performs a significant role in this theorizing about the direct causation of 'good' environments. Ideal climates, some argued, produce the most civilized societies. But hitherto unexpected changes in long-term trends and the occurrence of more frequent extreme weather events will subvert this steady-state model as climate zones drift away from their original residents. This chapter explores the history of beliefs and understandings of the effects of climate on people. This includes the long-held, if wholly undemonstrable, conviction deeply embedded within our culture that environments and most particularly climates define the essential nature of those who dwell within them. This has been the belief of generations of 'environmental determinists'. Evidence will be presented for the equally emphatic response of Modernists to annihilate climate difference, having unimaginatively absorbed the determinists' diagnosis, to destroy inferior climates in order to spare the climate-oppressed. Even 2500 years ago, Herodotus speculated on the implications of a changing climate. For him, the human frame responded physiologically in a wholly predictable way to each climate type.

'The eternal flux of nature': anthropo-geography and climate determinism

In 1884 Wladimir Köppen formalized the typology of global climate zones and regions according to three variables, the prevailing climate type and its seasonal fluctuations, precipitation and temperature, published as the Köppen–Geiger system (Figure 2.1) in the paper, 'The thermal zones of the earth according to duration of hot, moderate and cold periods and to the impact of heat on the organic world' (Köppen, 1884).

The painstaking analysis of the available data was accompanied by Köppen's opinions on the relative value of the various climate types to human advancement, in all respects. He was clear about the predominance of the

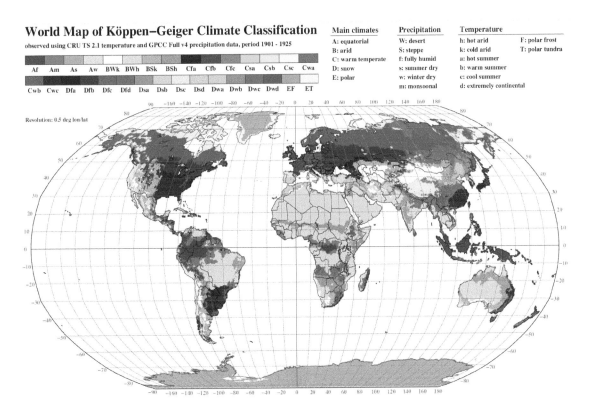

Figure 2.1

The thermal zones of the earth according to the duration of hot, moderate and cold periods and to the impact of heat on the organic world. Updated version of the original 1884 map to depict 1901–25 (Köppen 1884), reproduced from the paper by Rubel and Kottek (2010), courtesy of Professor Franz Rubel of the Climate Change and Infectious Diseases Department, Institut für Öffentliches Veterinärwesen, Veterinärmedizinische Universität Wien. See http://koeppen-geiger.vu-wien.ac.at (accessed 12 October 2016).

temperate climate in enabling the achievements of 'civilization' but allowed some anomalies (1884, p. 357):

> Therefore, the advantage of the temperate climates in comparison to the hot climates regarding the entrepreneurial spirit and the pursuit of greatness shows even more distinctly in America than in the Old World. However, the more general this phrase is stated, the more the singular exceptions must attract our interest, which are the places where we find grand constructions and rests of comparably high civilisations in the tropical lowlands.

Volken and Bronniman, Köppen's recent translators, felt it necessary to add the following Endnote E15 to their paper to firmly disassociate themselves from Köppen's beliefs: 'The original text has an attitude of climate determinism and racism. Environmental determinism was prominent in late 19th and early 20th century European science (see Hulme, 2011).'

Just two years (1882) before Köppen's paper appeared, Friedrich Ratzel (1844–1904) published his work *Anthropo-Geographie oder Grundzüge der Anwendung der Erdkunde auf die Geschichte,* (*Anthropo-geography: or Broad Application of Geography to History*). The notion of 'Anthropo-geography' was well received, not least in the United States. The propositions, including that of a fundamental belonging of people to the land type of their forbears, 'Lebensraum', a term later appropriated to sanction aggressive national expansionism, resonated with established classical themes, suggesting a diaspora of distinct cultures emanating from a small number of cultural epicentres, an idea explored by his immediate successors. We will consider the anthropo-geographer-psychiatrist Willy Hellpach momentarily. One should note the parallel lyrical and poetic tradition of what became known as 'regional geography', associated with Paul Vidal de la Blache (1845–1918). This is the more creative, literary geography emphasizing the particular qualities of 'place', which was displaced by the new enthusiasm for systematic quantitative geographical investigation in the 1960s.

In the 1920s and 1930s, Ellsworth Huntington (1876–1947) popularized climate determinism in the United States. Despite his tenuous, un-established post at Yale, he established a national reputation, publishing liberally through the Yale University Press. What could be more respectable? Yet the whole construct of environmental determinism became spectacularly and risibly discredited as having been indelibly connected into the belief systems of the various totalitarian regimes of the 1930s and 1940s as revelations emerged of the appalling genocides conducted to 'ethnically cleanse' both home and overseas territories. He was certainly not an isolated figure in the US, at least not until the mid 1930s when he subscribed to the eugenics movement. Somewhat earlier, Edwin Grant Dexter, Professor of Education at the University of Illinois, pursued climate- and weather-driven human and societal difference, publishing in 1904, *Weather Influences: An Empirical Study of the Mental and Physiological Effects of Definite Meteorological Conditions*. But this is a benign anthology, gentle in tone and interesting particularly in its correlations of monthly weather data against mortality statistics and police sickness absences in New York (Figure 31, p. 181). Dexter concluded that 'sickness and death are generally more prevalent during the winter and early spring months, though the latter (death) is tremendously increased by the intensely hot spells of the summer' (ibid. p. 197). In his chapter 'Weather influences in literature', he evoked 'skyey influences' (ibid. p. 41) under which 'the more sensitive and susceptible … become living barometers', evoking Byron (whom Dexter maintains became more religious in outlook in Mediterranean sunshine), Rousseau, Shelley, Beethoven, Goethe and Walt Whitman allegedly 'drinking the sun's rays in at every pore'. He then referred to 'Racial Traits', the observed effects of such empirical laws governing the influence of climate upon the mass of people, notwithstanding maverick individuals, views from a university professor potentially providing fuel for an impending interest in making controlled artificial environments.

Huntington may have been yet more influential in inadvertently providing a seemingly more academically respectable rationale for the commercially driven imperative to deliver artificial environments at a global scale. His contemporary, the American geographer Ellen Semple, authored the 1911 *Influences of Geographic Environment on the Basis of Ratzel's System of Anthropo-Geography*, dedicated to her tutor, who was indeed Friedrich Ratzel. Semple wrote, not at all unreasonably (Preface):

> The eternal flux of Nature runs through anthropo-geography, and warns against precipitate or rigid conclusions. But its laws are nonetheless well-founded because they do not lend themselves to mathematical finality of statement. For this reason the writer speaks of geographic factors and influences, shuns the word geographic determinant, and speaks with extreme caution of geographic control.

However, the contents of Semple's (1911) Ch. XVII, 'The Influences of Climate Upon Man',[1] proceeded to belie the cautionary note and to reinstate the determinist case with vigour:

> Importance of climatic influences – Climate in the interplay of geographic factors – Its direct and indirect effects – Climate determines the habitable area of the earth – Effect of climate upon relief and hence upon man – Man's adaptability to climatic extremes – Isothermal lines in anthropo-geography – Historical effects of compressed isotherms – Historical effects of slight climatic differences – Their influence upon distribution of immigration – Temperature and race temperament – Complexity of this problem – Monotonous climatic conditions – Effects of Arctic cold – Effect of monotonous heat – The tropics as goals of migration – The problem of acclimatization – Historical importance of the temperate zone – Contrast of the seasons – Duration of the seasons – Effect of long winters and long summers – Zones of culture – Temperate zone as cradle of civilization.

She insisted on survival in climatic hardship as the driver of human endeavour, the case being (Ch. XVII):

> … that his (Man's) later evolution depended far more upon the powers which she (Nature) developed within him. These have no limit, so far as our experience shows; but their growth is painful, reluctant. Therefore they

develop only where Nature subjects man to compulsion, forces him to earn his daily bread, and thereby something more than bread.

More specifically (Ch. 27):

If, as many ethnologists maintain, the blond Teutons of the north are a bleached out branch of the brunette Mediterranean race, this contrast in temperament is due to climate ...

with the surprising deduction that (Ch. XVII):

Where man has remained in the Tropics, with few exceptions he has suffered arrested development. His nursery has kept him a child ... without the respite conferred by a bracing winter season, and its combination with the high degree of humidity prevailing over most of the Torrid Zone.

These propositions were depicted in aggressive and overtly racist advertising for domestic and workplace air-conditioning in the interwar and immediately post-war periods. The machines were marketed with the message that they would prevent descent into the behaviours of the torrid zone by manufacturing the transformative temperate zone, wherever the market had woken to its climate-determined destiny. In fact Semple was particularly anxious about colonists succumbing to hot humid climates (1911, p. 1435): 'Transfer to the Tropics tends to relax the mental and moral fiber, induces indolence, self-indulgences and various excesses which lower the physical tone.'

Such deterioration, of course, being the novelist Somerset Maugham's habitual plotline observed wryly by indigenous local powerbrokers, well-adapted to their hot humid climates.[2]

Peet (1985, pp. 309–333) suggests that 'Environmental determinism was geography's entry into modern science'. This geographic version of social Darwinism intended to explain in Whiggish terms the inevitability of Europeans' aggressive imperialistic ambitions and the inability of the conquered to resist: as a direct consequence of climate difference.

Livingstone (2002) has identified a 'genealogy' of what he calls a 'moral climatology', a persistent inclination in the Western elite to use climate as a 'vehicle for moralistic construals of global space'. He describes a long history behind this 'tradition of tropical censure' from Aristotle, to Falconer, to late 19th century Scottish missionaries. It is ironic, he observes ruefully, that earlier generations depicted Paradise as a tropical environment. Livingstone cites the two sceptics amongst the discoverers of the mosquito transmission of malaria at the close of the 19th century at the London School of Tropical Medicine: Manson and Sambon. They were persuaded of human adaptive capacity, arguing that tropical 'lethargy' was the result of parasitism, not climate. They were in a minority, Livingstone argues, the 'belief in the authenticity of moral climatology underscored the application of germ theory'. Others in the 1890s, for example Ralph Abercromby, warned of demoralization, degeneracy, depravity and debility latent in the tropical climate zone. What then might be the ramifications of the fringes of the temperate climate zones morphing in this direction through the coming century?

The objective here is not to traduce the reputations of once honoured geographers and anthropologists. (Others are doing this work.) Rather, it is in detecting any propensity towards a belief that cool temperate conditions delivered the most 'civilized' populations, that 'mental and moral worth derived … from climatic circumstances' (Livingstone, 2002, p. 168). Might a 'moral' imperative have incubated within this worldview to transform the daily environments of those less fortunate inhabitants of warmer tropical and torrid climate zones into temperate environments, effecting some change for the better in the indigent populations and protecting transplanted individuals from the North? Did the apparently continuous dissemination of essentially the same idea reach the key figures inventing the means for making artificial environments and the emerging industry hoping to deliver it at a global scale? Or was it really 'all about the money'? Later in this chapter the early career of Willis Havilland Carrier, the alleged inventor of 'artificial weather', is explored for any clues as to an incipient 'moral climatological' position. The more recent discovery that the world's climate regions are shifting adds a certain piquancy to this notion of environmental determinism. This will undermine the credence of climate stereotypes. The market for indoor 'artificial weather' may be changing by the decade. Artificial weather is very energy intensive and a global imperative now exists to radically reduce energy consumption and CO_2-related emissions from buildings. This means, in turn, that the dependence on mechanical systems for environmental control (heating and cooling) must be radically reduced. Could it be that the influence of climate on a society actually does matter if one can no longer justify its annihilation?

Evocations of 'perfect' climates

Environmental determinism is an ancient idea with an unbroken continuity, persisting still today, as the struggle endures to enable everyone on the globe to join the chosen temperate region dwellers. Its classical origins infuse the writings of Parmenides, Strabo, Diodorus and Hippocrates of Kos from 460–370BC, reprised for clients and their constructors by Vitruvius. Hayes and Nimis' (2013, p. x) comment on Hippocrates' 4th century BC 'tendentious assertions about Asians' in the second section of their new translation of Hippocrates' *Airs, Waters, Places*:

> … negative stereotypes about Asians gave this text a prominent place in discussions of environmental determinism in the early modern period, where similar arguments were used to differentiate and hierarchize races of humans.

The assertions, in that they follow the medical treatise, acquire the gloss of apparent medical substantiation. Hayes and Nimis derive their observations on the continuity of determinist ideas from the broader commentaries provided by McCoskey (2012) and Thomas (2000). Thomas identifies Hippocrates' second section as an exposition on the divide between Europe and Asia. W.H.S. Jones (1923, p. 66) comments in his translation of Hippocrates, on which Hayes and Nimis base their work:

> In tone it is strikingly dogmatic, conclusions are being enunciated without the evidence upon which they are based … it is scarcely medical at all, but

rather ethnographical. It bears a close resemblance to Herodotus, but lacks the graceful *bonhomie* which is so characteristic of the latter writer.

Jones was not over-enamoured with the integrity of his subject author.[3] For Hippocrates, the correct environment was key to civilization:

> For where the changes of the season are most frequent and most sharply contrasted, there you will find the greatest diversity in physique, in character, and in constitution. Diversity of environmental experience is [the] most important factor that creates the most robust and progressive constitutions ...

Ironically, the most benign environments (arguably the body-temperature atmosphere of Palo Alto, California) do not of necessity yield only the most vigorous and civilized inhabitants. Hippocrates continues (Jones, 1923, p. 135):

> ... if the situation be favourable as regards the seasons, there the inhabitants are fleshy, ill-articulated, moist, lazy, and generally cowardly in character. Slackness and sleepiness can be observed in them, and as far as the arts are concerned they are thick-witted and neither subtle or sharp.

This is rather surprising, harsh even, but earlier in the text (ibid. p. 115) Hippocrates is absolutely explicit:

> ...with regard to the lack of spirit and of courage among the inhabitants, the chief reason why Asiatics are less warlike and more gentle in character than Europeans is the uniformity of the seasons, which show no violent changes either towards heat or towards cold but are equable. For there occur no mental shocks nor violent physical change, which are more likely to steel the temper and import to it a fierce compassion then is a monotonous sameness. For it is changes of all things that rouse the temper of man and prevent its stagnation. For these reasons, I think, Asians are feeble.

Where does this place the finely controlled, unvarying artificial environment? Thirty years earlier in 1893 the Rev. Tozer was interrogating the Greek geographer Strabo's 17 books of the *Geographia* (Strabo, 63BC – approx. AD19). Tozer knew that current debates were being fuelled in part by the exhumation of observations by the ancient geographers and writes: 'the editor of Strabo so often finds himself treading on the still warm ashes of modern controversies, into the service of which his author's remarks have been pressed'.

In his introductory comments Tozer was fearful of his work 'becoming a textbook of disputed questions', still a potential hazard. Tozer was a little sceptical of Strabo's intellect; he detects Stoic leanings, suspecting that he 'believes in providence rather than the gods'. The 17 books are short on statistical evidence. Strabo believes stable government is 'the primary agent for advancing civilisation', the Pax Romana, as Hume argued over 1700 years later. However, this can only emerge in benign conditions, the prevailing climate assuming particular importance in determining these. In Book 2, 'Prolegomena', Chapter 8, 'Superiority of Europe over the other continents', II.5.26, Tozer (p. 87) reports on the original text:

... he [Strabo] describes the advantages which the continent of Europe in general possesses in this respect, as compared with Asia and Africa, especially in its temperate climate, its equal distribution into mountains and plains, which supplied respectively a warlike and a peaceful element to the population, and its furnishing its occupants with the necessaries of life rather than superfluities and luxuries.

Tozer includes a reconstruction of Strabo's map of the inhabited world.[4] He summarizes Strabo's account of its extent (ibid. p. 80). It is confined entirely to the northern hemisphere, an irregular oblong reminiscent of the shape of the Chlamys, the Greek mantle, usually a double square, and attenuated east–west, its greatest length corresponding to a 'parallel', a line of Latitude.

Figure 2.2 shows a map marked by the current author superimposing Strabo's temperate climate envelope (that within which advanced civilization could develop, a considerably warmer temperate climate than Köppen would accept) with two epicentres of optimal conditions, Rome and Baetica. The blue

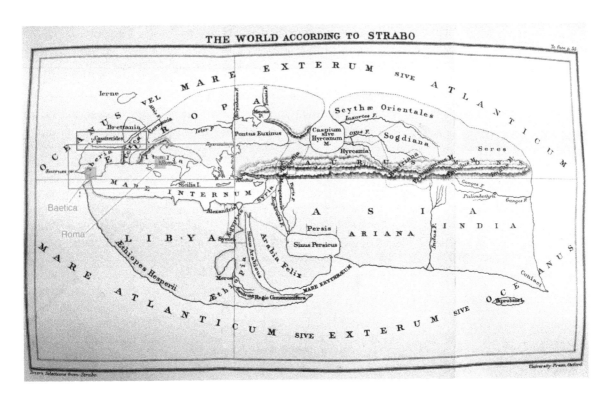

Figure 2.2

Map reconstructing 'The world according to Strabo' from Tozer (1893) marked by the author to show – very crudely – Strabo's Temperate envelope within which he argues that advanced civilization can develop a warmer Temperate climate than Köppen would accept, with the epicentres of optimal conditions, Rome and Baetica. The blue outline depicts the much truncated, more northerly and cooler later 20th century equivalent envelope, the domain perhaps of the culpable 'blond blue eyed Anglo-Saxon bankers' as recorded by Maureen Dowd, OP-ED columnist *New York Times* 'Blue Eyed Greed?' (28 March 2009): 'At a press conference Thursday in Brasilia with Prime Minister Gordon Brown of Britain, President Lula observed "This crisis was caused by the irrational behavior of white people with blue eyes, who before the crisis appeared to know everything and now demonstrate that they know nothing," charged the brown-eyed, bearded socialist president...'.

outline depicts the much truncated, more northerly and cooler equivalent preferred envelope adopted by the later 20th century.

In 1733, Dr Arbuthnot (Chapter IV, p. xxxi), mentioned in Chapter 1, marvelled at 'the great Limits sufferable by Human Bodies ... it is wonderful to observe one Creature, that is Mankind, indigenous to so many Climates'. He comments (ibid. Conclusion, Chapter IV, p. xxxi) that animals left to their own devices choose the countries and climates most adapted to their constitution:

> ...and so, perhaps would a Human Creature, if he were left to his own choice, but he is in Society, and under Government, and subsequent to Passion, to which he sacrificeth the greatest Blessings of Health, and Life itself.

Arbuthnot anticipates the notion of a 'social practice'. He continues with a theme to recur in Elsworth Huntington's commentaries on environmentally determined behaviours and in the ridiculing of Huntington in the 1950s by Manley (1958, p. 105): 'there are likewise other Causes of Migrations of great Herds of Mankind, which is their swarming so as to overstock the Countries which they inhabit'.[5]

William Falconer MD FRS (1781), offered a more scientific theory of the causation of difference in humankind as a respected medical doctor, winner of the Fothergillian Medal of the Medical Society of London. In Chapter II 'on the effects of heat on the human body', we learn that, under heat:

> ... the bile however must be excepted, which is considerably increased in quantity, and as some think rendered more acrimonious in quality ... The disposition of the body and juices to putrefaction is also much augmented.

In Chapter IV Falconer proposes that hot climate dwellers have a more passionate temper, a more amorous disposition (but sadly shallow and base) and a more vindictive disposition. He quotes Roman architect, engineer, theorist and writer Marcus Vitruvius Pollio, known as Vitruvius, practising in the first century BC and after him, 'Hoffman' (sic),[6] ascribing 'the cowardice of the people of hot climates to the small proportion of blood in their bodies ... But this is little more than vague hypothesis'.

Whereas in Chapter III, different outcomes are offered in cooler climates: 'Cold on the contrary, in similar circumstances, corrugates or wrinkles the cuticle, and causes the cutaneous/papillae to contract; and to retire deeper into the skin.' This involuntary response is compounded: 'The secretion of the bile is diminished by cold, and its quality rendered less acrimonious', and unsurprisingly, the body's 'humours less disposed to putrefaction'. Furthermore, 'cold, by blunting the power of feeling, in the manner above described, tends to greatly diminish the sensibility of the system in general'.

As a consequence (ibid. Book I, Chapter XVI):

> The intermediate climates have always been esteemed, both in ancient and modern times, to be the most favourable to human nature. Galen observes the great superiority of the inhabitants of the temperate zones over those both of the torrid and frigid; which appear both in manners and bodily accomplishments, and still more remarkably in the intellectual faculties.[7]

He cites Aristotle's agreement with this sentiment, given as a footnote to Problem 13, that extremes of temperature are unfavourable to the powers of the mind, as well as to those of the body and that a temperate region is useful to both (Ross, 1924).[8] He admits some advanced civilizations confusingly cross varying climate types, the Greeks happily straddled hot and temperate zones but, after reflection, he decides they, 'maintain a good administration of the affairs of government … this is probably the cause of that superiority'.

However, in general, Falconer argued that Europeans do not prosper in hot climates, their body weight melts down and is excreted, they become debilitated like the natives and their bile enters into their systems inclining them to indolence (ibid. Book 1, p. 12). There is just one problem with 'the inhabitants of moderate climates', but it is a major problem, 'fickleness … remarkably instanced in political matters' (ibid. Section V, p. 20).

Falconer is intrigued by a different variant of climate, that which has come to be known as the Continental climate, a type explored further in Chapter 8. Falconer's Chapter VII is titled, 'Of Climates subject to a Great Variety of Temperature'. He wrote: 'But another kind of climate yet remains to be taken notice of, that wherein the heat and cold prevail alternately, during the different seasons, to a great degree'.

He complained that 'modern' writers had not dealt with this climate type, 'but [it] did not escape the observation of Hippocrates', who, Falconer records, thought that in such a climate the shape and character of the people are much less uniform than in either a hot or cold country. Falconer defines the territory: 'subject to such vicissitudes as a great part of North America and that vast continent between Asia and Europe, called Siberia and Tartary'.

Falconer believes the populations benefit greatly from the environmental diversity (Ch. VII Of Climates Subject to Great Variety of Temperature):

> … the variations of climate make men much more active in the affairs of life, more brave and resolute in their conduct, and more austere and rugged, though more upright and just in their behavior.

In short, all the attributes of people in a temperate climate, but more so. This is a remarkably prescient observation insofar as temperate climates may tend towards the Continental through the coming century.

Richard Kirwan (1787), phlogistinist, had a keen interest in the science of meteorology, invigorated by the availability of new equipment: 'However desirous the ancients might have been of cultivating this science; the want of instruments necessarily denied them all access to it.'[9]

In Kirwan's view, Drebbel, 'inventor of the thermometer', and Torricelli were the founders of meteorology. He calls for a 'conspiracy' of nations to collect data worldwide but believes that he has discovered a fundamental principle that the Mean Annual Temperature of the Standard Situation, in every latitude, ranges from 84°F at the Equator to 31°F at the Pole, at 50° Latitude it is 52.9°F, quite predictably but with, perhaps, an increasing variability as one approaches the Poles. The data cited by Kirwan is fascinating. The mean temperature in Cairo for 1761, recorded by a not wholly reliable Mr Niebur, was 71.4°F with 'greatest heat' of 101°F (38.4°C) in June and July; Pekin enjoyed an annual mean over five years of 55.5°F, the greatest heat being recorded as 98°F, but on 25 July 1773 the heat arose to 110°F (41.75°C) under Southerly winds, more like contemporary extreme conditions (Kirwan, 1787, p. 93). In London, data recorded at the Royal Society between 1772–80 yielded an 'annual temperature' of 52°F, with monthly averages of 63.22°F in June, 66.3°F in July and 65.85°F in August, varying by only 2°F in the eight year period. He comments on the stability of the summer temperatures in London but bemoans the failure to grow grapes there successfully, as summers in Paris are a little warmer. Chapter XVI 'Comparison of the Temperature of London with that of other noted Places', presents a comparative table in which the annual mean variation, the January mean and the July mean for London are all indexed as 1000 (Kirwan, 1787, p. 112). Sadly, he does not give the algorithm for determining his unique and vivid index but one seems to 'feel' the differences in reading the table. They are not barometric pressures. Paris achieves 1037 in July, Pekin 1067 and Madras 1565. Diurnal variation in June in London was 12°F, 10°F in July and 15°F in August, offering plenty of scope for night time ventilation cooling. This is not at all the case now within the relatively recently identified urban heat island, as will be revealed in Chapter 4.

Ellsworth Huntington (1919, Chapter XIII) wrote *World-Power and Evolution*. He was writing in despair, as were many others, at the close of the First World War. As the full scale of the battlefield horrors was becoming known in the United States, he looked to the theory of climate determinism for a retrospective explanation of the disaster and some direction as to how to deal with the aftermath:

> This brings us to the question of where Germany stands in reference to climate. We may say at once that the German climate is much superior to that of her enemies who live east and west of her. This is illustrated in Figures 26A–B [Figure 2.3], two maps of energy and civilization, which I have described elsewhere … we are merely showing that in her relation to the other belligerents, Germany's energy has been in accord with what would be expected on the basis of her climate.

The implication being that the crystallization of a surfeit in national energy into open warfare in Europe may well recur. By way of evidence, Huntingdon reminds us that it was down to nervous energy furnishing 'an example of the aggressiveness which caused Rome to wield so wide an empire'. He is very preoccupied throughout the text with the mapping of health and business output against climate.

In Chapter V, 'Climate and Health', Huntington uses words now familiar to us, quoting Professor Fisher an 'expert in food' and Dr Fisk an 'expert in drink', authors of *How to Live* (New York, 1917):

> Air is the first necessity of life. We may live without food for days and without water for hours; but we cannot live without air more than a few minutes. Our air supply is therefore of more importance than our food or water supply, and good ventilation becomes the first rule of hygiene.

This could have been written in the early 18th century or even in the 4th century BC. Huntington provides a yet more authoritative sounding, if patronizing, gender-specific explanation:

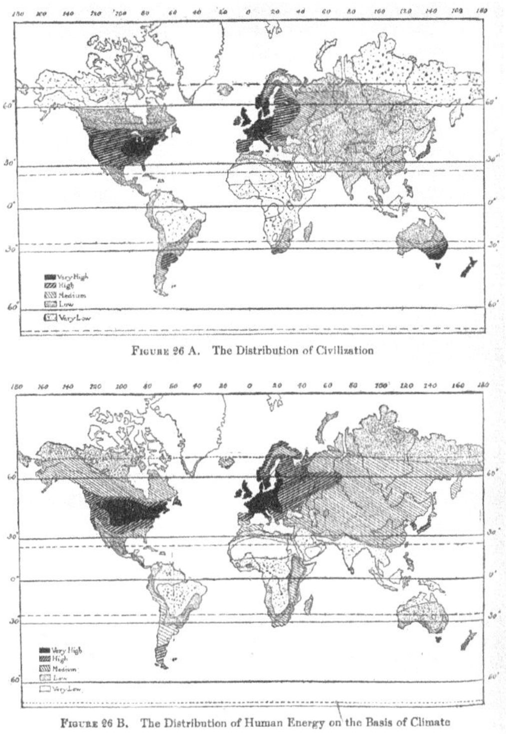

Figure 26 A. The Distribution of Civilization

Figure 26 B. The Distribution of Human Energy on the Basis of Climate

Figure 2.3
'The Distribution of Civilization' and 'The Distribution of Human Energy on the Basis of Climate', Figures 26A and 26B from Chapter XIII 'Germany and her Neighbours', in Huntington, E. (1919) *World-Power and Evolution*. New Haven, CT: Yale University Press.

The air varies constantly in temperature, humidity, movement, electrification, percentage of carbon dioxide, dustiness, and many other characteristics … coupled with all this is the fact that the human body is far more sensitive to the air than to any other feature of environment. A person who without a tremor can eat raw oysters, hot soup, cold lobster salad, frozen ice cream, and hot coffee, will become uncomfortable in a minute if air at a temperature of 65° instead of 70° blows on her neck.

Our interest in this kind of statement, all unevidenced, thereby implying that it is all self-evident, not least in its casual gender stereotyping, is whether there is a meaningful connection to the emergent air-conditioners' determination to develop stable artificial indoor environments and then make the market for them. Appendix E addresses 'Proper Air in Houses' and Appendix F 'Directions for Ventilation and Humidification'. Humidity is presented as the key variable in comfort and health. Humidity control distinguishes air conditioning from mechanical ventilation. Huntington prescribes curtains in absorbent material with their hems dangling in trays of water. Many distinguished collaborators were credited in the expanded argument for the predominance of climate presented in his 1922 book *Climatic Changes*, not least Professor Chamberlin at the University of Chicago, various Harvard professors and a Professor Reid from Johns Hopkins.[10] Humphrey's 1920 work, *Physics of the Air*, is a key reference to support his theories on how and why climate changes.

Early 20th century science on changing climates

Humphreys was Professor of Meteorological Physics at the United States Weather Bureau in Washington, DC. His 1920 book *Physics of the Air* is assembled from articles published in the *Journal of the Franklin Institute*. There is absolutely no speculation in any of the papers on the possible effect of climate, weather or air on human development and behaviour. Huntington was being selective, he focused on the very last chapter of Part IV, 'Other Factors of Climatic Control' which speculated on 'How Sunspots may Change Earth Temperatures' and later in the book on the 'Influence of Carbon Dioxide on Temperatures'. Sunspots figure prominently in 'Climatic Changes'. Humphreys correlated sunspot and volcanic activity against surface temperatures from records of sunspot numbers maintained since 1749 available from the Smithsonian Institute. He plotted sunspot activity S against volcanic ash 'pyroheliometric P' against temperatures T, taken from Köppen in his Figure 193 and observed (p. 597):

> It is probable that the sun-spot effect is not directly proportional to the actual number of spots, but, however this may be, the direct combination of the curves P and S gives the resultant P + S, which, as a glance at the figures shows, actually parallels the curve of temperatures, T, with remarkable fidelity.

And, commenting on the magnitude of temperature change (p. 602):

> The actual temperature range from sun-spot maximum to sun-spot minimum varies, roughly, from 0.5°C to 1°C, or possibly more, while the effect of volcanic dust appears to be fully as on rare occasions even much greater. In some ways, and in respect to many things, a range of average temperatures of even 1°C is well-nigh negligible, and, therefore, however important the results may seem to the scientist, the ultra-utilitarian would be justified in asking, 'What of it?' Much of it, in a distinctly practical as well as in a purely scientific sense, as is true of every fact of nature. For instance, during the summer, or growing season, a change of 0.5°C produces a latitude shift of the isotherms by fully 80 miles.

It is not too difficult to imagine the confidence of the above quote catching Huntington's imagination. Here was the mechanism by which climates changed, thereby changing the development of the evolving strands of humankind. Humphreys explained that increased sunspot activity results in increased ultra-violet radiation becoming entrapped in the earth's atmosphere, converting some oxygen to ozone, although there was no observational evidence to support this conclusion in 1920. He also speculated that CO_2 comprises a mere 0.03% of the atmosphere and if concentrated into an outer atmospheric layer it would be some 40cm thick (ibid. p. 607), much as Surgeon-Major Billings, whom we will meet in Part 1 Chapter 6, relates in the late 1880s. This would be sufficient, Humphreys says, to produce full absorption, any increase could not therefore significantly affect global surface temperatures. There follows (ibid. p. 607) what one speculates might have become a disproportionately impactful passage, notwithstanding its odd punctuation, in validating the beliefs of later climate change sceptics:

> Assuming that the present amount of carbon dioxide in the isothermal region absorbs 1 per cent, of the solar radiation and 10 per cent, of the outgoing earth radiation (values that seem to be, roughly, of the correct order), and using equation A, page 571, it will be seen, if the experiments here referred to and the assumptions are substantially correct, that doubling or even multiplying by several-fold the present amount of carbon dioxide, which would leave the absorption of solar radiation practically un-changed, and increase the absorption of terrestrial radiation at most to only 14 per cent, could increase the intensity of the radiation received at the surface of the earth about half of 1 per cent., and, therefore, the average temperature by no more than about 1.3°C. Similarly, reducing the carbon dioxide by one-half could decrease the temperature by no more than approximately the same amount, 1.3°C.

Physics of the Air was directed at trainee pilots and apparently much used in inducting the coming Second World War generation of airmen and women. Former airmen in the US administration, including at least one president, remembering Humphreys, might well have been sceptical in later decades in the face of European lobbyists' concerns about the effects on global mean temperature of increases in anthropogenic CO_2 emissions. Humphreys (ibid. p. 569) adds unambiguously:

> Suffice it to anticipate here the general conclusion that while variations in the amounts of carbon dioxide in the atmosphere may have somewhat modified our climates, it probably never was the controlling or even an important factor in the production of any one of the great climatic changes of the past, nor can be of any great climatic change the future may possibly bring.

On page 14 he cites Willis Carrier's 1912 paper on humidity, 'Saturation Benefit' in the *Transactions of American Society of Mechanical Engineers* (vol. 33, p. 1005), an indication of Carrier's parallel scholarly status.

Confusion over climatological causes of health

Clarence A. Mills, Professor of Experimental Medicine at the University of Cincinnati and a practising hospital doctor, cast climate determinism as 'medical climatology'. He proposed a set of climatological causations for the various states of health he had been collecting since the First World War, published as *Living with the Weather* and *Medical Climatology* but brought together with greatest impact in his much cited but, one suspects, now rarely read, later 1942 publication *Climate Makes the Man* (Mills, 1944; originally published 1942). He refers warmly to Huntington and sympathises with the latter's attempts to persuade the sceptical about the predominance of climate in human affairs. It is almost entirely anecdotal and self-referential, and was very popular. Mills used paper-based 'social media', publishing news-bite synopses of the book's arguments in newspapers. *Climate Makes the Man,* in the June 1943 edition of *The American Mercury* starts:

Ellsworth Huntington first put forward the well-supported thesis that climate does indeed influence and direct the course of human progress. The writer's contribution to the subject has been to explore how climate accomplishes these effects – to detail the mechanism of its action on body and mind.[11]

In his book Mills explains that the reader must think of the body as a machine, giving off unused energy as waste heat, requiring an effective cooling system (p. 145):

Tropical lowlands everywhere are blanketed by a continuous moist heat which makes an active life impossible. Natives of such regions are slugish or lazy not as much from choice as from necessity; if allowed greater ease of heat loss, they soon become more active.

They need to consume more glucose, whereas: 'The coolness of temperate lands offers man his chance for a most active existence … vitality runs high in all its aspects.'

But there are risks in this over-vitality, the prospects of heart disease and nervous collapse. Falling barometric pressure triggers domestic arguments and children conceived in winter and spring will be more successful, 'as Huntington and Peterson have shown', with potentially extraordinary consequences amongst the readership. He concludes by warning of mental sluggishness as heat builds up.

Mills does not express unreserved enthusiasm for air-conditioning and here our argument stutters. However, in Chapter 15, 'Made-to-Order Indoor Climates', he commends air-conditioning in tropical industrial environments, he cites a case in Manila, but continues:

Great strides have been made in the mechanics of air-conditioning, but the over-enthusiastic hopes people once had for it have faded considerably as experience has emphasized its handicaps. Chief disadvantage is the sharp contrast between air conditions indoors and out … everyone has experienced the disagreeable shock of going from a cooled building out into the choking air of a severely hot day. People have collapsed from the sudden change.

Furthermore such conditions encourage infection. He evokes the more recently identified phenomenon of 'adaptive comfort' through comparison of preferred theatre air-conditioning set points across the USA: 'Fortunately there exists a surprisingly superior method, reflective radiant conditioning.'

He had a test room set up with radiant heating and cooling panels, concluding for readers that this is a much healthier and effective form of 'conditioning' interiors. Many contemporary hospital engineers may well agree with that.

Mills argued that the Great Depression had been caused by a series of unusually warm and stormless seasons that sapped human energy so that government aid could not counteract 'the depressive effect of the weather'. At least one contemporary economist working in the field of climate change and its potential impacts now cites both Huntington and Mills without irony or qualification. In 2011, H.-M. Fussel, Guest Scientist at the Potsdam Institute for Climate Impact Research re-enters this ideologically charged territory by writing that:

... the importance of climate for human welfare has long been recognized [Ritter (1852), Huntington (1915), Mills (1942), Lamb (1982), Diamond (1999), Sachs (2003)], most research in the last decades has attempted to explain the wide divergence in wealth by cultural and political factors [Landes (1998), Acemoglu *et al.* (2001), Rodrik *et al.* (2004), Engerman and Sokoloff (2005)]. Increasing interest in the potential impacts of anthropogenic climate change, however, has led to a renaissance of research on the influence of climatic factors on economic prosperity.

Fussel maps climate against his definition of 'habitability' and then against predicted economic development (and potential economic consequences), as generated by a global economic model 'G-Econ+', and projects these mappings to 2080. The correlations are striking; Huntington and his many predecessors would not be at all surprised. As a member of the academic community he qualifies the outputs (Fussel, 2011):

... the current analysis does not consider demographic and technological developments that may aggravate or alleviate the changes in climatic favourability estimated here. For example, the availability of air conditioning has certainly influenced this relationship, and the potential availability of an effective and affordable malaria vaccine would further do so.

He continues, provocatively:

Another worthwhile topic for future research is the combination of estimated changes in habitability with projections of future demographic and/or economic development. Even a cursory analysis of Figure 8 suggests that many of the areas where climate change is expected to decrease habitability coincide with regions that are currently experiencing significant population growth.

Certainly two eminent geographers rather mourned the suppression of environmental determinism at the end of the Second World War. Sewell and Kates, reporting in a survey paper on late 1960s thinking on climate and human endeavour as members of the National Science Foundation Task Group on Human Dimensions of the Atmosphere, wrote revealingly: 'In retrospect the retreat from environmentalism appears as a retreat from grand theory in geography.'

They quote the geographer Rostlund, 'environmentalism was not disproved, only disapproved', and graph the declining numbers of papers on the subject from 1916 to 1967 with a brief uplift through 1939–45 (Sewell *et al.*, 1968). Nonetheless, as Huntington expanded his argument to insist that physical stature, productive variation, even cranial form derived ultimately from climate (Chapter 10 'The Origin of New Types among Men' of his 1919 *World-Power and Evolution*), the support of academic Geographers receded. Meyer catalogues this decline of support for Huntington's theory. He cites (Meyer, 2000, p. 172) Preston James of the University of Michigan writing in 1941 (Meining, 1971):

> ... the significance of the elements of the environment is determined by the nature of the people ... No climate ... should be described as inherently favorable or unfavorable except in terms of specific human cultures.

More recently, Fleming (1998) mercilessly ridicules Huntington, 'the unrestrained and undisciplined geographic determinist, eugenicist, and popular writer'. Fleming's researches do reveal that in 1903, Huntington joined the Harvard physiographer William Morris Davis's expedition to the Caspian and Aral seas to excavate the remains of an Aryan civilization which appeared to have thrived but was subsequently extinguished by the climate-change induced desiccation of the post-glacial sea around which their civilization had advanced. Perhaps this study provided Huntington's template for the inevitable effects of the climate-driven cycle of civilization, its 'mainspring'. Huntington was importing and supplementing the propositions of European climate determinists. In terms of our interest in the pre-disposing causes of the artificial environment, his most important contribution may have been the collation of various factory productivity studies purporting to demonstrate that an ambient air temperature of 20°C at 60% RH yielded the optimal productive indoor environment. Huntington's most immediate German co-combatant was Willy Hellpach, a social psychologist, admired by the National Socialists but whose reputation was rehabilitated to enable a productive post-war academic career.[12] Jarzombek (2000, p. 120) is less kind, noting that Hellpach's supervisor had been Wundt, author of *The Pathological in Modern Art* (1910), proposing that art had to become an antidote to nervousness in modern times but that only certain races were capable of shedding their anxieties and therefore useful in the 'modernization' of Germany.

After 1945 it appears that respectable academics simply avoided discussion of the possibility of difference based on climate, certainly in formulating public policy advice. Perhaps the potential availability of air-conditioning, mechanical cooling and heating had, in their minds, negated the issue. But, in effect, they had joined a long dissenting tradition of dissociating human behaviour in every instance from environmental influence. For example, David Hume's 1741 essay 'Of national characters' (Hume, 1994, p. 83), argued ferociously that a kind of 'social contagion' determined the evolving nature of societies, not external physical phenomena:

> If we run over the whole globe, or revolve all the annals of history, we shall discover every where signs of a sympathy or contagion of manners, none of the influence of air and climate.

He presents nine arguments:

1. Highly evolved government determines and spreads the national character (he cites China).
2. Contiguous landscapes of small independently governed cities or provinces yield just as much diversity in manners as transcontinental comparisons giving an example as relevant now as then: 'I believe no one attributes the difference in manners in WAPPING and St. JAMES, to a difference of air or climate'.
3. National characters change abruptly at political boundaries.

4. Distinct sets of people, i.e. Jesuits, Armenians and Jews preserve their distinct character wherever they may have settled.[13]
5. Co-location of nationals does not dissolve national character (he cites the 'bravery of the Turks' living alongside the 'cowardice of the modern Greeks', a surprising inconsistency in the unfolding argument).
6. Colonials preserve their home nation's characteristic manners.
7. National characteristics change over time (here he returns to condemn the modern Greek and other unfortunate national groups).
8. Contagion of manners across neighbouring friendly states.
9. There is manifestly a wide variation in manners across the same nation (p. 83):

> … and in this particular the ENGLISH are the most remarkable of any people….Nor is this to be ascribed to the mutability and uncertainty of their climate, or to any other physical causes; since all these causes take place in the neighbouring country of SCOTLAND, without having the same effect.

Notwithstanding Hume's pungent observations, the consensus view was that the influence of climate did matter, so that 170 years later, by the early 20th century, the consistently held view was firmly that climate could either expedite or seriously undermine 'civilizing' tendencies. There had been attempts at manufacturing completely artificial environments through the second half of the 19th century, a disastrous sealed, bellows-driven hospital at York, UK but not until the first decade of the 20th century were the necessary constituent technologies of electrically driven and controlled fans, ducts, dampers and condensers integrated into complete environmental systems, named 'air-conditioning' by its new industry, a term credited to Stuart Cramer, a textile engineer, in 1906 (Wampler, 1949, p. 13).

Billington and Roberts catalogue evidence of early attempts to air-condition public buildings. They refer to a mechanically cooled theatre in Cologne, Germany using a steam operated chiller running on well water, circulating pre-cooled ammonia brine through substantial coils, completed in 1903 and published in 1904 by J. Musmacher in the May issue of *Ice and Refrigeration*. This was published in Volume 26, indicating this was by no means untrodden territory. The technique sounds strangely reminiscent of contemporary work on thermal storage systems which we encounter in Chapter 5.[14]

Was there a conscious intent to neutralize climate, to make a level playing field, a flat cool temperate environment for thinking, working, governing, learning and dwelling worldwide? Was air-conditioning envisioned by its inventors to liberate the reach of civilization globally, to give populations across Köppen's more extreme climate zones good quality odourless air, enhanced personal cleanliness, free of body odours, in fact the suppression of the physiological response of perspiration? By accident or design, the new technology facilitated the contemporary cultural phenomenon of cleanliness. The new young clerical class, smartly attired, was no longer rained upon by dust and smuts, clearly a significant cause of unhappiness, as portrayed in real estate advertisements for the first all glass office buildings in Chicago, extolling the prodigious availability of natural light, so much light that the spitting gas and oil lamps could be kept unlit.

Fabricating climate: 'the joyous spitefulness' of cold air

Willis Carrier's contribution to the invention and monetizing of air-conditioning technology is reasonably thoroughly recorded, although Carrier's archives have become difficult to access in recent years. Margaret Ingels' (1952) biography is the 'General Staff History' of the Carrier Corporation's origins and growth, sent out by Carrier to important clients. Ingels, the officially sanctioned biographer, was the first female mechanical engineering graduate of the University of Kentucky in 1916. Ironically she had hoped to be an architect, but in 1917 joined Carrier, became a research engineer, and developed into an important spokesperson for the company and for women in engineering. Gail Cooper provides a very much more detached account (Cooper 1998).

In 1902 Carrier invented a humidifying machine for the Sackett–Wilhelms Lithographic and Publishing Company after a series of hot, dry summers had impaired production. In 1903, it is claimed, he invented the principle of using a fine mist as the condensing medium in a mechanical ventilation scheme attempting to humidify an industrial process, the process which interested the meteorologist Humphreys.

The financial drama of the company's creation in the very difficult commercial environment of 1915, when it was widely thought Germany would win the European war and thereby destroy the European market for American products, seems to have conditioned company policy thereafter. It was highly aggressive, with important consequences for the future built environment worldwide. In his forward to Ingels' book, the company's first and surprisingly young bank manager Cloud Wampler (1875–1973) of Harris Trust and Savings Bank, co-incidentally managing agent for Carrier's first premises, reported on the nature of Willis Carrier's abrupt departure from the Buffalo Forge Company. Carrier's air-conditioning team, but not Carrier, was made redundant by the owners, the Wendt brothers. He departed with his boss Irvine Lyle, his ingenious idea (which was securely patented in 1904), Lyle's brother and L. Logan Lewis (both trained as engineers in Britain) and three other co-workers. They raised between them the very substantial capital of US $32 600, equivalent to £6936 at the exchange rate of $4.7 in 1915 (equivalent in worth today to £1 910 000, using average earnings in 2013). This was a colossal personal risk.[15]

Although Wampler writes, 'what they accomplished was one of the finest examples of the workings of the American free-enterprise system', he added 'Fear was everywhere' from 1915. The business faltered in 1928, running at a loss, merging in 1930 with Brunswick-Kroeschell and New York Heating and Ventilating Corporation facing similar difficulties to gear up their combined borrowing capacity, losing money for five further years.[16] Wampler remained a close financial adviser to Carrier until becoming president of the corporation in 1942. This long period of financial stress seems to have induced in the corporation an exceptional ruthlessness in marketing air-conditioning.

The heating, ventilating, air-conditioning business was highly aggressive, briefing against rivals' equipment, launching gratuitous patent infringement actions and hostile takeover bids. The Sturtevant 'heritage' website reports on tensions with Carrier:

> The Sturtevant people took a dim view of the flamboyant founder, Willis Carrier, whom they referred to (at least internally) as the "father of humidity",

a derisive slam at what they thought was excessive self-promotion and boastful claims of being the father of air conditioning.[17]

Carrier invented 'centrifugal refrigeration' in 1922 and its commercial applications proliferated, marketed as 'Manufactured Weather', oriented to make the Carrier Corporation 'Weathermakers to the World' by 1939. Architectural critic Reyner Banham (1969) observed how appropriate the reference to weather was to their work, which was based on close observation of the mechanics of atmospheric humidity.

Did Carrier and his peers understand the 'philosophical' context within which their work fell, environmental determinism, anthropo-geography and their increasingly sinister connotations? Did they have the intellectual equipment?

Although Lyle graduated from what was to become the State University of Kentucky with a practical degree, Carrier was Ivy League educated, unusual one presumes amongst his peers in the industry. Lyle reports that Carrier was charming, wholly persuasive, supremely confident, a Cornell University graduate (1901) with an MEE, granted the title Mechanical Engineer in 1901. Portrayed as a wholly single-minded inventor with little interest in, or understanding of, the wider social and cultural implications of his work, this is belied by the mandatory Primary Entrance Examination requirements for all applicants to Cornell, on top of which were required subject-based examinations. How broadly educated was Carrier? Was he equipped to reflect on the consequences of his work?

As described in the Cornell Register for 1896/7 and 1898/9 the requirements were fearsome: for the Primary Examinations students needed preparation in Shakespeare's *As You Like It*, Eliot's *Silas Marner*, Macauley, Milton, Scott, De Quincey's *Flight of a Tartar Tribe*, Sir Roger de Coverley and others; histories of Greece and Rome in Greek and Latin; in Hygiene, Hutchinson, Huxley, Jenkins; elementary French or German, 400 pages from at least three authors; a lot of mathematics from the binomial theorem (a probability distribution) to quadratic equations, so that students were 'required to perform all operations even when somewhat complex with rapidity, accuracy and to solve practical problems readily and completely'.[18]

The Register continues:

For the College of Civil Engineering the following subjects are required for admission: English, Geography, Physiology and Hygiene, History [the student must offer two of the four following divisions in History: (a) American, (b) English, (c) Grecian, (d) Roman], Plane Geometry, Elementary Algebra, Solid Geometry, Advanced Algebra and Plane and Spherical Trigonometry and either Elementary French or Elementary German.

Ingels records that Carrier took the option to spend the fourth year of the mechanical engineering course specializing in electrical engineering at Sibley College, a newer foundation within the Cornell campus with a vocational outlook.

Our as yet incomplete argument drives us to have a particular interest in Carrier's knowledge of contemporary Geography and climate studies. Significantly, the New York State Weather Bureau was based in the College of Civil Engineering. Ebenezer Tousey Turner was the Lecturer in Meteorology in the College. Carrier's transcendental experience in the drizzle on the railway station platform may have derived from his exposure to Cornell climatology teaching whilst a student.

Banham (1969, p. 180) reported that the provision of air-conditioning did not materially affect office rental values after the allegedly first installation of 1928 in the Milam Building in San Antonio, Texas. The loss of lettable plan area to accommodate ductwork was insufficiently compensated for in higher rent.[19] Carrier persevered and by 1921 Carrier had opened an office in London. The company had certainly established a base in India by 1932 and CIBSE's history group reports that installation drawings survive for the Palace of Jodhpur.

The Carrier Corporation reports:

> In 1943, Carrier's advertising and public relations manager, Walter A. Bowe, spoke before the American Management Association. He noted that large government contracts had created a seller's market, but this would swiftly reverse when World War II ended. Bowe suggested that there was still much to be 'sold' to consumers in wartime, including a company's name, reputation and contribution to the war effort. When the 'buyer's market' returned, he said, 'those who have established themselves in the minds of potential customers will reap the benefits.' In fact, Bowe added, 'the demand for air conditioning, both for comfort and as a production tool, will be great. And we have resolved not to lose touch with those who will be our customers when "V" day comes.'[20]

Banham explains how Carrier's miniaturized domestic air conditioner made it possible to proliferate the post-war suburban development of Levitt's thermally insubstantial single storey house types across virtually all North American climate regions. The rapid and widespread adoption of the post-war suburban house meant a significant proportion of the population was entrapped in houses designed and constructed with little resilience and with a new dependence on mechanical cooling.

Banham (1969) concludes very succinctly:

> ... while European modern architects had been trying to devise a style that would 'civilize technology', US engineers had devised a technology that would make the modern style of architecture habitable by civilized human beings.

There is a strange resonance here with the environmental determinists' arguments discussed above. Banham (1969, p. 162) continues:

> In the process they had come within an ace of ... making architecture culturally obsolete, at least in the senses in which the word 'architecture' had been traditionally understood, the sense in which Le Corbusier had written *Vers une Architecture* ... Not that the inventors of air-conditioning knew this, or were interested in the outcome.

But is this really true, was it just about the money? One can only speculate. Carrier is obstinately silent about the broader socio-cultural implications of the epidemic of artificial 'weather-making' he unleashed. Ingels proudly records

Carrier's unsentimental research strategy, more reflection may have evoked effete tropical traits for Carrier and his circle: 'I fish only for edible fish, and hunt only for edible game – even in the laboratory.'

By 'edible fish', Carrier meant technologies that would gel expeditiously with consumers' aspirations and sell. There was already a substantial literature speculating on consumer aspirations, first amongst which was Thorstein Veblen's (1889, p. 21) grim proposition that base pecuniary emulation drove the will to consume: 'this emulation shapes the methods and selects the objects of expenditure for personal comfort and decent livelihood.'

Veblen did allow for other incentives for acquisition apart from pecuniary emulation: 'The desire for added comfort and security from want is present as a motive at every stage of the process of accumulation in a modern industrial community.'

The notion of what is acceptable as sufficient is highly influenced by pecuniary emulation. Emulation was probably the backbone of the domestic air-conditioning market, supported by ruthless advertising campaigns depicting successful families gathered around their new air-conditioner whilst served by 'failed' individuals, gardeners for example, glanced through the window beyond, labouring for them in the summer heat (advertising the Mueller Climatrol Type 910, 1955). Mexicans were depicted in sombreros sleeping through the middle of the day unproductively (Carrier advertisement in the *Saturday Evening Post*, 1949) whilst businessmen in blue ties prosper in their temperate microclimates next door to heavily perspiring contemporaries in yellow ties struggling in ambient temperatures (Figure 2.4).

Figure 2.4
Post-war advertisement for air-conditioning with a political flavour. Courtesy of the Advertising Archive.

Conditioned air becomes 'the environment'

By the mid 20th century Potter (1954), co-Professor of American History at Stanford University, identified 'abundance' as the all-encompassing context for what had become American consumer society in his Charles R. Walgren Foundation Lectures. He devotes Chapter VIII to (p. xx):

> ... the impact of abundance in shaping one of the major institutions which guide the individual in the orientation of his values and thus ultimately play a basic part in the formation of the personality: this is national advertising, with all that its existence implies concerning the transition from a society oriented to production to one oriented to consumption.

Potter doesn't mention personal investment in air-conditioning, perhaps 1954 was on the cusp of the explosion in sales?

The 'reception history' of air-conditioning machines is clearly very much more complicated. Arsenault (1984) records the reception offered to this Northern technology in the Southern states of the USA.[21] He suggests that Carrier's 1922 invention of the lighter, smaller, more efficient centrifugal compressor, the new refrigerant Carrene and subsequently Freon in 1931 replacing dangerous ammonia gas, a threat to the survival of the Brunswick-Kroeschell company to be absorbed into Carrier, enabled the manufacture of domestic-scaled air-conditioners and mass experience of 'comfort cooling'.[22] Arsenault records that the first air-conditioned environments outside industrial process plants were made in three hugely popular cinemas in Dallas in 1924.[23] Arsenault (1984, p. 598) suggests that the accepted wisdom of environmental determinism set lower expectations of human endeavour and behaviour in the Southern states:

This was the age of Walter Prescott Webb, Ellsworth Huntington, and Ulrich Bonnell Phillips, when the link between climate and culture was often thought to be a simple relationship of cause and effect. The southern climate, in particular, was credited with producing everything from plantation slavery to the southern drawl.

Arsenault evokes the presence of the unmodified Southern climate, and the determinist belief that it was the fundamental cause of Southern aberrance, to be suppressed if at all possible (p. 598):

> 'Let us begin by discussing the weather, for that has been the chief agency in making the South distinctive,' was the opening line of Phillips's 1929 classic *Life and Labor in the Old South Boston*. According to Phillips, the hot, humid southern climate 'fostered the cultivation of the staple crops, which promoted the plantation system ... which not only gave rise to chattel slavery but created a lasting race problem'.

He continues:

> Some southerners will praise air conditioning and wonder out loud how they ever lived without it. Others will argue that the South is going to hell ... As one Florida woman recently remarked, 'I hate air conditioning; it's a damnfool

invention of the Yankees. If they don't like it hot, they can move back up North where they belong.'

He concludes, without obvious irony:

Air conditioning has changed the southern way of life, influencing everything from architecture to sleeping habits. Most important, it has contributed to the erosion of several regional traditions: cultural isolation, agrarianism, poverty, romanticism, historical consciousness, an orientation towards non-technological folk culture, a preoccupation with kinship, neighborliness, a strong sense of place, and a relatively slow pace of life. The net result has been a dramatic decline in regional distinctiveness.[24]

As Arbuthnot predicted in 1733, 'Too great and sudden Refrigeration by Ventilation, may be dangerous'.

Further north in the US, sceptics baulked at prominent examples of sealed artificial environments. James Fergusson (2006) notes the prominent objectors to Frank Lloyd Wright's Johnson Wax Building: H.L. Mencken, Sinclair Lewis and, somewhat surprisingly, J.K. Galbraith. Less surprisingly, a visit by Henry Miller went sour:

Henry Miller visited the new Johnson building during a tour of America in 1940; an account of his journey, entitled 'The Air-Conditioned Nightmare', was published in 1945. 'This place is flawless—deathlike,' he wrote. 'Man has no chance to create once inside this mausoleum. Down with Frank Lloyd Wright!'[25]

But to some extent the avante-garde's objection to artificial environments was subsumed in a condescending unhappiness at the presence of new technologies in everyday existence and disdain for 'the perceived complacency of a surging middle class full of Babbits who smugly added foolish comforts to their way of life', as Marsha Ackermann records (2002). The dehumanizing influence of technology threatened 'to banish from American life all that was natural, passionate and spontaneous' (ibid.).

By 2006 Fergusson could write that 83% of US households contained one or more air-conditioning units, that the Carrier Corporation employed 45 000 people in 172 countries, Chinese households owned more than 100 million residential air-conditioners, triple the number five years before. The Chinese *People's Daily* reported air-conditioning had accounted for 15% of annual power consumption annually and up to 40% in summer. In the US one-third of all electricity produced was driving air-conditioning, some 8% of the world's energy (Ferguson, 2006).

Other commentators have expressed surprise at the absence of discussion about the impact of air-conditioning. Nash (1999, p. 58) cites the *Time* magazine correspondent Frank Trippert: 'scholars have been aware of the social implications of the automobile and television ... but [avoided] charting and diagnosing all the changes brought about by air-conditioning.'

Nash speculates that this neglect is the consequence of a consensus reaction against Huntington and Walter Prescott Webb's environmental determinism, tacit approval of the diaspora of cool temperateness.[26] Nash reveals that Federal policies endorsed the use of air-conditioning. From 1950 the Internal Revenue Service offered special tax deductions, allowances, and tax credits for homeowners who installed air-conditioning. Physicians could prescribe air-conditioning as a medical necessity which would entitle the patient to additional benefits from the Internal Revenue Service. More extraordinarily, pressure from the Federal Housing Administration (FHA) on mortgage lenders persuaded them to penalize borrowers who did not install air-conditioning in the hotter states. Nash adds that in 1959: 'the National Weather Service began issuing a "discomfort index", a composite of heat and humidity that represented subtle government support for climate control and the air-conditioning industry.'

He concludes, not wholly unconvincingly, 'The cold war and air-conditioning shaped the broad context of western growth', referring here specifically to economic growth in the western states of the USA. The few sceptical commentaries on the proliferation of air-conditioning have elicited a strong defence of the technology (Diamond, 2013). In 2013, the economist Arthur Diamond identified the non-believers as Cooper (1998), Ackermann (2002) and Cox (2010). He suggests that air-conditioning is fundamental to progressive socio-economic improvement, empowering the lower socio-economic classes so that, he argues, it is fundamentally reactionary and undemocratic to restrict its availability:

> … a major benefit of air conditioning is that it increases people's sense of choice and control over their lives … it is a likely candidate to be what Rawls (1971) called a 'basic good,' one that is needed to achieve almost any life plan, e.g., health, longevity, food, clothing, and shelter. Air conditioning is similarly a basic good for many popular and defensible life plans … these episodes suggest that air conditioners might be another example of Schumpeter's claim that the glory of capitalism is that it brings within the reach of ordinary people, goods and services that were previously available only to the rich.

Perhaps, in addition to what Diamond regards as a basic human entitlement, air-conditioning can soothe the psychological and physical stress arising from environments created by the capitalist system in full flight by cleansing the air itself. Reyner Banham (1969, p. 31) made an intriguing observation:

> … the incidence of compulsive hand-washing in the early literature of psychoanalysis suggests that atmospheric pollutants may have corroded the minds, as well as the bodies, of those who had to endure these conditions.

Banham's aside suggests another potential driver for the interest in sealed, artificially conditioned, all glass buildings: a disinclination to be in contact with damp brick masonry which might harbour real or imagined pathogens. A disinclination which might, in some instances, be symptomatic of Obsessive Compulsive Disorder (OCD). Paul Scheerbart, author of innumerable pieces on an imaginary world reconstructed in glass, 'Glasarchitektur', and his circle including Bruno Taut, fantasized about the Backsteinbakillus, a fictional microbe that thrived in damp brickwork, the name evoking the mediaeval brick architecture of the Hanseatic ports for greater ironic effect, a refined Architecture of punctured masonry envelopes which will be encountered in Chapter 4.

G.E. Berrios of the Department of Psychiatry at the University of Cambridge provides an account of evolving 19th century interpretations of Obsessive Compulsive Disorder (OCD). Initially, OCD was considered a 'partial insanity' or monomania up to the 1850s (he identifies the 1853–54 debate of the Societe Medic-Psychologique in Paris as the significant moment of sea-change in the understanding of OCD). It then was interpreted as a variant of the then newly-formed notion of psychosis caused by a dysfunctional nervous system. From the 1890s, OCD was identified as a neurosis proper, deriving from 'volitional, intellectual or emotional impairment', *delire emotif* (Berrios, 1989, pp. 283–295).

He explains, surprisingly to the layman, that contemporary clinical description still benefits from the 27 cases reported by Legrand du Saulle (1830–86). OCD seemed to affect the higher social classes and to be more frequent in fastidious and rigid personalities, particularly, but not exclusively, those of women. Tension was temporarily released after the compulsive behaviour was enacted. Early onset was frequently observed, many patients were under 30. In the advanced stages patients become housebound. Onset could be correlated with the descent of a disturbing zeitgeist upon the nation, a time of heightened social anxiety. Interest revived in the subject in the late 1980s but, very intriguingly, Berrios states, 'its nosological status, natural history and classification remain unclear'. Penzel explains how 'sufferers', 'patients' or perhaps 'clients' reconfigure their familiar spaces to cope with their condition. This description by Penzel has an architectural resonance:

> In an attempt to keep clean and minimize compulsions, some sufferers will create two different worlds for themselves; one clean, and one dirty. When contaminated, they can move freely about their dirty world and touch and do anything, since everything in it is already contaminated. Nothing in it has to be cleaned or avoided … this dirty world usually takes in most of the outside world, and can also include parts of their home or work areas … they may also be able to live freely in their clean world, as long as they themselves are clean when they enter it, and also stay that way. The clean world is usually a much more restricted area than the dirty one, and is often limited to special places at home or at work … the two worlds may exist side-by-side like parallel universes that are never allowed to meet.[27]

The glass tower: purity and poetry

This description of parallel environments is strangely evocative of the threshold of a glass tower on a plaza set into the chaos of a city. Close study of these moments of entry and transition, the thresholds, particularly in the earliest glass buildings, may reveal them to be yet more charged even than Christopher Alexander suggested, as balm to a 'modern' condition (Alexander *et al.*, 1977). Rapoport (1990, Chapter 3) describes a male patient with OCD for whom entry into the home required a highly complex procedure achieving entry in 74 steps. It is not a ritual. Rapoport describes it as a 'tic' or hiccup of the mind but cannot explain the circumstances of sudden onset. Could particular reconfigurations of the environment trigger or ease the condition? One might even ask if there is evidence of sufferers' desire to change their physical environment to ease their inner compulsive 'impulsion', to enlist their architects as agents of change, consciously or otherwise. Might such change exploit new clean environmental technologies (i.e. electricity in 1890s Europe, air-conditioning in 1940s America) to make interior environments accommodate their clients' secret procedures for releasing tension?[28]

The conundrum of glass architecture remains opaque to the author and his colleagues but another potential insight might arise from Mary Douglas's (1966) explanation of the huge significance of water in the attempt to achieve 'purity'. Douglas quotes from Mircea Eliade's 1958 *Patterns in Comparative Religion* (p. 200, translated from *Traite d'Histoire des Religions*, 1949): 'In water everything is "dissolved", every "form" is broken up, everything that has happened ceases to exist ... water possesses this power of purifying, of regenerating, of giving new birth.'

This evokes a reading, bizarre as it might sound, of the all-glass building perceived as a vessel of water, a translucent sac, completely sealed from the contamination and danger of the city beyond. There may be a more literal provenance for the universal practice of building icy crystalline towers for the commanders of industry and commerce, constantly replenishing the air they breathe: Nietzsche's presentation of the freezing air of the icy mountaintop as the ideal environment for the exclusively male figure in complete control of his destiny in *Thus Spake Zarathustra* (translated by Common, 1977):

> The grave-diggers dig for themselves diseases. Beneath old rubbish rest bad vapours. One should not stir up the marsh. One should live on mountains. With blessed nostrils do I again breathe mountain freedom. Freed at last is my nose from the smell of all human hubbub!

And, later in the text:

> The hour in which I frost and freeze, which asketh and asketh and asketh: 'Who hath sufficient courage for it? Who is to be master of the world? Who is going to say: *Thus* shall ye flow, ye great and small streams!'

Gaston Bachelard (1971, p. 48) discusses the extreme importance of pure air in this 'world conception':

> For a true Nietzschean, the sense of smell must give the happy *certainty* of an odorless air, the confirmation of immense happiness, the blessed consciousness of feeling nothing. It guarantees the total absence of odours.

Bachelard explains that the 'scent instinct' is given to the 'Ubermensch' so that 'he can flee at the slightest sign of impurity'. Bachelard continues:

> *Pure air is a consciousness of the free moment*, of a moment which opens a future ... *coolness* is the truly *tonic* quality in air, the quality that gives the joy of breathing, that *endows immobile air with dynamism* ... it corresponds with to one of the most important principles of Nietzsche's cosmology: cold, the cold of heights, of glaciers, of absolute winds ... thanks to cold, the air acquires *aggressive virtues*, it takes on that *'joyous spitefulness'* that

awakens the will to power, a will to react coldly, in the supreme freedom of coldness, with a cold will.

The Nietzschean 'world conception' appears to have so informed the ethos of early business education that Santiago Iñiguez de Ozoño, the Dean of Spain's IE Business School could respond, in a debate about the transformation of the business education ethos globally one year after the 2008 banking crisis:

> It is a different sort of leadership than the one which has grown in the past decade. It is not charismatic leadership, but teamwork. We will also see in the future many institutions getting rid of this spirit of elitism or arrogance which has contributed to create this atmosphere of overconfidence. They [believed that they] were actually beyond any controls or rules – that Nietzschean moral of the super-masters.[29]

Did the graduating classes of the earlier ethos literally recreate Nietzsche's optimum environment in their new workplaces in which to enact his 'world conception'?

Would the Glasarchitektur circle of Bruno Taut (born 1880), Paul Scheerbart (born 1863), Mies van der Rohe (born 1886) and their acolytes have encountered Nietzsche at school? Whether or not his work was included in their school curricula, Pehnt (1973) emphasizes that, '*Thus spake Zarathustra* was a canonical text for the Expressionists', Taut quoting passages on his sketches, a general sympathy for the beneficial properties of high mountain environments. Taut publishing *Alpine Architecture* in 1919.

In Basle in January and March 1872 the philosopher had delivered scathing public lectures on the newly 'mechanized' German educational system which had become centrally controlled by the Prussian State later in 1872. He was appalled by the increasing number of more vocational 'Realschulen' at the expense of the elite academic Gymnasien in a system driven by economic dogma towards 'bread-winning' and the acquisition of vocational skills (Reiter and Welmon, 2015; Young, 2010).[30]

Scheerbart's Polish father was probably too poor to place his son in a Gymnasien; Taut attended the Baugewerkschule but eventually entered the Technical University at Charlottenburg; Mies van der Rohe was sent to the Cathedral school in Aachen, winning a scholarship to the Gewerbeschule (Schulze, 1985, Chapter 1). The intellectual member of the group had had the least academically rigorous education. We have seen that Carrier was compelled to read extraordinarily broadly compared to more recent higher education practice with consequences we can only speculate upon. Much reliance has been placed in the Anglophone world on a literal reading of Scheerbart's envisioning of an all-glass built environment, interpreted as the sanctioning of history for the abandonment of heavier weight, referential Architecture (1971). His confederate Bruno Taut was pre-occupied with the actual making of buildings with much more glassy envelopes than customary out of a useful connection with the glassmaking firm Puhl and Wagner, but perhaps the richness of Scheerbart's Berliner humour has defied complete translation. This is an intense, even cruel, humour experienced through the undaunted unravelling of every scenario within a complicated joke about the absurdity of all humankind, in which the listener may also be implicated to his/her disadvantage, regardless of their manifestly growing discomfort and receding friendship.

Scheerbart's 1914 novel *The Gray Cloth and Ten Percent White: A Ladies Novel*, recently translated by John Stuart in 2001, is certainly not a straightforward propagandizing tool for Glass architecture. Newly married to Clara, the central character, a priggish Swiss architect and a former archaeologist, Edgar Krug, travels the world in an airship attempting to win commissions for vast structures entirely of coloured glass. His condition for marriage is that Clara only ever wears a particular shade of grey to not despoil the glass interiors. Clara's friends are openly hostile. Does Krug enjoy an obsessive condition and does the dress fulfil a tension-releasing role within each new clean, shiny, smooth scheme? Glass is universally applied regardless of context, not least climate. Clara is very sceptical, hovering above the Pole in an airship she writes to her friend Miss Amanda: 'God only knows why Edgar still wants to have glass architecture here. I do not know. Indeed!'

Her husband's fame is such that a film is made of their lives, she watches as an actress delivers the overtly ludicrous line:

> I am prepared to wear gray clothing for my entire life. I am doing this because I love glass architecture so much that I would never wish to create competition with a colorful outfit.

One large project is for the Mediterranean Motor-ship Society:

> … the ships as they arrived at the port, were to enter a huge realm of glass. And large glass walls were therefore fastened to all sides of the harbor. These walls had very little ornament and revealed large panes of single-colored glass.

In fact they were double glass walls with an electrically lit cavity, an arrangement Henri Sauvage actually achieved at the Decre department store in Nantes in 1931. Krug hates brick buildings:

> 'It must, indeed, be really excruciating,' said one woman, 'to travel over a large brick city.' 'That,' replied Mr. Krug, 'I never do. My air chauffeur zealously studies the maps in order to avoid all brick sites along the way … I also do not like to hear about the residents of brick houses … it pains me that architects today still build with bricks. No, these architects can no longer gain any expect.'

Dietrich Neumann (1992) reproduces a sketch of Mies van der Rohe admiring his model of a glass tower for the Friedrichstrasse competition sketch by a visiting student, Sergius Ruegenberg, later to become an assistant in Mies' office but ultimately an enthusiast for a quite different Expressionist Architecture as one of Hans Scharoun's more important collaborators (Figure 2.5). The spirit in which the sketch was undertaken is unclear but it is difficult not to notice that Ruegenberg has drawn the sun beating down on the transparent model. In fact he has taken an orange crayon and coloured the sun as a burning disc, the

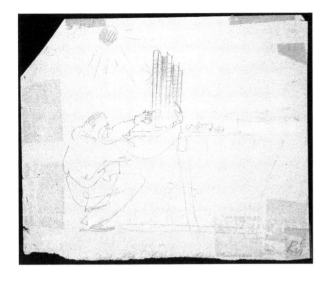

only colour on the sketch. It is clear that Mies van der Rohe had no viable plan
for making an inhabitable interior within the tower.

Neumann (1992) reports an account given in a 1933 anthology of private office
buildings by Seeger that the all-glass façade Mies proposed, coupled with double
height floors, was intended to deliver the Prussian building regulations' natural
lighting requirement at the back of what was an illegally deep plan.[31] In *ABC
Beitrage zum Bauen* (1925), a sketch by Mart Stam of the subsequent Curvilinear
scheme shows a challenging but very elegant structure of circular columns
without tapered heads supporting thin flat slabs with edge upstands. There is no
equivalent environmental strategy. In Mies' own account he dwells on the play
of light and shade on the sinuous reflective façade of his model.[32]

There is nothing remotely ironic in Taut's vision for the StadtKrone, his
proposal for a symmetrized city centre complex which will stabilize it against
the chaotic sprawl fuelled by the modern industrial economy, although this is
written from the catastrophic circumstance of 1919 Berlin, four years after his
close friend Scheerbart's premature death. Altenmueller and Mindrup (2009)
provide a useful translation of Taut's original intention:[33]

> The light of the sun penetrates this crystal house, which reigns above the
> entire city like a sparkling diamond. [This house] sparkles in the sun as a sign
> of the highest serenity and peace of mind. In its space, the lonely wanderer
> discovers the pure bliss of the building art. While climbing up the stairs to the
> upper platform, he looks to the city at his feet and beyond to the sun rising and
> setting, towards which this city and its heart are so strongly directed.

Le Corbusier's *Une Ville Contemparaine* of 1921–22 was realized in what
Banham referred to as 'German-style glass towers'. His fascination with glass
persisted through the Second World War. Drawing 19 in *Concerning Town
Planning* of 1946 is annotated, 'Seen through the glass wall of the dwelling …
Nature is entered in the lease' (the drawing originally appearing in *La Maison des
Hommes*, Plon, 1942). Le Corbusier's proposals for sustaining a sealed artificial

environment in the Continental climate of Moscow and the temperate environment of Paris have been closely scrutinized by many architectural and cultural scholars: Frederick Starr, Anatole Kopp, Catherine Cooke, Tim Benton, Kenneth Frampton, Stanislaus Von Moos and more recently by William Curtis, Jean-Luc Cohen and, of course, others.[34] But for our particular project, to start to understand 20th century designers' drift to all glass hermetically sealed artificial environments, there are some particular points to note in these histories.

Although Le Corbusier had access to very competent structural design in concrete through his cousin, competent enough to innovate and lead at an international level, the team's understanding of environmental physics and engineering – the ability to understand and influence thermal, ventilation, acoustic and lighting conditions – was very much less sophisticated. This is most clearly evidenced in the illustration of a fan in *Vers une Architecture* which was, by 1923, completely archaic. Large scale mechanical ventilation installations in the United States were being driven by centripetal Sirocco type fans, the Detroit-based American Blower Company (ABC) having acquired the US licence from its Belfast inventor already in 1909.[35] The diagram of 'machines for air extract' in *Precisions* (1930) suggests that an immense and very old-fashioned fan on the roof of a Unité block is capable of pushing air back into a black box, the 'usine' which is removing 'carbonic acid' whilst a 'usine hermique chauffeuse frigerantes' circulates air through a 'doppelfassade'. This sketch reprises the failed attempt to incorporate the system into the Central Union of Consumer Co-operatives building, the Tsentrosoyuz 1929–36 in Moscow. In that scheme his rather older ventilation collaborator, the acoustician Gustave Lyon (1857–1936) proposed his patented positively pressurized raised floor plenum to distribute fresh safe air for occupants through a myriad perforations in the floor plane, an idea illustrated by Dr Reid in the early 1840s and actually built in the Isolation Ward at Johns Hopkins Hospital in Baltimore from 1875, which is considered in Chapter 6.

The 'doppelfassade', a 'mur neutralisant de verre ou de pierre', added by Le Corbusier borrows from Constructivist buildings in Moscow such as Ilia Golosov's 1926 Zuev Workers' Club and, as Benton shows, from the incorporation of large windows in Swiss and German buildings in the mountains.[36] In Le Corbusier's application of the idea, the cavity void is contiguous through both glazed and stone-faced façade elements. The proposed construction details for the second scheme of 1928–30 show an implausible assembly of simple steel angles and angle stops holding the glass panes in place (Allen Brookes, 1982). It would have leaked air prodigiously across the sometimes spectacular indoor–outdoor temperature gradients. Le Corbusier (1930, p. 64, Figures. 44–46) reproduces a different order of window section, an American break-pressed frame. He wrote expansively to justify this, employing a diagram of the dissected chest cavity:

What is the basis of life? Breathing … I propose only one house for all countries, the house of exact breathing … I set up the factory for the production of exact air … I produce air at 18°C humidified according to seasonal needs … I blow this air through carefully laid conduits … air that has been breathed in and out returns to the factory … there it goes over a bath of potassium where it loses its carbon … through an ozonifier … to compressors that cool it, if it has been heated too much in the lungs of the occupants … how does your air coming at 18°C keep its temperature? Answer: it is the neutralising walls in glass.

This was an essentially 19th century miasmatist's view of what was needed to sustain life.

The Soviet commissioning clients of Tsentrosoyuz led by Vyacheslav Molotov were wholly sceptical. Molotov was famously humourless and completely ruthless. The British Ambassador Sir Archibald Clark Kerr later described 'the governessy presence of that boot-faced Molotov' whenever he met Stalin in his reports to the Foreign Office in early 1942 (Gillies, 1999, p. 126).[37] Le Corbusier appealed to ABC, in the belief that they would commend the idea and furnish the mechanical equipment to make it work.[38, 39] Although it is generally thought that ABC were contemptuous of the idea and tried to sell air-conditioning directly to the Soviet authorities, their response was courteous and detailed, although delayed by their efforts to extract a competent translation in French to send. ABC explains that the idea is misconceived, that the heat transfer through the outer panes will be so dominant that the energy demand would be an order of magnitude higher. The need for cooling will place a yet greater demand on energy. The same applies today. Disastrous solar gains rendered Le Corbusier's Salvation Army Hostel in Paris uninhabitable, the double façade glazing details are dated 1929 (Allen Brookes, 1982).

The Prefet de Paris condemned the Salvation Army Hostel and the Paris police enforced a deadline of three months in which to deliver acceptable summer conditions, an appalling situation for an architect to be in and very embarrassing with the benefactor, Princesse de Polignac. An acceptable solution was delivered just in time by a combination of sun-shading, 'Brise-soleil' configured by Xenakis, and opening windows (Benton, 1987). A recent paper by Mingotti (2011) on the fundamental physics of double glass façades shows that in Moscow, as in many other locations and climates, there are fundamental problems with the idea when the various energy balances are calculated. The great concern is that a continuing pre-occupation with sealed all-glass buildings is still sanctioned by the propogandizing of high Modernism of the 1910s and 1920s, a truly Whiggish phenomenon.

Le Corbusier's next attacks on an extreme climate, the hot dry heat followed by the monsoons of Gujurat, for the Sarabhai family, notably the Millowners' Association building, do not attempt to incorporate any environment-making technology. Perhaps the memories of his environmentally malfunctioning buildings in Moscow and Paris were far too painful. Instead the architecture is used to fashion the internal environments within the extensive environmental penumbra developed between inside and outside. The clients certainly had access to the technology: we have established that the family concern, the Calico Mills, were humidity controlled.[40] The 1954 Cultural Centre at Pali is defended by a flooded roof, 'all precautions are taken against the excessive temperature', visitors are encouraged in the evening and night-time (Boesiger and Girsberger, 1967, p. 242).

Returning to the West, Henrik Schoenefeldt's inspection of the United Nations building archive reveals a voluminous and largely one-way correspondence between Le Corbusier and the UN architectural team. Le Corbusier did not understand that Carrier technology had advanced to the point where the UN tower could be turned into a block of ice, at will. It is tempting to conclude that the American sphere of building pursued the sealed glazed air-conditioned type with its superior technology and infinite supply of energy whilst the European equivalent, largely in thrall to Le Corbusier, pursued architectural form and detail to fashion its interior environments. This book argues that it is the latter tendency in which by far the greatest benefit can be found.

On reflection, could it be concluded that 'history' has conspired to annihilate climate, to insist on a temperate environment or risk certain descent into enervated chaos? The former President of Wolfson College at the University of Cambridge, the eminent historian Gordon Johnston, declared to the author in a stressful moment, his faith in the 'cock-up' theory of history, the definitive antidote to Whiggish historicism. The emergence of the artificially-induced cool temperate climate and the lightweight building envelopes (sustained by mechanical services technology) appear to have been the outcome of an unconnected series of collisions between quasi-mystical visions, classically inspired prejudices towards regional populations, the relentless marketing of technological innovations by capitalistic enterprises with little declared interest in their wider socio-cultural consequences and the convergence of the business model for air-conditioning and what this 'freed' architects to create (built form that could, in principle, and often was, oblivious to the climate). This comes at a high cost: environmentally, the associated CO_2 emissions are immense; physiologically, adaptation and resilience may decline; culturally from disconnection with the natural world and psychologically, embracing incarceration in an unnaturally hygienic environment – fostering a belief in the control over nature at a global scale.

Anthropogenic contribution to a changing climate

More recently, the Inter governmental Panel on Climate Change (IPCC, 2014, p. 119) described the notion of climate as the 'history of weather':

> Climate in a narrow sense is usually defined as the average weather, or more rigorously, as the statistical description in terms of the mean and variability of relevant quantities over a period of time ranging from months to thousands or millions of years.[41]

Atmospheric scientists regularly remind us that the changing climate is the product of anthropogenic contribution, in the form of greenhouse gas emissions, superimposed on a natural variability within the climate. Hans Graf, the University of Cambridge atmospheric physicist, explains that there are no structures to this natural variability, just a continuum of one mean global temperature that varies. There is compelling evidence that 1400–1900 was cooler, particularly during the 17th century Dutch Golden Age and the early 19th century whilst the 20th century was relatively warmer. The synchronization of the two trends is difficult to identify and impossible to predict with full confidence, they constitute a chaotic system. The United Nations Framework Convention on Climate Change (UNFCCC) emphasizes this coupling of natural and man-made causation:

Note that in its Article 1, UNFCCC defines climate change as, 'a change of climate which is attributed directly or indirectly to human activity that alters the composition of the global atmosphere and which is in addition to natural climate variability observed over comparable time period'. The UNFCCC thus makes a distinction between climate change attributable to human activities altering the atmospheric composition and climate variability attributable to natural causes.[42]

Judgement as to the most likely alignment of the phasing of the two causes is critical to the predicted overall amplitude of change. Although historic data provides evidence for future trajectories, the pattern of variability may change. The climate predictions presented in this book, which follow through the text, mesh these cycles by identifying the models with closest fit to the pattern of variation of the preceding 25 years. The outcomes for England differ from those of the Hadley model, as shown in Chapter 9: they are less extreme. One potentially very important consequence is that adaptation measures would be effective for longer.

That the relative concentration of CO_2, a relatively minor constituent of the atmosphere, might significantly affect global climates, had already been proposed in the late 19th century by Svante Arrhenius (1896) in a paper given to the Stockholm Physical Society.[43] It scaled up the 'greenhouse effect' theory to explain the energy balance of the earth's atmosphere. This was based on work that evolved over the preceding century attempting to describe how glasshouses induced warmer environments. Arrhenius devised a model to predict the flows of energy within the atmosphere. By varying the notional concentrations of CO_2, he could venture predictions in the event of the

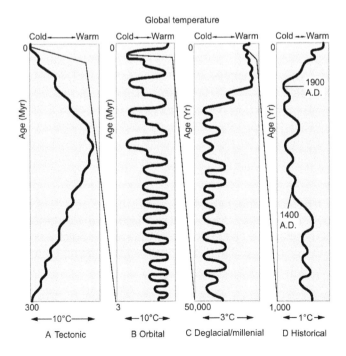

Figure 2.6
Long-term climate variability over 300 million years reproduced from www.pages-igbp.org/products/pages-magazine?site=1 (accessed 17 June 2015). Originally from Ruddiman, W.F. (2001) *Earth's Climate: Past and Future.* New York: W.H. Freeman & Sons.

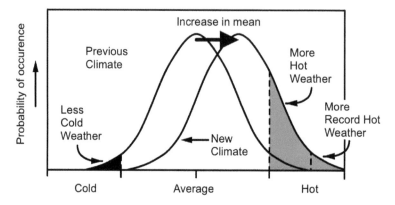

Figure 2.7
The IPCC proposition that small increases in mean annual temperature will increase the propensity of extreme heat events. Reproduced from www.unep.org/climatechange/Publications/Publication/tabid/429/language/en-US/Default.aspx?ID=6306, sourced 6 June 2015.

concentration rising.[44] This seemed inevitable already at the commencement of the 20th century in that annual global combustion of coal in 1904 was estimated to have risen by 44% in only eight years. By 1908 (p. 56) Arrhenius reported:

> If therefore the percentage of carbon dioxide is doubled, the absorption of plants would also be doubled. If at the same time the temperature rises by four degrees, the vitality will increase in the ratio 1:1.5 so that the doubling of the carbon dioxide percentage will lead to an increase in the absorption of carbonic acid approximately in the ratio of 1:3 … an increase of the carbon dioxide percentage to double its amount may hence be able to raise the intensity of vegetable life and the intensity of the inorganic chemical reactions threefold.

Arrhenius explained that the carbonic acid content of the air must increase at a constant rate as the consumption of fossil fuels continues to increase. In 1908 he calculated the CO_2 by plants to be 13 000 million tons, 15 times that from burning coal in 1906, 1/50 of the CO_2 in the atmosphere. He forecasted 'the possibility of enormous plant growth' and raised the possibility of a new Carboniferous period. Nonetheless, he suggested there is 'good mixed with evil', and reaches a very surprising similar conclusion to the post-IPCC world (ibid. p. 63):

> By the influence of the increasing percentage of carbonic acid in the atmosphere, we may hope to enjoy ages with more equable and better climates, especially as regards the colder regions of the earth, ages when the earth will bring forth much more abundant crops than at present, for the benefit of a rapidly propagating mankind.

By the late 1950s the eminent meteorologist H.H. Lamb could publish dense comparative historical climate data to demonstrate climate shifts, some 'very quick'. He wrote: 'Our attitude to climate "normal" must clearly change.

1901–30 and still more 1921–50 were *highly* abnormal periods' (Lamb, 1959, pp. 299–318).

Lamb reviews nine theories of which CO_2 concentration is but one, coupled with volcanic ash, others being solar activity and changes in the earth's surface.

Recorded atmospheric concentration of CO_2 equivalent (CO_2e) has risen from 315ppm to 400ppm since 1955 and a rise of 1°C in global mean annual temperatures is recorded since 1900. It is now generally thought by climate scientists that a 2°C rise in global mean temperature will lead to catastrophic and irreversible climate change effects. A 2°C rise is correlated with 450ppm of CO_2e (IPCC, 2014, p. 151). On 9 May 2013, data collected at the Scripps Institute of Oceanography at Mauna Loa revealed that 400ppm was surpassed for the first time since measurement began in 1958, data corroborated independently by the National Oceanic and Atmospheric Administration (NOAA).

The book will now explore the implications for buildings, climate by climate as they morph and shift across the globe. Academics in Vienna have recast

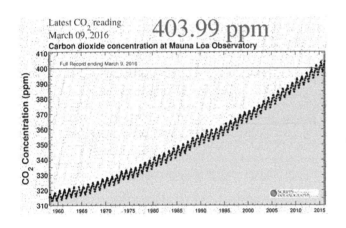

Figure 2.8

The Keeling Curve: Monthly average carbon dioxide concentration measured at Mauna Loa, CO_2 reading 403.67, accessed 2 May 2015. Named after Charles David Keeling, now maintained by his son Ralph Keeling. Curve reproduced with grateful thanks to Ralph and the Scripps Institute CO_2 Program. Ref. Atmospheric CO_2 concentrations (ppm) derived from in situ air measurements at Mauna Loa Observatory, Hawaii: Latitude 19.5°N Longitude 155.6°W Elevation 3397m. Source: R.F. Keeling, S.J. Walker, S.C. Piper and A.F. Bollenbacher, Scripps CO_2 Program. See http://scrippsco2.ucsd.edu (accessed 12 October 2016).

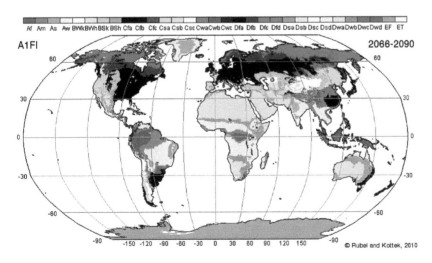

Figure 2.9

Köppen recast with IPCC 4th Report model predictions to 2076–2100. Reproduced from the paper by Rubel and Kottek (2010), by courtesy of Professor Franz Rubel of the Climate Change and Infectious Diseases Department, Institut für Öffentliches Veterinärwesen, Veterinärmedizinische Universität Wien. See http://koeppen-geiger.vu-wien.ac.at (accessed 12 October 2016).

Köppen's map to depict the distribution of climate zones within the period 2076–2100, predicted by the IPCC's fourth round of modelling (Figure 2.9).

This is an extraordinary prospect. The implications for the design and redesign of buildings will be explored particularly in the Continental climates within the northern hemisphere. The prospects are considered for low carbon environmental design strategies across the globe and the sort of architecture which could develop (Mordaunt Crook, 2003, p. 60):

And then there was climate: we have records of a documented sequence of very bad weather in the 1840s and 1850s. Ruskinian polchromy was as much an answer to Dickensian smog as a reaction against Neo-Classical purity.

Notes

1. Available online: www.gutenberg.org/files/15293/15293-8.txt (accessed 30 October 2013).
2. For example see *Rain* in Maugham, W.S. *Collected short stories, Vol. 1*, Vintage Classics Kindle edition, an ironic depiction of climate determinism in which a bigoted missionary condemns his junior colleagues' 'pitiful lack of energy' in the heat and humidity of Pago Pago.
3. The ten years of translation work was weaved through duties as Bursar and Steward at St. Catherine's College Cambridge (guardian of the College finances, controller of the wine cellar and compiler of the 'placement' at feasts).
4. The map is derived from: Meurer, P. (2004) The Strabo Illustratus Atlas , ed. P. Haas, D. Le Bail and F. Weissert, publ. Antiquariat Gebr. Haas, Bedburg-Hau; Le Bail & Weissert, Paris. Reproduction of 'COSMOGRAPHIA UNIVERSALE, a C19th rebind of 191 prints on 164 paper units assembled in the first half of the 1570s including a set of 11 historical maps drawn from Strabo's texts to illustrate an Italian edition of Strabo's Geographia.
5. Professor Hulme (2011) quotes from Gordon Manley (1944 and 1958).
6. Falconer may be referring to Johann Christoph Homann, professor at the University of Göttingen, pioneer of medical geography and deviser of the new discipline of 'Geosophy' – a more scientific, Christian, less mystical 'Geomancy' – author of (1725) *Medicinae Cum Geosophia Nexu*, cartographer of the 'unsettled lands' of North America, indicating their relative healthiness and therefore the likely productivity of settlers and transplantees.
7. Galen of Pergamon, c.129–216/17, was a pre-eminent physician and subscriber to the humoral theory after Hippocrates, hence the interest in bile, see www.sciencemuseum.org.uk/broughttolife/people/galen.aspx (accessed 10 October 2013).
8. Footnote 13 is dense and sets out a much debated conundrum about the substantiality of the universal (1042a26-31) involving the status of living organisms, as Mary Louise Gill explains in her paper 'Logic and Metaphysics in Aristotle and Early Modern Philosophy', *Synthese*, **96**(3), 379–397. See www.jstor.org/stable/20117819 (accessed 7 November 2015). The footnote is not central to the argument but no less illuminating in our context and, possibly, disproportionately significant.
9. https://archive.org/stream/estimateoftemper00kirwiala#page/n5/mode/2up (accessed 28 October 2013).
10. Raymond Holder Wheeler (1943) 'The Effect of Climate on Human Behavior in History', *Transactions of the Kansas Academy of Science* 46, 33–51 provides a survey of support for environmental determinism across North America. He identifies four parallel schools of determinists, one clustered around Clarence Mills whom we will meet shortly. Wheeler's position is abundantly clear, 'In cooler climates man is more vigorous, more aggressive, more persistent, stronger physically, larger, braver in battle, healthier, and less prone to sexual indulgence. In warm climates man is more

timid, smaller, physically weaker and less courageous but more inclined to physical pleasures, more effeminate, lazier, and less aggressive.' By now familiar sentiments. See www.jstor.org/stable/3624926 (accessed 7 November 2015).

11. 'Climate Makes the Man', *The American Mercury*, June 1943, 735–740. See www.unz.org/Pub/AmMercury-1943jun-00735?View=PDFPages (accessed 25 June 2015).

12. Sociologist Nico Stehr has written on Hellpach's 1938 essay 'Kultur und Klima' illuminating the relationship between Nazi ideology and climatic determinism, see Stehr (1996), and for the original essay, Hellpach (1938). See also Egbert Klautke (2012) *History of European Ideas: Defining the Volk: Willy Hellpach's Völkerpsychologie between National Socialism and Liberal Democracy, 1934–1954*, History of European Ideas. DOI: 10.1080/03044181.2012.735086, available digitally from UCL SSEES.

13. Judaic perceptions of climate difference are summarized in Altmann, A. (2005) The Treasure Trove: Judah Halevi's Theory of Climates. *Aleph* 5, 215–246. Much of what Halevi reports resonates with the largely Christian context of the theories described here but a very interesting project awaits. See www.jstor.org/stable/40385835 (accessed 6 November 2015).

14. Billington, N.S., Roberts, B.M. (undated web resource) *The Story of Comfort Air-conditioning*. Pergamon Press, International Series on Building Environmental Engineering Vol. 1. See www.hevac-heritage.org/electronicbooks/comfortAC/8-CAC2.pdf (accessed 14 February 2012). See also, Arnold, D. (1999) The evolution of modern office buildings and air conditioning. *ASHRAE Journal*, June 1999, 40–54.

15. Officer, L.H. and Williamson, S.H. (2013) Purchasing Power of British Pounds from 1245 to Present, *Measuring Worth*. See www.measuringworth.com/ppoweruk/ (accessed 29 October 2013).

16. Wampler, C. (1949) *Dr. Willis H. Carrier: Father of Air Conditioning (Newcomen Address)*. Newcomen Society of England, American Branch, New York (rare pamphlet in author's collection).

17. See www.sturtevantfan.com/BFS.html (accessed 30 October 2013).

18. The Cornell University Library Division of Rare and Manuscript Collections, Willis Haviland Carrier's Deceased Alumnus records in Alumni Box 20. Course catalogues referred to as 'Announcements' or 'Registers' from this period have been scanned and are available at: http://ecommons.cornell.edu/handle/1813/22306 (accessed 12 October 2016).

19. Banham credits the origin of the idea of making deep plan office floors to a Chicago real-estate agent, the omission of the plan re-entrants that had hitherto provided air and light, exploiting the latest air-conditioning and fluorescent lighting technology, as published in September 1949. Banham traces the viability of the glass office slab to this moment. The agent was George R. Bailey, who, as a partner in Turner, Bailey, Barley and Zol, 25 years later wrote a piece reproduced in the *Cornell University Preservation News* (Volume 14, Number 9, 1 September 1974) condemning the spot preservation orders being put on historic office buildings in Chicago, not least the Marquette Building, as being 'tremendously unfair', to give a flavour of the man, real estate to the core.

20. Quoted at www.willliscarrier.com/m/1941-1950.php (accessed 27 October 2013).

21. http://links.jstor.org/sici?sici=0022-4642%28198411%2950%3A4%3C597%3ATEOTLH%3E2.0.CO%3B2-W (accessed 12 October 2016).

22. Arsenault (1984) credits Stuart W. Cramer, textile engineer from Charlotte, North Carolina and I.H. Hardeman with the introduction of air-conditioning to the South, writing, 'Cramer actually coined the term "air conditioning" in 1906. Hardeman, a graduate of the Georgia Institute of Technology working under Carrier, convinced his boss that air conditioning would eventually "Revolutionize the textile industry...". At Hardeman's urging Carrier published an article in *Textile World* in April 1906 describing the benefits air conditioning could bring to the textile industry. By the time the article appeared, Hardeman had already sold a primitive air-conditioning system to the Chronicle Cotton Mills of Belmont, North Carolina. After visiting the Chronicle Mills in the summer of 1906, Carrier added several important refinements to his system. It was during this visit that he discovered the principle of "dew point control," later an

axiom of air-conditioning technology. By 1908 air conditioning had been installed in several North Carolina cotton mills and by 1911 in rayon mills, where temperature and humidity problems were especially acute.'

23. Arsenault cites 'Home-Made Weather', *Literary Digest,* CIV, p. 28. (15 February, 1930.)

24. Arsenault locates the additional sources: Ingels (1952, pp. 1–19); 'Blow, Cool Air', *Time*, LXXX, p. 42 (27 July, 1962); Carrier Corporation, *The Two Faces of Janus: The Story of Carrier Corporation* (1977, p. 1–28); and Friedman, *The Air-Conditioned Century*, pp. 25–26.

25. James Fergusson, 'A brief history of air-conditioning', *Prospect Magazine*, 24 September 2006. See www.prospectmagazine.co.uk/magazine/abriefhistoryofairconditioning/#.UozmhY6I0WY (accessed 20 November 2013).

26. Prescott Webb, 1888–1963, distinguished historian of Texas, proposed a pragmatic model of westwards expansion involving necessary technological advances, evolving soil conditions, and hot arid climates. His papers are held in the Briscoe Centre for American History at the University of Texas Austin. See www.lib.utexas.edu/taro/utcah/00344/cah-00344.html (accessed 27 July 2015).

27. F. Penzel, *Expert Opinion: OCD and Contamination*, International OCD Foundation. See https://iocdf.org/expert-opinions/expert-opinion-contamination (accessed 30 September 2016).

28. The author notes that Historians of Psychology associated with the Department of History and Philosophy of Science (HPS) at Cambridge have found this possibility intriguing and suspect there may be an interesting if challenging cross-disciplinary research exercise to be undertaken (members of the HPS 'Coffee with Scientists' seminar group).

29. Santiago Iñiguez de Ozoño, Dean of Spain's IE Business School, *The Economist*, 1 December 2009. Dean's debate: 'How do business schools remain relevant in today's changing world?' *Economist Online* extra, available at www.economist.com/node/15006681 (accessed 12 October 2016).

30. Reiter, P. and Wellmon, C. (2015) How the Philologist became a Physician of Modernity: Nietzsche's Lectures on German Education. *Representations* 131. Available at https://chadwellmon.files.wordpress.com/2015/06/rep131_04_reitter.pdf (accessed 27 July 2015).

31. Seeger, H. (1933) *Burohauser der privaten Wirtschaft,* p. 14, as quoted in Dietrich (1992).

32. Article by L. Mies van der Rohe (pp. 213–214) in Taut, B. (1920–22) *Fruhlicht Eine Folge fur die Verwicklichung des neuen Baugedankens*, Ullstein Berlin Frankfurt/M Wien.

33. Originally published as Taut, B. (1919) *Die Stadtktrone. Jena:* Diederichs Verlag. Available at http://onlinelibrary.wiley.com/doi/10.1111/j.1531-314X.2009.01035.x/full (accessed 12 October 2016).

34. Kopp, A. (1970) *Town and Revolution: Soviet Architecture and City Planning*, transl. Thomas E. Burton, Thames and Hudson, London; Cohen, J-L. (1992) Le Corbusier and the mystique of the USSR: theories and projects for Moscow, 1928–1936, transl. K. Hylton, Princeton University Press, Princeton, NJ.

35. For the history of Sirocco fans see, American Blower History webarchive.

36. Clearly visible in the photograph taken by Richard Pare in 1999, exhibited at 'Building the Revolution: Soviet Art and Architecture 1915–1935', Royal Academy, 2011.

37. Gillies quotes from Kerr's dispatch to Anthony Eden. Foreign Office paper FO 800/300/9.

38. With very grateful thanks to Rosa Urbano Gutiérrez for sharing her scans of the response from the American Blower Company to Le Corbusier made in the Fondation Le Corbusier, Paris. See her article on the progress of the mur neutralisant prototype: Rosa Urbano Gutiérrez (2013). Le pan de verre scientifique: Le Corbusier and the Saint-Gobain glass laboratory experiments (1931–32). *Architectural Research Quarterly* 17, 63–72. doi: 10.1017/S1359135513000365.

39. ABC was by then a substantial business competing with Carrier and Sturtevant and very recently absorbed into 'American Sanitary and Radiator' (now 'American Standard') for $4 million.

40. An account of the Calico Mills under the stewardship of Ambalal Sarabhai is available at www.dnaindia.com/india/1375259/report-calico-mills-passes-into-history-for-rs270-crore (accessed 3 August 2013). The plant was fuelled by natural gas and was subject to a workplace study focusing on humidities in the various loom sheds, see Millar, E.I., *The Ahmedabad Experiment Revisited, Work Organization in an Indian Weaving Shed, 1953– 1970*. See http://moderntimesworkplace.com/archives/ericsess/sessvol2/E_Millar.pdf (accessed 3 August 2013).

41. The IPCC definition continues, 'The classical period for averaging these variables is 30 years, as defined by the World Meteorological Organization. The relevant quantities are most often surface variables such as temperature, precipitation and wind. Climate in a wider sense is the state, including a statistical description, of the climate system.'

42. The United Nations Framework Convention on Climate Change (UNFCCC or FCCC) 1992 (unfccc.int): 'The Parties to this Convention, Acknowledging that change in the Earth's climate and its adverse effects are a common concern of humankind, Concerned that human activities have been substantially increasing the atmospheric concentrations of greenhouse gases, **that these increases enhance the natural greenhouse effect**, and that this will result on average in an additional warming of the Earth's surface and atmosphere and may adversely affect natural ecosystems and humankind.'

43. Reprised for the general reader in Arrhenius, S. (1908) *Worlds in the Making: the Evolution of the Universe*. Translated by Borns, H. London: Harper and Brothers.

44. Fleming provides a useful overview of Arrhenius' contribution in Fleming, J.R. (1998) *Historical Perspectives on Climate Change. New York:* Oxford University Press.

3
Temperate climates

Figure 3.1
'Foret au bord de l'eau avec une charrette au premier plan', a Dutch Golden Age 'Campagna' painting by Jan Hackaert (1629–85) with figures in the foreground by Jan Lingelbach (80 × 100cm). Hackaert captures the atmospherics of the cool Temperate climate even whilst purportedly portraying an Italian river landscape (author's collection).

'If there is one thing I hate,' said father, 'it's unnecessary fresh air.' He gave his hard-case scowl at the assembled Cottrells, who had the passion for draughts common among intellectuals, and enjoyed himself going round their sitting room shutting all the windows … it would have been so funny if anyone had believed it.

(E. Arnot Robertson, 1933, *Ordinary Families*, Chapter VII)

To mend the air when it is unhealthy or corrupted, is a work scarce thought possible to be done by any human contrivance; unless by appealing the wrath of Heaven by prayers and supplications, which, like the nail driven by the Consul, have sometimes, as we read, put a stop to the most destructive contagions.

(Leoni, 1726, *The Architecture of L.B. Alberti*, p. 91)[1]

The joke in E. Arnot Robertson's novel is that the discomfitted father had been a merchant navy seaman acclimatized to sometimes spectacular environmental conditions. The 14th century architectural theorist Leon Battista Alberti (as translated by Leoni, above) speculated on the probable impossibility of artificially restoring 'vitiated' air to its original freshness. This 'work scarce thought possible' was eventually achieved by the inventors of air-conditioning (as discussed in Chapter 2), but Alberti continues to qualify his scepticism: 'not that I deny the possibility of amending a great many of those defects which proceed from the air, by curing the earth of exhaling noxious vapours.'

Alberti would tackle the source of the miasmas, requiring swamps to be drained at a vast scale, but Carrier focused on the eventual destination of the air. Indoor air quality (IAQ) is frequently raised to support decisions to pursue an artificial environment, perhaps another manifestation of the obsessive need for a notional cleanliness. It may be just about achievable to deliver truly 'clean' air but only at huge cost in energy and CO_2 emissions in present-day HEPA-filtered clean laboratories[2] and operating theatres. The employment of the very top end of air filtration technology captures particulate matter above a certain size but cannot intercept gases or odours. Disproportionate fan-power is required to propel useful quantities of air through the most effective filters, as they are dense obstructions. The US Government standard requires interception of particles 3 microns or greater in diameter, relatively large for a microorganism.

It is highly likely that the required energy input and associated CO_2 emissions will make buildings artificially conditioned to deliver this level of cleanliness difficult and expensive to own as the century progresses. They are not resilient and there is no less energy intensive contingency in their design if their demand for energy cannot be fulfilled. These are questions of concentration and dilution of contaminants, the sealing of buildings and the causes of respiratory disorder.[3] Can one make a viable, safe, naturally ventilated, free-running building in a temperate climate (buildings that do not rely on mechanically-operated services for heating and cooling)? This challenge is investigated in this chapter. Two complex higher education buildings are presented that were specifically designed to address this challenge.

What constitutes a temperate climate?

Köppen's distinctions involve denoting broad climate types and refining them with data for precipitation and temperature. From the more focused perspective of creating free-running, low carbon buildings, one might ask where across the globe do night-time temperatures reliably remain at or below 18°C. Under such a condition, one might be reasonably confident about the success of various natural cooling strategies. Figure 3.2 shows the rather surprising outcome of our analysis using 2014 IPCC (Intergovernmental Panel on Climate Change) model base data to map the occurrence of night-time temperatures in which 75% of the night-time peaks fell below 18°C between May and September 1996 and 2005.

One might then ask where across the globe do temperatures remain at or above 6°C? A well-insulated and airtight public building with a reliable occupancy might generally avoid additional heating in these conditions. Figure 3.3 depicts the outcome of this exercise for the period 1996–2005.

Figure 3.4 combines these low and high temperature characteristics, mapping the distribution of night-time temperatures at or below 18°C and peak temperatures at or below 38°C. Northwest Europe falls comfortably into this zone, the southern parts of Africa and South America, much of Australia and New Zealand, areas of southwest China. These areas will migrate north and south in the future due to anticipated changes in the climate and capture very much more territory as the rolling forward of Köppen's original map suggests. This is the field of play for much of what is described in this book.

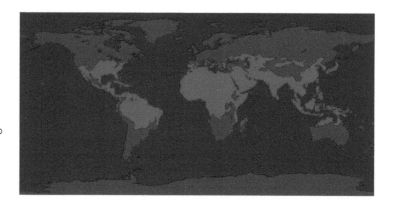

Figure 3.2
The recorded occurrence of current mean night-time temperatures between May and September 1996 and 2005: red zones show night time temperatures in which 75% of night-time peaks remained below or equal to 18°C yielding potentially beneficial environments for very low and zero energy passive night cooled buildings. Graf, Xiaoyong Yu and Short, University of Cambridge, unpublished.

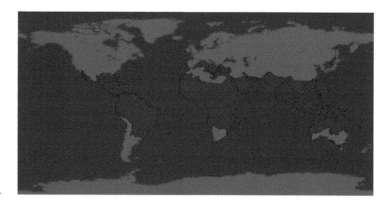

Figure 3.3
The recorded global distribution of temperatures remaining at or above 6°C through 1996–2005, suggesting very little heating should be required in the non-domestic stock. Graf, Xiaoyong Yu and Short, University of Cambridge, unpublished.

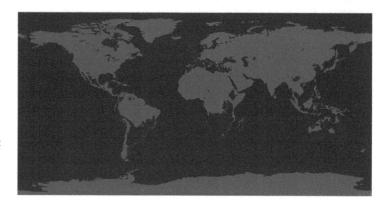

Figure 3.4
The potential field for zero carbon building: combining the low and high temperature constraints, blue zones show temperatures within range zero to 38°C and 50% of night-time minimum temperatures below 18°C from May to September through 1996–2005: beneficial environments for very low and zero energy passive night cooled buildings. Graf, Xiaoyong Yu and Short, University of Cambridge, unpublished.

The Queens Building, Leicester

A barrister, having toured the Queens Building for the School of Engineering and Manufacture at De Montfort University in Leicester and listened to a dry explanation of the building physics supporting the design, concluded audibly that the building was very quietly, very subversive. The Queens Building is not formed as the prevailing laboratory building type, it is its inverse. The prospective Head of the new school sketched out the type of building he required, a deep rectangular plan of double-loaded corridors flanked by inward-looking teaching rooms and laboratories, stacked on four floors with a full fifth floor plant room, encircling a large open atrium, all sealed and air-conditioned, exactly as the new engineering building he had recently visited in Sweden. When he saw the proposed plans he resigned. The design team only discovered this later from the Pro-Vice-Chancellor, who was determined to realize the building. The building is not hermetically sealed, far from it, and has only one air-conditioned room, a small tribology lab. Nor does the Queens Building contain the customary false ceilings and raised floors to conceal the rapidly evolving spaghetti of cables and ductwork. Most importantly, the physical forms of the technology installed in its laboratories, or rather their casings, do not dominate or drive the

Figure 3.5
View of the Queens Building across the disparate city landscape of Leicester, taken from the cover of the Royal Fine Art Commission publication, *Design Quality in Higher Education Buildings* (original photograph by Peter Cook).

design or the imagery in which the building is realized. It is not, for example, streamlined, it is a building in a specific place, not a moving vehicle. It bucks Louis Sullivan's (1896, pp. 32–34)[4] somewhat exasperated prescription for a modern non-domestic building:

> … above this an indefinite number of stories of offices piled tier upon tier, one tier just like another tier, one office just like all the other offices, an office being similar to a cell in honey comb, merely a compartment, nothing more, 5th, and last, at the top of this pile is placed a space or story that, as related to the life and usefulness of the structure, is purely physiological in its nature namely, the attic. In this the circulatory system completes itself and makes its grand turn, ascending and descending. The space is filled with tanks, pipes, valves, sheaves, and mechanical etcetera that supplement and complement the force originating plant hidden below ground in the cellar.

The Queens Building has no penthouse plant room across its top disconnecting the building below from the sky as the axonometric view (Chinese perspective) indicates (Figure 3.6).

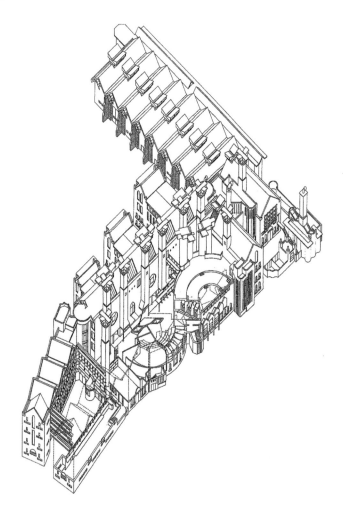

Figure 3.6
Axonometric view of the Queens Building. A new type of configuration for a large densely occupied public building, turned inside out, an un-modern collection of simple single ridge roofed components (author).

Design strategy

The environmental design strategies for the various parts of the building are based on the exposure of high thermal mass to the air volumes within, plan forms narrow enough to cross-ventilate and side light, deeper plan elements which are connected to the sky above exploiting passive stack ventilation and top lit, all to reduce the reliance on 'artificial weather', mechanical heating and cooling. Professor Randall Thomas (1996) describes these strategies as detailed, quantitative case studies in *Environmental Design*, to the extent of reworking the original stack-effect calculations. The detailed design of the amphitheatres, the most complex exercise, is reprised in Chapter 5. Here we are interested in how to 'give form' to an environmental intent, an intent not nearly so prescriptive, deterministic and exclusive as sceptics might believe.

The building construction is highly integrated and heavyweight with high thermal mass achieved through the use of load-bearing brick masonry. The rationale for high thermal mass is its ability to retain heat and coolth, contributing to stable radiant temperatures, the exchange between the skin and the surfaces visible to it. It becomes difficult to change the temperature within such a building as opposed to highly volatile very lightweight enclosures. There is an embodied energy penalty with these materials but it is essential to the low carbon intent to eliminate mechanical cooling, it has no separate steel or concrete frame, which can be deleted from the calculation, but is built up course by course (Figures 3.7 and 3.8). It can also be argued that although brick

Figure 3.7
First lift of the polychromatic load-bearing Queens Building brickwork, the more intense the polychromy, the less load being borne
(author).

Figure 3.8
The lift of brickwork completed to eaves level. The aspirating acoustic buttresses to the noise generating Mechanical laboratory. Photograph by Peter Cook.

masonry has a higher embodied energy, it is durable and has a longer service life than many other materials. Compared to other materials and structural approaches, the embodied energy in the overall life cycle compares relatively favourably. Structural timber is attractive in these terms but with little mass, perhaps phase-change materials can recover some of the beneficial effect of exposed mass.

The heavy load-bearing masonry interior is revealed to the interior air volume to exploit its thermal admittance to the maximum, so that almost all conventional applied finishes were omitted (Figure 3.9). This is quite a radical proposition, much as Hendrik Petrus Berlage (1856–1934) shocked Amsterdam's chattering class with the naked brick interiors of his Beurs (bourse) building (final design, 1898–1903), 'a gradual self-liberation from eclecticism',[5] and the full height vestibule of his Diamond Workers' Union building.[6] P.V. Jensen Klimt pursued similar exposure in his post-Backsteingotik Grundtvig Church, not least in the stair chamber descending to the crypt.[7]

The contractor of the Queens Building was thrown by the 'un-modern' way in which the building grew in construction, not as a discrete sequence of packages: frame, floors, cladding, services, ceilings … the rhythm of contemporary wholly subcontracted contracting, but as continuous and integrated assembly. The cost formula was very different and probably unprecedented in recent years, whole packages of work were omitted but the superstructure is entirely fair faced (i.e. no coatings, claddings or coverings added). Services (lighting, water and waste

Figure 3.9
The Queens Building concourse, intended to be the equivalent of MIT's Infinite Corridor in encouraging interaction. The auditoria volumes sit on pilotis to the left side of the concourse, there are no corridors, the concourse contains all the 16% balance space allowed by the Polytechnics and Colleges Funding Council, its exposed mass stabilizes internal temperatures (Peter Cook).

water, electrical, heating, ICT) accounted for 24% of the cost rather than the customary 40% or more for a laboratory building, permitting greater cost to be slewed towards the superstructure, the architecture. The design process had to be highly integrated, focused and quick, against the atomized co-existence of the various design professions. The 'dynamic thermal modellers' at Leicester Polytechnic (later to be De Montfort University) led by Professor Lomas and the fluid flow mathematicians led by Professor Linden at the Department of Applied Maths and Theoretical Physics, University of Cambridge had to work in indecent haste to deliver results and their subsequent sensitivity studies but, above all, our joint interpretation of the data and its implications to maintain the programme of supplying information. They found their domain entirely intermeshed with all other aspects of the design. The acoustician was obliged to broker the likely reverberation time against the minimum useful exposure of thermal mass, the enthusiasm for natural light against the noise of the city centre. The mechanical and electrical engineers had to configure every pipe and cable; they are all clearly evident. The structural elements are all visible and entirely implicated in the 'vision'.

The provenance and the architectural response

In 1991, The *Times Higher Education Supplement* ran the story 'Polys start building work' (Brookman, 1991). Six projects were awarded £34 million, £10 million of which went to what was to become the Queens Building out of a field of 64 applications. Would the Higher Education Funding Council have sanctioned this experiment so wholeheartedly? As the design progressed, a dramatic change in Further and Higher Education in Britain was introduced by the Conservative government, a significant number of polytechnics acquired full university status. The 1992 Universities, as they became known, became publicly incorporated, with 'attitude' and a quite different view of what democratizing higher education meant to that held by the Russell Group of older, established research-based universities. The 1992 Universities acquired the physical assets they occupied from the county councils and city corporations, substantial urban estates in many cases. Leicester Polytechnic rebranded itself as De Montfort University. The Queens Building became an icon of this movement, quickly identified in 1993 by critic Hugh Pearman in the *Sunday Times*. Pearman wholly understood the ambition and wrote (Pearman, 1993), 'the polys got pre-fabricated sheds. Universities got buildings that were sometimes so individual they forgot what they were there for. This is exactly what happened in Leicester.'

The sheds were delivered in the very lightweight prefabricated CLASP system at Leicester Polytechnic, a system so ubiquitous in the Midlands and North (adapted for coal mining subsidence areas, the parallel SCHOLA system prevailed in the south) that a visitor might believe they were in one of more than 6000 similar sites (Figure 3.10). The prefabricated system is lightweight and lacks thermal mass, therefore it has virtually no resilience to excess heat or cold. Leicester Polytechnic's Fletcher Building was notorious for its summer overheating. The new building was highly characterized to its place and the spaces within it particularized, each type identifiable across its freestyle elevation, matching openings to activity behind (Figure 3.11).

Figure 3.10

The Queens Building from the northwest seen across the 1960s CLASP system accommodation for the Polytechnic, a radical reinterpretation of the needs of Further and Higher Education (Peter Cook). Ruskin (1893, vol. 2, p. 176): 'We must, however, herein note carefully what distinction there is between a healthy and a diseased love of change … consider the different ways in which change and monotony are presented to us in nature … change being most delightful after some prolongation of monotony, as light appears most brilliant after the eyes have been for some time closed.'

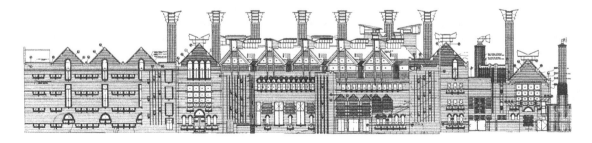

Figure 3.11

Northwest elevation to Mill Lane (Alan Short). Beresford Hope, 'The Skyline in Modern Domestic Buildings', in *RIBA Transactions xiv*, 1863–64, 103–115: 'I asked myself the question, what principles of beauty and convenience are there which give a charm to Gothic and yet can be grafted into Italian without detriment to its unity of style? My answer to myself was, high roofs and tower like chimneys.'

The Leicester Polytechnic campus certainly seemed to be in the damp hollow at the bottom of the hill which the old 'new university' occupied.

The environment of the Polytechnic was blighted by late 1950s demolitions which had led to absolutely no subsequent development. The city council's Chief Planning Officer Konrad Smigielski's vision in the 1950s for Leicester was a future of monorails, slab blocks and flyovers obliterating the past in order to challenge nearby Nottingham and Coventry as a co-recipient of the connectedness promised by the M1 motorway. It largely failed, preserving by default some of the city's late Victorian, Edwardian and inter-war commercial and industrial buildings. Leicester had been prosperous, growing rapidly between 1861 and 1891 from a population of 68 000 to 174 600, and to 227 000 by 1911, fuelled by the hosiery industry, employing mainly women, then the boot and shoe industry employing mainly men and the all-male heavy engineering industry in particular.[8] Between 1918 and 1939 the new city continued to grow despite national recession. Was there anything to remember, any residue of authenticity? Smigielski thought absolutely not. Was there any evidence of a meaningful built past at the site? The 1888 Ordnance Survey map reveals a dense mat of terraced houses interspersed randomly with factories evidently being expanded across homes, not back-to-back tenements but terraces in red brick dignified by gratuitous brick and cast stone embellishments, with gardens and alleys (Figure 3.12a). Photographer Peter Cook's long view of the Queens Building (Figure 3.5) as employed by the Royal Fine Art Commission, shows it set into its very fragmented context evident from the site plan (Figure 3.12b; Short and Chiddick, 1995).

The River Soar is canalized and Mill Street, bounding the north side of the Queens Building site, sways westwards towards it. Robert's city map of 1741 shows the lane clearly delineated through open fields, Spencer's 1879 record

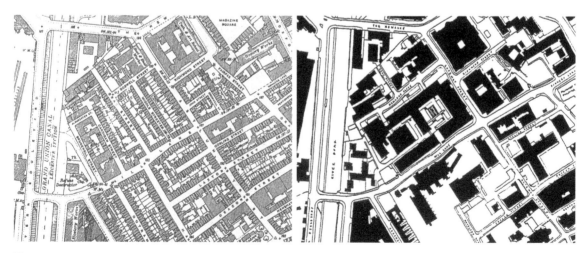

Figure 3.12a
The 1888 Ordnance Survey map of the western reaches of Leicester adjacent to the River Soar shows a dense urban community, and a prosperous one at the turn of the 20th century, largely demolished in the 1960s and left as a wasteland.

Figure 3.12b
The new building endeavours to mend the street and inform alignments of future buildings. Might an environmentally responsive architecture be more capable at mending the urban fabric?

shows Swann's Mills alongside the natural course of the river, gone by 1888 (Figure 3.12a). The inflections survived the 1960s but were left abandoned in a vacuum. The plan for the Queens Building was articulated and flexed to reinforce what had emerged as an ancient route. Was this a clue to the capture of the *genius loci*?

The axonometric drawing (Figure 3.6) shows that the Queens Building is a loose aggregation of component buildings, each a single ridge-roofed prism in brick masonry presenting a gable, its single means of expression, to the world beyond, an archaic form. The fundamental form is clear in the Sepulchral Chapel of Lorsch (876–82) and is the basic unit comprising the urban landscapes of the Hansa towns extending across the north coast of Europe into the Low Countries (Rivoira, 1910). Norman Shaw redeployed it in his designs for houses at 180 and 185 Queens Gate, London and Ernest George's designs for houses at Collingham Gardens, London. Each prism has a masonry stack reconnecting its lower floor to the sky and is narrow enough in section to be effectively naturally lit and cross-, through- and stack-ventilated with no use of energy at all (Figure 3.13).

This plan-forming strategy is non-modern, it is shallow and potentially capable of extreme refinement in design and very flexible in its effective use, the strategy for forming the well-appointed English country house towards the end of the 19th century. This is the architecture documented by architectural historian Hermann Muthesius (1861–1927), which was not lost on Le Corbusier, the first of whose 'five diagrams' seems to represent this sprawling but disciplined English house planning strategy. The additional cost in superstructure is saved in the mechanical and electrical element – the former with a life of a century or more, the latter some 11 to 18 years. The 'free-planning' of the Queens Building and the author's rather later Chadwick Centre for Berkhamsted School drew heavily on Richard Norman Shaw's extraordinary ability to assemble elemental prisms into highly articulate configurations, for example the manner in which the gables at Cragside (1869–84) address the approaching visitor and the forest beyond (Figure 3.14 and Figure 3.15), but in particular the composition of the more bourgeois red brick Lowther Lodge (1872) alongside Kensington Gardens, now the Royal Geographical Society (Figure 3.16).[9] Yet more economical, ingenious and exhilarating was architect William Butterfield's rather earlier inspiration from a Hansa town packed into a very tight, square city site which he used in the ingenious design for All Saints Margaret Street, London in 1850 (Figure 3.17).

Here Shaw also appears to have been borrowing, in part, from the Backsteingotik of the Hanseatic ports, Stralsund, Lubeck and Wismar, the demolished stable block very reminiscent of the 15th century Archidikonat at St. Marien-Kirchhof in Wismar, an archetypal single-ridge roofed prism where the roof plane plunges below and through an arcaded parapet (Figure 3.18).

The buildings do not have large expanses of seamless glass but have sufficient glazing to both light their interiors and to offer prodigious free opening areas to deliver adequate air supply in hot conditions. It is essential to shade the glass. Ernest George demonstrates how to almost fully glaze a façade but defend it by maintaining the continuity of the envelope through it at the house he made for W.S. Gilbert (of Gilbert and Sullivan) on Harrington Gardens,

CRAGSIDE, NORTHUMBERLAND.
R. NORMAN SHAW. R.A.A. ARCHT

Figure 3.13
Richard Norman Shaw, view of/from Cragside, Rothbury,
Northumberland. Reproduced by courtesy of the Royal Academy.

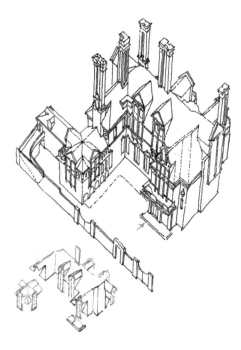

September 1993

Low energy gothic:
Leicester engineering school
Maxwell on Rationalism
Piers Gough hits the beach
Powell Moya's Wansbeck Hospital
Herzog in Hannover
Product: Energy and green issues
Security and ironmongery

Figure 3.14
A slice of the Mill Lane elevation of the Queens Building
taken from the cover of *Architecture Today,* issue 41
(original photograph by Peter Cook).

Figure 3.15
The gabled single ridged roof
developed further as an
environmental device at the
Chadwick Centre, Berkhamsted
School (original photograph by
Paul Riddle).

Figure 3.16
Author's sketch of Richard
Norman Shaw's Lowther Lodge
made during the conceptual
design of the Queens Building.
Shaw achieves a tight
configuration of gabled single
ridge-roof elements with a
Backsteingotik stable block
(bottom left), now demolished.

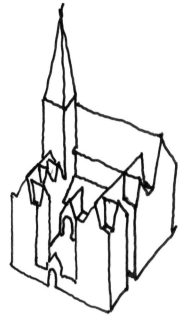

Figure 3.17
Author's sketch of Butterfield's massing of All Saints Margaret Street, an even tighter composition of single ridge-roofed prisms on a square site.

Figure 3.18
The 15th century Archidikonat at St. Marien-Kirchhof in Wismar (author).

London (1882–83) and the northeast facing elevation of No.7 Collingham Gardens, London (1885), his constructor partner Harold Peto's own house, incorporating very large areas of glazing within a fine mesh of stone and terracotta transoms and mullions, which the design for the Queens Building developed into deeper sunshading stone blades offering vertical defence to southwesterly and southeasterly low angle sun (Figures 3.19a and b; Grainger, 2011).

It seemed to be very much more productive to look to the open campus planning of the late 19th and early 20th century Ivy League institutions of North America, most particularly the Philadelphia Quaker architects Walter Cope (1860–1902) and John Stewardson (1858–96) authors of Bryn Mawr College, the University of Pennsylvania (usurpers of Frank Furness's post as campus architects in 1894), Washington University in St Louis and Princeton, rather than the post-war pattern of building at English Old Universities. This is characterized by 'cod quadrangles' with corridors rather than staircases. Here was an opportunity to look again at European collegiate architecture through these American designers' anxiety to deliver instant authenticity, strong characterization, uniqueness of place, exhilarating at its most audacious and exploding familiar collegiate elements across generous lawned landscapes, some by Olmsted, in fierce continental climates.

Figure 3.19a
Glazed gable at the Queens Building defended by brick and stone brise-soleil within the plane of the wall.

Figure 3.19b
A glazed gable façade in Collingham Gardens by Sir Earnest George and Peto (both photographs by author).

The 1992 Universities, confronting a similar dilemma in a milder climate, were asset rich but with a baser provenance to shake off in the viscous British academic class system. Perhaps even more was at stake as higher education identities were struck very quickly in the early 1990s in the aspirant minds of students and guardians.

The environmental rationale for the heart of the building, the Central Building, has been published many times (Figures 3.19 and 3.20).

To the south of an open concourse, the ground floor classrooms for up to 64 students are connected to the sky by 25m high stacks relieving the rooms of their single-sided situation. The classrooms are 8.4m deep with high ceilings of white painted double tee precast car park planks punctured by higher level windows with internal 'dormers' pushing up into the laboratory above. Lancet headed windows sit between the tee downstands to recover high level glazed area, the tees sit on padstones. The stacks and other heavy masonry spaces are lifted into the air on steel props (Figure 3.21) decongesting the main circulation floor from the traditional heavy footprint of a masonry building. This strategy is informed by French architect, theoretician and conservationist Viollet-le-Duc's vision for a hybrid masonry and iron public architecture at a prodigious scale, realized more modestly by Hector Guimard.

Perhaps the occlusion of the lower public floors of load-bearing institutional buildings with heavy masonry drove the now universal application of the light structural frame and all its unintended consequences for public buildings. The line of stacks through the concourse provides the engineering students and staff with an analogous experience to walking along Trinity Street in Cambridge. The fair face masonry wall they buttress is in calcium silicate brick, a pressed

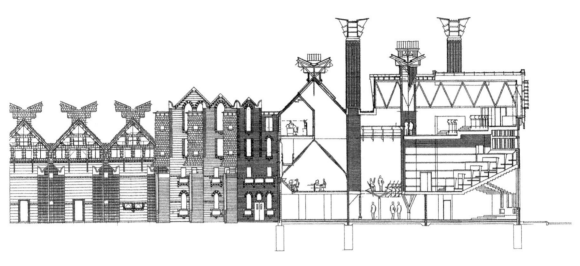

Figure 3.20
Cross-section through the Central Building.

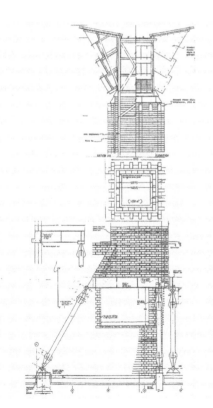

Figure 3.21
Detail of a 25m ventilation stack propped on steel supports to decongest the public concourse at ground level, a device invented by Viollet-le-Duc.

brick to give sharpness and accuracy, permitting startling colours – sienna hues are possible. (Or were possible at that time: the insurance industry subsequently killed this material due to a perceived risk. The bricks will crack if badly detailed. All the brickyards that produced these bricks are now closed.) The interior masonry is different to the exterior, the building develops insideness and outsideness. Controversially, the medium of outsideness is monochromatic, the mortar is the same colour as the brick to emphasize the plane of clay, a device used by George Aitchison (1825–1910) on Lord Leighton's studio building in Holland Park, London also as employed by Louis Sullivan. However the internal masonry walls are striped, the coloured courses being syncopated headers and stretchers in the Morse code developed by William Butterfield for Keble College in Oxford, ridiculed in its time, and highly evident in the Backsteingotik (Clarke, 1964). Butterfield's rhythms are very complex, jumping between alternating courses (Figure 3.22). This was a revelation to the Queens Building architects leading ultimately to the invention of a new brick bond for the School of Slavonic and East European Studies. The more intense the pattern at the Queens Building, the lower the structural load descending through the masonry plane.

Beyond the wall is the Mechatronics Laboratory, 50m in length, a galleried library. It connected to Electronics at the east end, Mechanical Engineering at the west end, in the hope that these constituencies will meet and develop robotic manufacturing machines and techniques. It is now an Institute of Energy and Sustainability. The double-height lab is top lit, therefore much of the double tee plank floor above is removed to admit light and the cut planks are propped by cranked 'spider frames'. Top light is shared with the concourse

Figure 3.22
Detail of the polychromatic brickwork at Keble College Oxford, the architecture of thermal mass (author).

through fire-resistant rose windows, set into the masonry fire compartment wall. None of the construction is shrouded in lightweight gypsum-based boards. At either end the lavatories are stacked in towers, stack vented through voids in the external wall formed by an inner octagonal plan placed within a circular outer plan, the scheme benefitted from scrutiny of William Burges' towers, notably the all header-bonded Melbury Road stair tower.

To the north of the concourse are two steeply raked amphitheatres, wholly naturally ventilated and passively cooled, thought to be an impossible aspiration in an urban environment in the guidance available at the time, glazed to intensify the acoustic tensions, the first water bath modelled building.[10] These were the prototypes for various subsequent free-running theatre auditoria as described in Chapter 5. The natural lighting for the design studios is based on the classic northlight studio form developed by Shaw for the newly wealthy exhibitors at the Royal Academy Summer Exhibition. The large north facing gables are cantilevered beyond the amphitheatres below, formed in timber and clad in cedar shingles after the discovery of the Coxhead twins' (Ernest and Almeric) architectural oeuvre in California in which the shingle skin appears to become quite plastic, folding and lifting over openings like drapery. Such plasticity is unrealized here, sadly, but it was clear that the shingles could take this level of flexing.

Circulation above ground level is directed along steel and lenslight walkways incorporating bench seats for informal supervision and for audiences awaiting an overdue lecturer. Although the device of the inhabited balustrade has been a recurring feature in contemporary architecture, Edward Cullinan's in particular, the geometry is derived from Gaudi's balcony seat cantilevered off the side of the El Caprichio villa 1883–85. The Funding Council allowed 16% of the total floor area as balancing space but thinking only in plan, omitted to specify a required height. The Queens Building possesses no corridors and 11% balancing space but much of it is 12.5m high. The rules were quickly amended.

Triangular prismatic trusses clear span over the Mechanical Laboratory, are clamped to a longitudinal building of smaller laboratories to the southwest and span onto a buttressed and a masonry diaphragm wall to the northeast (Figure 3.8). The buttresses are similarly cellular, in perforated brickwork to make sound-absorbent cells within the damper controlled air supply routes to the periodically noisy activity within (Figure 3.7). Internal heat loads can be high due to the equipment and numbers of people. Air is exhausted through gravity-opening, rain-defying ridge rooflights. The buttresses are substantial, they brace a travelling crane and take the two-way sliding, folding connectors absorbing the thermal movement of the roof trusses, which is considerable. The brickwork is intensely polychromatic, the buttresses step back in homage to the bright buttressed basilican reading room of the Furness Library at the University of Pennsylvania. At the opposite end of the building is the tight trapezoidal courtyard of the Electronics Labs. Heat gains within were predicted to be very high, 60% of the lightweight white elevations to the courtyard open to cross-ventilate vigorously as required. Lightshelves attempt to even out the distribution of light. They transpired to be highly efficient dust collectors. The external masonry elevations are pierced by collections of windows climbing across the floors in raised enclosures to throw light further across the section, others below form an almost continuous chain of top-hung windows behind the heat generating electronic equipment on benches.

Frank Furness's muscular transformations of even the most extreme European architectural inventions and reinventions provided a permanent benchmark for the expressive tenor of the building. They provided a broad backdrop to the problem of making two engine test cells for a tenth of the customary cost. The intense problem of noise reduction to 35dB at the public pavement alongside, a condition of the planning consent, was solved by triple layering the envelope with no connecting ties, the serpentine 'crinkle-crankle' brick masonry outer wall (Figure 3.23), as formalized by Thomas Jefferson at the University of Virginia, cantilevers vertically some 6m, the inner blockwork leaves adopt a more rectilinear geometry but wholly detached from the sinusoidal skin, a physical architectural solution to a problem of environmental physics. Monitored in use, its performance satisfied the local authority's requirements.

Supply air enters at roof level into large brick attenuator boxes, the central stack exhausts the ambient air, engine exhaust is taken up to the top of the stack. An electricity substation sits alongside in a miniature Battersea power station.

This is a recipe for a low carbon building, surprising but effective, developing some unique identity for its new occupants in a very new institution. Its reception is worth reviewing briefly if only to document our continuing surprise at the survival of the Late Modern as the default idiom for public architecture

Figure 3.23
The brick masonry crinkle-crankle envelope containing the noise within the engine test cells (author).

even now being imported furiously into China and constructed at a staggering scale, currently some 20 billion m^2 with a further 20 billion to come.

The Queens Building: Reception history

The building was received in a very particular way in the UK. At the time, most architects were committed to early modern architecture and residual loyalty to late modernism, and displayed a general anxiety about the mysteries of 'sustainability' which was perceived to be beyond the control of the construction industry. Its reception was quite different abroad, particularly in the US, where it was universally well received and lauded as offering a tangible glimpse into a better future. Perhaps it is fundamentally an American building.

In the UK, the *Architects Journal* published early drawings of the scheme, unsanctioned, in its January review of the forthcoming year, 1990. The acerbic critic Martin Pawley reviewed forthcoming projects in a highly entertaining piece and concluded:

> Does any project here embody this same charismatic image power [he is referring to a smokey glass BMW showroom in Basingstoke much filmed for television advertisements]? Perhaps in its way only Peake Short and Partners' monstrous Leicester School of Engineering and Manufacture, with its lavish ambition to bend twenty-first century technology to the discipline of the medieval street. In a sense this was done by Norman Foster at Willis Faber Dumas nearly 20 years ago, but now the trick is to be repeated, and this time in section and elevation as well as plan. If it is built, the result will surely become as great an international symbol as the Albert Memorial.

He coins a new style identifier 'High Tech Ormolu' in an attempt to interpret the elevation or perhaps the plan. Read in haste it is an ambiguous piece but from the proponent of ephemeral architecture it should no doubt be interpreted negatively.[11] The New York design magazine *Metropolis* headlined the arrival of the Queens Building as (Young, 1994):

> Something strange is happening in English Architecture. Two of the country's newest, most forward-thinking buildings are also aesthetically pleasing, functionally complex, and most important, environmentally responsible. What's more, they are university buildings.

The other building was the 'Constable Terrace dorm' at East Anglia, ironically a proto-Passiv Haus design, sealed and mechanically ventilated, the very inverse of the Queens Building strategy and questionable in its resilience in the hotter weather of a changing climate. *Metropolis* continues: 'Though on the cutting edge of environmentally sensitive architecture, it is not a glass and steel concoction by one of Britain's vaunted high-tech architects … instead it is a surefooted, red brick and cedar shingled Post-Modern fantasy'.

One is tempted to quote more from this laudatory piece (Young, 1994):

> Happily … the Queens Building is as bold and visionary as the architects who conceived it. A truly holistic structure, it is designed from the inside out. And, it breathes, drawing fresh air in through its panoply of odd shaped

windows, perforated brickwork and beehive of louvers that adorn the outside walls. Stale air is exhaled through ventilators concealed in the ceiling's ridge lights and gargantuan brick chimney stacks.

The first formal publication of the completed scheme was in the UK journal *Architecture Today*. This was a matter of fact presentation of general arrangement drawings for construction, with a crayoned part elevation on the cover giving some impression of its redness (Figure 3.24; Short, 1991).

Shortly after the completion of the Queens Building, the *Sunday Times* critic Hugh Pearman visited the building and toured Leicester. His piece (Pearman, 1993) delighted in the apparent confrontation of the old new university (University of Leicester) by the upstart polytechnic turned 'new, new university' (De Montfort University). Musing on the *de facto* if unintentional confrontation, not least with its own Engineering Building designed by James Stirling and James Gowan 30 years earlier (perhaps the most iconic post-war piece of architecture in Britain for Stirling's generation). Pearman wrote of the Queens Building: 'it is perfectly clear that the *building*, looming up unexpectedly in its slightly wrong-side-of-the-tracks bit of town, knows what it's up to. Its self-appointed task is to gobsmack.'

He continues:

The easy reaction to this thing would be to condemn it as a time-expired piece of post-modern camp. At first glance its picturesque jumble of forms recalls some of the stumbling efforts of architects at the start of the 1980's to come to terms with public hostility. Conventional office blocks or shopping centres were done up as Camelot to disguise their bulk and function, at least for a time. Such cynicism finds no place here. Odd though it may seem, this is a rigorously functional building, reflecting in its shape everything that is

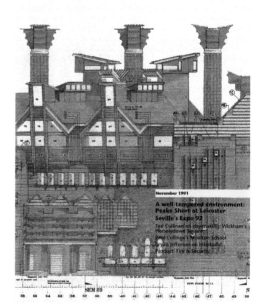

Figure 3.24
The first outing of the emerging design for the Queens Building.

happening within, from lecture theatres to laboratories and design studios. It is as if a number of smaller buildings, each utterly different in form, had been subsumed into a whole, with all the lumps and protuberances that such an approach implies. It is an approach that Stirling was thoroughly familiar with.

His picture of a collection of buildings and the capture of the spaces between is vivid. He concludes by identifying it as 'New Romantic', over-rich, identifying the resonances with Shaw and Furness, praising the truly inspired structural engineering by Stephen Morley and deciding:

> For the students, there is no sense of the second-rate prevailing here. It has, rather, something of the ritualistic, slightly theatrical quality you might associate with student life on the older universities … the building is clever and satisfying.

For the architect, this was the whole point of the exercise. It was a Finalist in the Royal Fine Art Commission Building of the Year competition, it won the Green Building of the Year Award, eliciting the bizarre headline 'Medieval building wins green award' (Nicholson-Lord, 1995). Others, campaigning on different territory seized on the apparent 'Gothicness', even though as Jencks pointed out, it was Queen Anne in idiom if anything. Historian and critic Alan Powers (1995, pp. 42–45) urged readers of *Perspectives* to think beyond the crude and artificial disjunction between Classical and Gothic tastes:

> The coincidence of the newly revealed St Pancras and the crypto-Gothic green Queens Building suggests that the Gothic needs to be reconsidered in the current pluralist mixture of British Architecture … [its] energy-conscious design, with solid walls enclosing big atrium spaces is a pure example of 'form follows function', and exposes how a functional image rather than real function controls the design of standard modern buildings.

He asks, 'Is Gothic the archetype of natural, green architecture?' A most interesting observation. He doesn't then mention German Romanticism, Caspar David Friedrich or Karl Friedrich Schinkel in this piece, perhaps because it would seem so absurd to relate their ideas to an industrial site in Leicester, but is that as absurd a prospect as imagining the tight courtyards of the ancient universities in an institution starting its career at the other end of the league table? *Architecture Today* reviewed the completed building at length, architect Peter Rickaby wrote an informed summary of its green aspirations and attributes (Swenarton and Rickaby, 1993). Further pieces appeared, citing its unfamiliar idiom in an entirely matter-of-fact way, 'Green Gothic' in the AIA's journal, the new president devoting a lot of her New York inaugural speech to the building (Davies, 1995). *Newsweek* was universally supportive (Menagh, 1995), and it featured in the November issue of *Archicree* (Lefevre, 1995).

Critic Charles Jencks announced that the building had a very special place in the architecture of the time. Jencks (1995, pp. 97–99) wrote:

> I do not believe any historical labels do justice to the synthetic thinking behind the building which borrows as much from the contemporary pavilion planning of Frank Gehry as the tradition of thin industrial structures.

Perhaps the most important commentary is not by a figure in the architectural orbit but by the founder and chief scientist of the Rocky Mountain Institute, physicist Amory Lovins and the economist Ernst von Weizacher (Lovins *et al.* 1997, pp. 24–25), citing the Queens Building as one of their '20 Examples of Revolutionising Energy Production' in their open letter to the Club of Rome. This is a recognition of the radical low-energy strategy and solution, addressing the concern to radically reduce energy consumption. Implicit in their praise is that such solutions are economically and practically viable. This is a solution that can be implemented today.

Architecture Today's 100th issue in July 1999, selected its ten top buildings of the 1990s. The co-editor Mark Swenarton concluded:

> ... the project that more than any other embodied the new thinking was the Queens Building ... [it] combined a radical proposition about energy – with an exuberance that made full use of the tectonic potential of passive environmental control ... what is most commendable is the fact that these architects were prepared to take on both the environmental and the architectural challenge in the same project, rather than ducking out of the one in order to pursue the other more easily.[12]

In his book, *Ecological Architecture*, James Steele (2005) concludes his chapter on the Queens Building in a similar vein, but celebrating the new freedoms of expression brought by post Post-Modernity here pressed into the service of a more ecologically sound world:

> The tall ventilation towers, which are the main breathing apparatus of the building, recall those on the ovens that used to make those bricks, reinforcing local memory. These towers have an equally essential purpose, not only replacing dependence on costly and energy-intensive mechanical systems, but also demonstrating to a skeptical global public that this dependence can be broken through a return to traditional methods augmented by appropriate new technologies.

The Lanchester Library

The Lanchester Library and learning resource centre at Coventry University transfers the thin-building low-energy design strategy of the Queens Building into a deep plan building. The deep plan type is much preferred by developers, one could say it has become a 'social practice', although the rationale behind the preference is becoming increasingly unstable. The Lanchester Library is a large, bulky building in a major city centre site containing 350 000 books. It sits on reclaimed industrial land on an appreciable slope, the former site of Morris Motor's first industrial-scale car factory. It is wholly naturally ventilated and passively cooled and it works well, as established by a rigorous post-occupancy monitoring exercise by another relatively disinterested research team (Figure 3.25). It was inexpensive to build, £17 million for 110 000ft^2 (10 200m^2) and much scrutinized and published globally, particularly in China (Krausse *et al.*, 2007; Short, 2004; Short *et al.*, 2004).

Figure 3.25
Aerial view of the Lanchester Library in an
aggressive city centre environment (author).

Whereas the Queens Building ventilation strategy is essentially 'edge-in/centre-out' in Professor Kevin Lomas's terms, the Lanchester Library is inverted, 'centre-in/edge-out' (Ji and Lomas, 1999; Lomas, 2007). The Chief Librarian's brief required a highly sustainable design by contemporary standards but, counterintuitively, insisted that the 110 000ft^2 take the form of an open, square plan unimpeded by stairs, toilets or service cores. This immediately implied a very deep plan building, a completely baffling prospect to the design research team. A plan this deep, 50m square, was presumed to be impossible to light naturally. Therefore it would generate continuous lighting heat loads. It would be impossible to thoroughly cross- and through-ventilate such large floor plates to dissipate substantial heat gains from the predicted 2500 users (Cook *et al.*, 1999a and b).

Six of the seven preliminary schemes proposed placed a single large atrium somewhere in the plan. In this much favoured atrium plan type, the principle beneficiaries are immediately adjacent to the perimeter of the atrium, but a simple Fresnal Square calculation (developed at the Centre for Built Form and Land Use Studies, University of Cambridge) showed that good natural light and ventilation was significantly less well distributed across the plan in terms of reader places served than that of regularly distributed smaller lightwells, given a beneficial ratio of lightwell height to depth. The less well-lit spaces between lightwells could harbour books but never far from reader stations (Figure 3.26).

Contemporary librarians generally think orthogonal arrangements of book stacks give maximum current and future flexibility. Sadly for architects, librarians believe that radial panoptical plans are less efficient in achieving the density of stored books required. The design creates a variant of a 'mat-building' plan (Figure 3.27; Sarkis *et al.*, 2001).

Mat-buildings are non-hierarchical, loosely organized on a recurring grid, and tend to be, but not necessarily, asymmetric, low rise and ragged at the edges. Sarkis *et al.* (2001) identify Candilis–Josic–Woods' Free University of Berlin (1963–73) and Le Corbusier's Venice Hospital (1964–65) scheme as exemplars. The origin of the term is credited to Alison Smithson in her September 1974 *Architectural*

GROUND FLOOR PLAN

Figure 3.26
Ground Floor plan.
1. Corner lightwells (air in).
2. Library issue desks.
3. Rooms joined directly to the air supply.
4. Main public stair.
5. Escape stair.
6. Main entrance.
7. Entrance arcade required by the university.
8. Research group annex.
9. Bookshop.
10. Café.
11. Energy Centre for this part of the campus.
12. Reprographics workshop.
13. Lettable shop.
14. Bibliographic services.

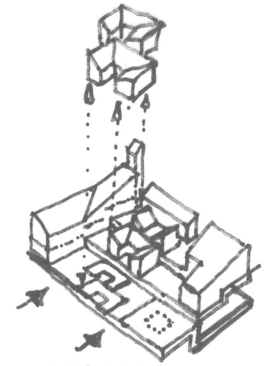

CLARE HALL
CAMBRIDGE
RALPH ERSKINE

Figure 3.27
A mat-building. Author's sketch of Ralph Erskine's 1967 Clare Hall
Cambridge. Neutral planes onto which are placed archaic type
buildings: courtyards, L- and U-shaped villas, pavilions and a tower.

Design article, 'How to Recognise and Read Mat-Building: Mainstream Architecture as it has Developed Towards the Mat-Building'.[13] The Lanchester Library is a very much more formal variant but, and perhaps this is the most important mat building characteristic, the plan form can be replicated and tessellated *ad infinitum* within its own adjacency rules even with the infrastructure to support naturally driven ventilation. Each quadrant of the library footprint is punctured by a lightwell that admits both fresh air and daylight into the building. Surprisingly, but necessarily, it is possible to part-cellularize the floor plans to predetermined rules by connecting spaces directly to the air inlets and allowing the passage of stale air out of these spaces at high level (Figures 3.28 and 3.29).

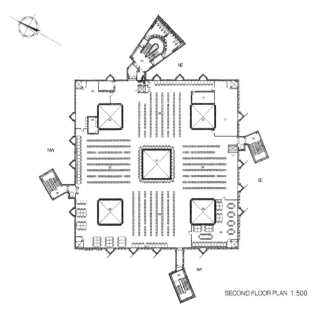

SECOND FLOOR PLAN 1:500

Figure 3.28
Level 2 plan.
1. Central atrium (air out).
2. Lightwells (air in).
3. Book stacks orthogonally arranged.
4. Main public stair.
5. Escape stair.
6. Silent Study room.
7. Seminar Room.
8. Subject Librarian's Office.
9. Photocopy Room.
10. Group Study Room.
11. Lift.

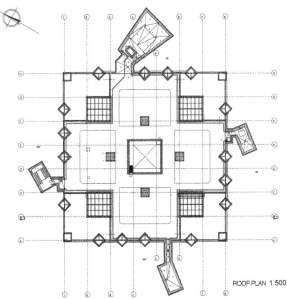

ROOF PLAN 1:500

Figure 3.29
Lanchester Library: Roof plan
showing stacks and atrium.

Larger teaching spaces span from the lightwells to the perimeter stacks, thus providing dedicated, but controlled, ventilation to these areas. For other spaces, acoustically treated transfer ducts permit the passage of air through intermediate spaces. More specifically, air is supplied to the four lightwells (Figure 3.30) from below through a 1.5m-high plenum of precast concrete planks top and bottom, the mass is important, located below the ground floor construction and supplied by intake air from each principal orientation. The plenum is compartmented but connected via dampered openings so that pressure differences created by the various wind directions can be equilibriated before supply into the occupied interior.

Delivering the building physics

Preheating coils occupy the whole base of each lightwell to temper the air in winter. Internal heat gains and the natural pressure differences between low level entry and high level exhausts draw air up through the supply lightwells and into each of the four high performance dampers. These devices were specifically developed by Landis Staeffer for the project and are designed to be airtight when closed.[14]

The floor-to-ceiling dimension of 4m and the complete absence of suspended ceilings (as at the Queens Building) assists the natural buoyancy-driven displacement strategy by maintaining warmer, 'vitiated' air above head height. The concrete floor structures are fully exposed to the air volume within to provide the necessary thermal storage, particularly for night ventilation induced coolth by lowering the radiant temperature of the soffit in summer, a very important component of the passive cooling.

In winter, the warmed air at each floor rises towards the soffit and passes beneath the soffit to the points of lower pressure at the perimeter and in the centre of the plan. The steel beams are castellated, punctured by circular openings, so as to provide for free movement of air across the soffit. There are

Figure 3.30
Lanchester Library: Air supply strategy.
1. Fresh air inlet.
2. Fresh air supply plenum.
3. Trench heating.
4. Heater batteries.
5. Lightwells providing ventilation and daylight.
6. Building energy management system-controlled louvres.
7. Ventilated void.
8. Retractable translucent blinds.
9. Well insulated roof.

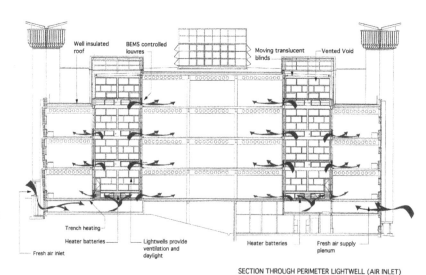

SECTION THROUGH PERIMETER LIGHTWELL (AIR INLET)

■ Fresh Air intake

no concrete downstands to obstruct the flows of air. The air is exhausted through 20 stacks at the perimeter, substantial towers 2.4m in diameter, the minimum free area required for the volume served (Figure 3.31).

The necessary order of magnitude required to move such a huge air volume several times hourly obliges these elements to become architecturally charged objects. A central lightwell is also configured to exhaust air from the lower three floors. Ground, first and second floors all connect into the same stacks, but computational fluid dynamics (CFD) analysis revealed a potential risk of backflow into the upper floor if it shared the same outflow routes as the lower floors (Figures 3.32a and b). The plenum fed, distributed void idea would certainly work more comfortably at three storeys instead of four but that is quite feasible and we will explore the three storey mat as a hospital design strategy in Chapter 6.

Research underpinning practice
Modelling exercises are frequently post-rationalizing compliance exercises. However, for this design research team, parallel modelling using different techniques is used to reveal unsuspected anomalies over extended periods of time, the principle complementary techniques being digital and physical although there is a strong belief in simple calculation of heat transfer and ventilation strength. As a result of the modelling analysis, the exhaust from the upper floor was decoupled from the lower floors and four additional stacks were provided to exhaust the upper floor only, just in time to catch the steel frame package. All the stack terminations are maintained at the same height to prevent U-tube effects driving flows down windward stacks and up those in the lee. In fact downdraughting was measured *in situ* by colleagues at the BP Institute for Multi-phase Fluid Flow at the University of Cambridge and it may be that this much feared and unanticipated mechanism is responsible in part for achieving reliable performance.

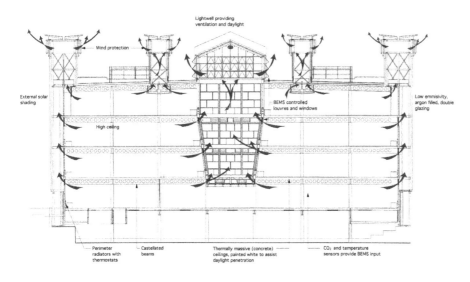

SECTION THROUGH CENTRAL ATRIUM (AIR OUTLET)
■ Warm Exhaust air out

Figure 3.31
Lanchester Library: Air exhaust strategy.
1. Carbon dioxide and temperature sensors providing input to the building energy management system.
2. Perimeter radiators with thermostats.
3. Exhaust dampers at a high level on each floor.
4. Castellated beams.
5. Thermally massive (concrete) ceilings, painted white to assist daylight penetration.
6. Lightwell providing ventilation and daylight with solar shading.
7. Stack termination with wind protection.
8. Building energy management system-controlled windows.
9. Dedicated stacks to the third floor.

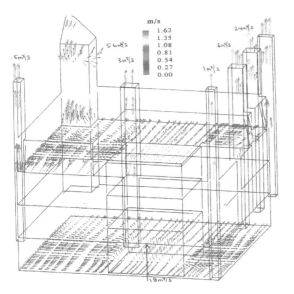

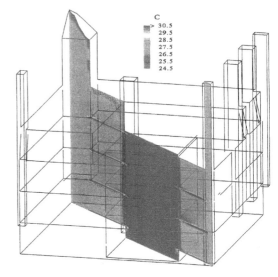

Figure 3.32a
Lanchester Library: Computational Fluid Dynamics analysis of airflow through shared stacks (Professor Malcolm Cook, formerly of De Montfort University).

Figure 3.32b
Collapse of stack pressure after a period, warmed air bleeds back into the upper floor (Professor Malcolm Cook, formerly of De Montfort University).

The stack terminations are derived from a sketch drawn for us by Professor Tom Lawson, Emeritus Professor of Aeronautical Engineering at Bristol University, proposing parallel racks of split tubes offset half a module at each layer to catch the wind on any face and generate so much turbulence that all force was spent. Warmed air would continue to rise driven by relatively much smaller pressure differences and theoretically the potential for downdraughting diffused.

The resulting stack tops appear to resonate with the design of car radiators, not least those of the mid 20th century Rolls-Royces, earning immediate local empathy in this UK motor city (Figure 3.33).

As the pressure differences are relatively so small in passive buoyancy-driven ventilation systems, the external wind pressures have the capability to drastically disrupt the proposed regime. A physical 1:200 scale model of the design set into its urban context, a matrix of blocks arranged across a 2m diameter table and a detailed model of an exhaust termination were tested in the Welsh School of Architecture wind tunnel under the eight standard wind directions. Into its exterior surface were drilled 157 pressure taps to align with intended intake and exhaust positions. Air was driven across the model at 8m/ sec, simulating a considerable wind. The experiment demonstrated a robust and successful ventilation strategy within its own terms. The results are expressed as a 'mean wind pressure coefficient' Cp, the ratio of the pressure recorded at the point on the model relative to the prevailing pressure. Negative pressures imply suction off the surface, drawing air out at that point or positive which suggests air will be drawn in at that point. They vary from +1.5 to –2.0. The output data is made usable by calculating the mean of the pressures recorded over time giving the 'mean point pressure' P, in Pascals.[15] The modellers explain that the intention is to maximize the wind pressure difference

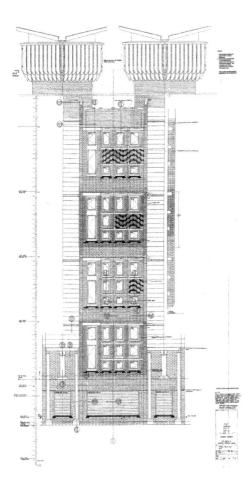

Figure 3.33
Lanchester Library: Bay of elevation
showing stack terminations.

between the low level inlets and the high level exhausts, a robust positive wind pressure difference to drive buoyancy flow, the natural stack effect, up through the building. A negative wind pressure gradient will suppress naturally driven flows and, beyond a certain level, flow will reverse, thought to be unacceptable at the time. The modellers suggest that at a temperature difference between inside and outside of 15°C in still conditions, a winter scenario, a reliable 15 Pascal buoyancy pressure could be expected. However, in the summer condition of a mere 3°C difference, a much lower and more fragile 3 Pascal pressure difference might be generated. Northwesterly winds on the northerly façades induce the least helpful situations on those façades, but boost the pressure difference for façades in the lee. This suggests, and as the BP Institute team detected, reverse flow will occur between stacks on the façade with higher positive pressure coefficients and those opposite. The pressure differences generated are relatively minute in comparison with fan-driven systems. There is no mechanical heat recovery, the system is so distributed (as it is in the Queens Building) that it could not be economical using existing technologies. Ironically, the unintended mixing regime will preheat the supply air through the outgoing exhausted air. Pressure gradients for each stack were measured for the eight wind directions. Stack 17 on the northwest façade achieved a positive gradient of 1.5 under a northwesterly wind and will tend to downdraught in this scenario. The colder outside air has a long way to travel down the insulated stack before striking an occupant and so there is no history of cold draughts being reported. Would we have designed the building differently to exploit reversing flows so that the building becomes a kind of pumping engine above a certain wind speed? Probably, the next generation of this environmental design approach remains to be invented.

The university very deliberately defined the building as a 'Library and Learning Resource Centre'. Small group teaching is permitted on the main library floors. There is an ebb and flow of self-suppressed noisiness. Talking, mindful of others, is permitted across the floors, except in designated quiet areas, so that active supervisions can take place. Although the building was designed for 2000 visitors, it immediately attracted 6000. It became the daily base for many students in the intervals between organized teaching events across the campus. It is strangely silent, there is no background noise from mechanical ventilation. Might the Coventry students derive similar but healthier solace from their library as Thomas Mann's character Hanno Buddenbrook did from his Form Room on one hellish Monday at school, entirely unprepared to recite Ovid from memory (Mann, 1901, p. 567):

> The first lesson, Herr Ballerstedt's class in religious instruction, was comparatively harmless. He could see, by the vibration of the little strips of paper over the ventilator next the ceiling, that warm air was streaming in, and the gas, too, did its share to heat the room. He could actually stretch out here and feel his stiffened limbs slowly thawing. The heat mounted to his head: it was very pleasant, but not quite healthful; it made his ears buzz and his eyes heavy.

Architectural apparatus

The simple box-like bulk of the building, its corners occupied by vent stacks and risers and therefore unglazed, reduces the palette of the elevations to vertical bays between diagonal stacks. They are articulated only in the attempt to shade the glazing within the depth of the masonry build up of the elevations. The recurring architectural problem evokes Louis Sullivan's ever prescient question (1896):

> Problem: How shall we impart to this sterile pile, this crude, harsh, brutal agglomeration, this stark, staring exclamation of eternal strife, the graciousness of these higher forms of sensibility and culture that rest on the lower and fiercer passions? How shall we proclaim from the dizzy height of this strange, weird, modern housetop the peaceful evangel of sentiment, of beauty, the cult of a higher life?

Sullivan advised:

> ... throughout the indefinite number of typical office tiers, we take our cue from the individual cell, which requires a window with its separating pier, its sill and lintel, and we, without more ado, make them look all alike because they are all alike.

From our 'environmental' interest we might take issue with the advice that all windows should be the same in a large public building because they are the same. Their situations vary and they can be tuned to respond. This text, so literally and gratefully seized upon by generations of compliant architects resilient to Sullivan's exasperation at the commercial arena and his deep sense of irony, may have caused the loss of many opportunities. The Lanchester Library elevations are tuned to the exigencies of each orientation in their mute way and then more locally to the condition behind. Figure 3.33 shows some of the variants in filling and opening voids within the brick and stone shading matrix. The north elevations are further tuned so that the free glazed area grows and shrinks as the elevation passes in front of the atria and bookstacks behind. The southwesterly and southeasterly elevations are quite different in character, being heavily modelled in cast brick and stone to form the brise-soleil within the depth of the deep wall. The window arrays are configured so that the sun will kiss the stone window transoms and cills on 21 June (Figures 3.34 and 3.35).

Regulations and governance
The ventilation strategy for this building develops a design entirely at odds with the prescriptive model of compliance with the Fire Regulations set out in Part B of the Approved Documents comprising the UK Building Regulations at that time. This raises a series of issues and challenges surrounding how our buildings are regulated and whether regulations and other forms of governance help or hinder the desired transition to a low-carbon society.

Figure 3.36 shows a naturally ventilated building with floors compartmented and discharging smoke into a fire-glazed atrium to be vented out of the top of

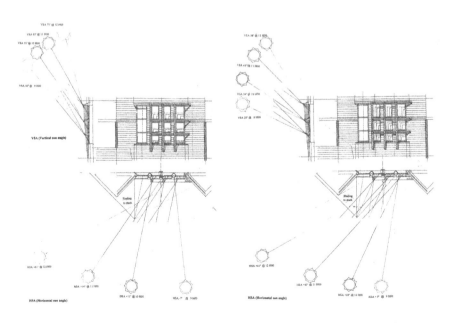

Figure 3.34
Lanchester Library: The passage of the sun across a bay of the east elevation through 09.00 a.m. to 12.00 p.m. on 22 June and 21 March/September. The dashed line indicates the extent to which the proposed vertical shading attached to the stacks will cast shadows onto the facade.

Figure 3.35
Lanchester Library: Detail of a stack, a termination and the brick and stone grid of a glazed bay (*Brick* Bulletin, Summer 2001 issue, original photograph taken by Paul White).

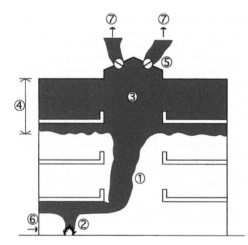

Figure 3.36
The principles of smoke venting an atrium building. Compartmentation: an atrium open to all floors.
1. Atrium volume contiguous with adjacent floors.
2. Seat of a fire, with smoke discharged into the surrounding space. The stack effect drives smoke generally upward to the atrium head.
3. The upper part of the atrium and adjacent floor act as a smoke reservoir.
4. Depth of smoke layer.
5. Controllable vents at the atrium head.
6. An air intake to a localized area around the fire.
7. An air outlet for the building as a whole.

the building and the Lanchester Library as a huge single volume or 'compartment' of 40 000m^3 with floor plates of 2230m^2, whereas the maximum compartment size described in the then current Approved Document Part B of the UK Building Regulations (DETR, 1995) was 800m^2 on a single level. The environmental strategy of the design would appear to be fundamentally impossible to realize.

In the built Lanchester Library, five lightwells puncture the floorplates, the glazing is not fire-rated and, at each level, rows of opening lights freely admit air supply to the floors. However, the regulations allow for alternative approaches to demonstrate compliance, as they do now in some US states.

(The situation in Illinois is considered in Chapter 8.) An alternative approach to demonstrate compliance is potentially time-consuming, risk-laden and the outcome is unpredictable. Understandably, this is anathema to the project management industry.

This issue is extremely important to the future realization of large, very low-energy, low carbon non-domestic buildings. Therefore, the author (Short *et al.*, 2006) captured the negotiations and the evidence presented to make the case for the Lanchester Library and the School of Slavonic and East European Studies in a research paper. This evidence can be used by clients and designers in the future to help overcome the hurdle of alternative routes to compliance.

Another form of governance is building insurance, which adds significant complication. Although the preservation of property and contents is not paramount in the legislation relating to fire and buildings, it is very much in the minds of insurers whose requirements for the issue of cover may exceed those of regulatory compliance. Should an insurer perceive an additional risk to a project, the premiums quoted will, of course, be higher and/or cover will be restricted. The restriction on cover could be particularly significant with respect to, for example, a library's book stock.[16]

These risks, or at least the perception of them, are extremely important determinants in the campaign to achieve a very significantly lower carbon built environment. It is not quite clear how actively the insurance industry, clearly alarmed at increasing risk from extreme weather events, is supporting measures to achieve the national carbon reduction target. Researcher Andrew Dlugolecki calculated a notional one-off 2–4% increase in annual premiums was required to cover this emerging field of risk.[17] Insurers may not realize the unintended consequences of their approach to risk on project managers' behaviours and may not care. The 2009 Chartered Insurance Institute report on climate change expresses useful sentiments but is nonetheless wholly candid in its Executive Summary:

> Climate change may affect liability risk. Insurers' clients may be exposed to claims that they did not reasonably foresee the possibility of weather-related damage (direct-link). Insurers should identify high-hazard categories, e.g. construction industry professions, and require policyholders to undertake regular climate change risk assessments, both on a business-wide basis and for specific projects … Insurers must engage in construction projects early to ensure that planning and design addresses these risks.

Perhaps this is the way forward, early cooperation based on full disclosure to contain Professional Indemnity Insurance premiums (Dlugolecki, 2009).[18]

Under the section 'Commercial Property' in Chapter 18, 'Sustainability as a principle in risk management of climate change', the authors record:

> Commercial property is generally exposed to almost identical issues in relation to the physical effect of increased risks from climate change as domestic property, but in addition the properties tend to have further aspects due to their intended use. Finding sustainable solutions, for both the building of new commercial property and the repair of existing stock damaged by an insured event, to assist in the reduction of energy usage are key here,

because commercial buildings tend to have much higher energy requirements than homes.

More encouragingly, Chapter 17 mentions passive air conditioning systems:

> … where the use of "chimneys" to pull cool air into the building, using the convectional draw created by the waste heat in the building, make good sense and require little or no energy to operate. Examples of similar projects become more common as new ideas are tried out, but the important element here is that post-loss repairs will require the specialist knowledge of these new techniques which may be both difficult to secure and expensive. Loss resiliency and severity is also untested.

The use of passive air ventilation is an additional challenge to the implementation of a nationwide adaptation strategy, which may not be fully appreciated by policy-makers (ibid. Chapter 18, p. 6). In their recommendations the authors take their constituency back into the cold shower: 'Insurers will need to take account in their pricing and claims management activities of changes in risk caused by the use of sustainable techniques and materials.'

The collapse of an environmental design strategy at an advanced stage in a design and construction programme would have potentially awful time and budget implications for all involved, including investors. This possibility will be perceived as a significant risk by an increasingly risk averse industry. The fire safety exercise described here took more than 18 months to achieve a resolution, an extraordinarily stressful experience, as did the later negotiation to achieve the SSEES building.

A further complication is an unintended consequence of the requirements of central government's Modernization Agenda for the Fire Rescue Service. This now requires Integrated Risk Management Plans from brigades, which are likely to delay the moment fire-fighting commences on-site, as the intent is to save firefighters' lives.[19] Formerly, strategic design judgements were made on the basis of the likely behaviour of a fire 10–15 minutes into its progress. It is now more likely to be 20–25 minutes and at a T^2 growth rate the fire will be considerably more advanced.[20]

Figure 3.37 shows a notional code-compliant Lanchester Library scheme, divided into 800m² compartments, four per floor with one-hour enclosures between them including the lightwells, wholly negating the possibility of a naturally ventilated scheme and adding some £800 000 (5%) to a £17 million project. Figure 3.38 shows how it could have been configured to be sprinklered throughout with mechanical smoke ventilation (Figure 3.38), adding in this case rather more at some £926 000.

Sprinklers are a concern in naturally smoke-vented buildings. The smoke temperature needs to remain significantly above ambient, some 30°C, or buoyancy will stall and smoke will stagnate and build rapidly. Of course, in a conventional compartmented building, sprinklers remove heat and reduce the possibility of flash-over temperature being reached.

The built design, described in Figure 3.39, was based in part on the guidance in BS 5588 Part 7 using its criteria for demonstrating adequate fire safety performance.

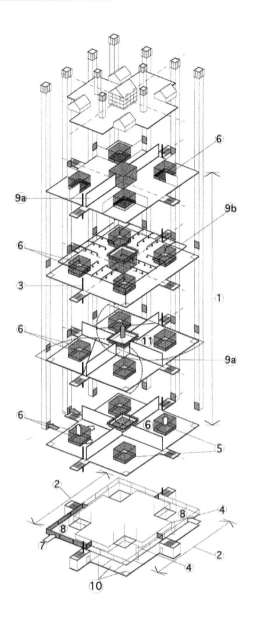

Figure 3.37
Lanchester Library: The code (1997) compliant option.
1. Height of the top floor below 18m, implying a 60-minute fire resistance between compartments.
2. Minimum of 50% of the building perimeter immediately available for a high-reach fire appliance.
3. Dry rising mains (DRM), all areas of a floor reached within 45m of DRM landing valves.
4. The basement requires natural vents equal to 2.5% of the floor area (27.5 m^2 free area), or a mechanical smoke extract to achieve a minimum of ten air changes/hour.
5. 1-hour fire-rated and smoke-sealed dampers at all lightwells to fail shut.
6. Lightwell-glazed side walls installed to achieve a 1-hour fire-rated assembly of frames, seals and glazing.
7. Plenum air intakes are rated to 1-hour to fail shut.
8. Plenum sub-compartmented, connected via 1-hour fire-rated dampers to fail shut.
9a. Each floor plate is subdivided into compartments at 800m^2 or less; or (9b) alternatively, to smoke extract and sprinkler mechanically, requiring provision of stand-by power.
10. Three additional plant rooms for the generator, sprinkler pumps and mechanical ventilation consuming the usable floor area, if the latter strategy is pursued.
11. Slot extract at the perimeter of the lightwell/atrium to provide mechanical ventilation if this is thought to be preferable to a fire-rating-glazed atria to achieve 1-hour fire resistance.

It was accepted that students and staff would become familiar with the layout through induction classes. It was agreed that the lightwells/atria would not be fire rated, could be part open but would be smoke sealed. The top of the atria/lightwells would be extended to form smoke reservoirs and a pattern of fail open/closed damper positions would be agreed. In all approaches an adequate fire and smoke detection system was *de riguer*. Building Control and the Coventry Fire Officers would not accept CIBSE guidance based calculations of smoke clearance, hence the need for CFD analysis to investigate potential solutions. The detailed specification of the CFD model is described in the research paper. Three fires were modelled, a 1MW steady state and a 1MW growing fire in an open plan area on the ground floor, 1 and 2MW fires steady

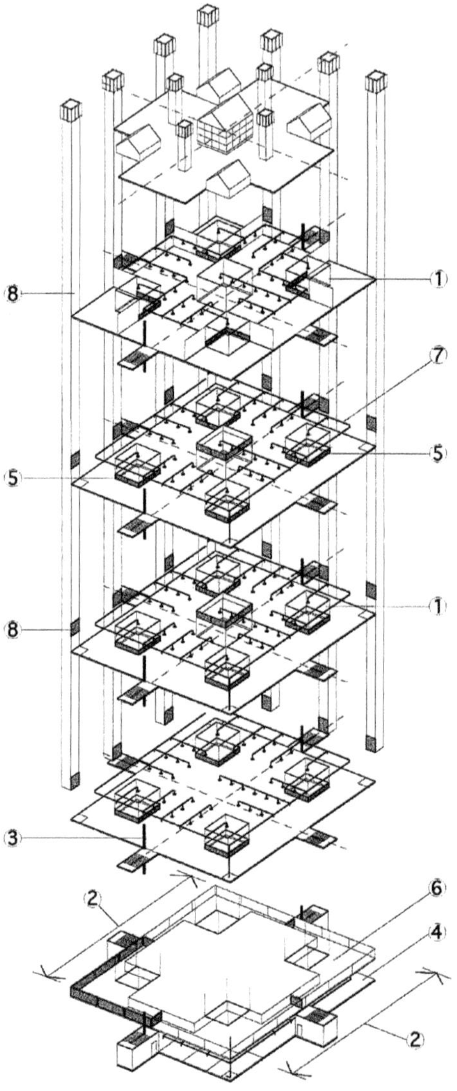

Figure 3.38
Lanchester Library: The sprinkler option, alternative fire strategy to BS 5588 Part 7.
1. Lightwell walls of the smoke-retardant construction to remain unbreached for an agreed period until the first fire brigade intervention.
2. Fire-fighting access as required by the Approved Documents.
3. Dry rising mains provided.
4. Ventilation of the basement to achieve ten air changes/hour.
5. Smoke-sealed dampers at all lightwells fail shut if smoke ventilation is mechanical: those connected to fire floor are open whilst all others are closed if natural smoke ventilation is proposed.
6. Plenum uncompartmented in terms of fire resistance.
7. Full sprinkler system to all floors is fed from the plant room on Level 1.
8. Dampers to perimeter stacks fail shut if smoke venting is mechanical: those on the first floor fail open if natural smoke driven ventilation is employed.

state at 15 and 17 minutes adjacent to a lightwell and a 5MW steady state fire in the centre of the ground floor to determine the likelihood and timing of flash-over in an open-plan area, expected at 500–550°C, thought to depict the most hazardous combination of circumstances. Figure 3.40 shows the simulation under this latter circumstance. It is evident that the building configuration is draining smoke as fast as it is produced. Light smoke appears in the atrium at 30–40°C at the first floor level.

Based on the original fire specification, a fire of this size would develop after approximately 21 minutes. At this time safe evacuation would have been achieved and the fire brigade should have been tackling the fire for some minutes.

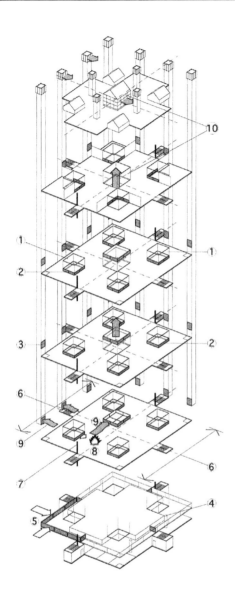

Figure 3.39
Lanchester Library: The final built fire strategy as brokered with Building Control and the Fire Brigade 1996–99.
1. Lightwell/atria not fire-rated but smoke-retardant for an agreed period, equivalent to the expected first intervention by the fire brigade.
2. Smoke-sealed dampers, but lightwells on the fire floor fail open, all other dampers fail shut.
3. Smoke-sealed dampers to perimeter stacks fail open, all others fail shut.
4. The 'basement' requires natural ventilation free to the area to achieve a minimum of 2.5% of the floor area (27.5 m²).
5. Plenum air intakes remain open to supply airflow to the fire affected floor.
6. A minimum of 50% of the building perimeter is immediately available to a high-reach fire appliance.
7. Dry rising mains (DRM): all areas of the floor reached within 45m of DRM landing valves.
8. A potential fire on the lower floor.
9. Smoke exhausted via day-to-day natural ventilation scheme.
10. A smoke reservoir formed by the top of the lightwell.

Does it work?

A post-occupancy monitoring team lead by Birgit Krausse (Krausse *et al.*, 2006 and 2007) analysed data gathered in the building over the years 2004 and 2005, publishing the findings in a number of research publications. The authors record, 'The building has been a marked success'. The researchers were based in the Institute of Energy and Sustainable Development (IESD) at De Montfort University, the same institution that modelled the design but the results are peer reviewed by a respected journal in the field. The monitoring period included an unusually hot spell with outdoor temperatures exceeding 30°C, peaking at 35.4°C. Cook *et al.* (1999a) predicted that 'dry-resultant temperatures would always be below 28°C and that 27°C would be exceeded for only 11 hours of the year'. Dry Resultant Temperature (DRT) includes a radiant component. The

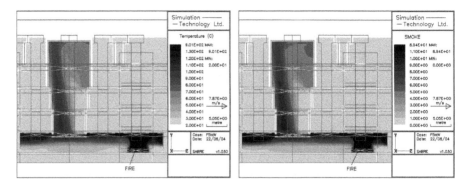

Figure 3.40
CFD simulations of smoke behaviour and temperatures in a vertical plane through the building deriving from a 5MW fire in the centre of the ground floor (Geoff Whittle in SABRE).

mass of the library is intended to suppress summer temperatures through beneficial radiant exchange between people and the exposed brick masonry and concrete construction. Moreover, refinement of the Building Energy Management System (BEMS) controls in service was expected to 'be capable of reducing internal temperatures even further'. The predictive analysis suggested that the building would satisfy the overheating criterion published by the Chartered Institution of Building Services Engineers (CIBSE), still current, that DRT should not exceed 28°C for more than 1% of the occupied hours wholly passively (CIBSE, 2005 and 2006) and should also satisfy the criterion that the dry resultant temperature should not exceed 25°C for more than 5% of the occupied year (CIBSE, 2002).

Krausse's team collected data recorded through the Library's BEMS sensors reinforced by additional data loggers installed for the monitoring exercise. The controls protocols are critical and involve imagining every predictable future environmental scenario. IESD drafted the essential controls protocols. Inevitably particular scenarios which actually occur later in the operating life of the building defy the technical imagination, hence the extreme importance of a meaningful year-long commissioning exercise in which controls can be adjusted as unanticipated scenarios unfold, all-night working in particular rooms for example. The ventilation regime for night-time cooling, potentially a very significant component of the passive cooling strategy, is based on a self-learning algorithm in the BEMS to 'predict' the likely passive cooling requirement for the next day. Theoretically this updates itself as the climate shifts. The BEMS is also configured to avoid over-cooling, a common problem triggering heating in the early hours of the pre-occupied day, even in hot spells, by monitoring slab temperature.

Unsurprisingly during the heating seasons the daytime indoor temperatures were dominated by the heating set points, programming and internal heat gains. Temperatures remained below 24°C during the daytime and decreased to approximately 21°C during the night, which is the minimum mid-week temperature set by the facility managers. 21°C is the set temperature for the lightwell supplies but it is not clear what drives this minimum, perhaps warranties on computing equipment on open floors and/or conservation of the book stock?

During the warmer periods of the year the internal temperatures are strongly influenced by those outside. However, individual hot days do not significantly raise the internal temperature, as is evident on the days around 1 July, 22 July and 6 September in 2004. Even during August 2004 and June 2005, prolonged periods of high ambient temperatures reaching 35°C, the internal temperatures only occasionally exceeded 25°C, a remarkable performance. The building is significantly more resilient than predicted (Figure 3.41).

However, the diurnal temperature swing was marked during this period, in excess of 9°C. Night-time ambient temperatures remained below 18°C, yielding a swing of up to 15°C, offering a reasonable passive night-time cooling potential. This is not available in the urban heat islands of temperate zone cities such as London.

Night ventilation cooling, although constrained by the set point, together with the exposed thermal mass, prevented the internal temperatures exceeding 26°C. During the entire two-year monitoring period Krausse *et al.* report the maximum internal temperature recorded was 26.4°C, measured on the top floor on 19 June 2005 when the external temperature had reached 35.4°C, a depression in temperature of 9°C achieved entirely passively, replacing a very substantial air-conditioning plant with the need for plant room space.

Temperatures stratify vertically so that the top floor with the lowest stack height is warmest, the ground floor coolest. The top floors of passively cooled buildings warrant careful design and certainly heavyweight roofs. Yet even on the third floor the CIBSE 2002 overheating criterion, less than 5% of occupied hours over 25°C, was met, with temperatures greater than 25°C only occurring during 3.8% of the hours of use. The internal temperatures never exceeded 27°C, rather less than the 11 hours Cook *et al.* predicted at the design stage.

The CO_2 concentrations for occupied periods during the six weeks investigated were typically between 400 and 500 ppm, with occasional peaks of up to around 700 ppm. The maximum CO_2 concentration recorded was 720 ppm, which is below the limit of 1000 ppm recommended for school buildings but suggesting air change rates may be allowed to increase.[21]

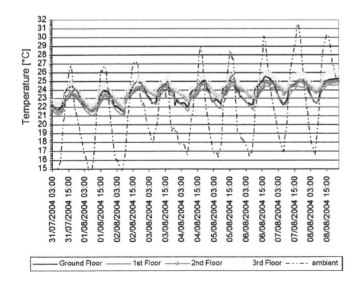

Figure 3.41
Measured results form the occupied building through an unusually hot spell in which external temperatures climbed to 31.5°C. Internal temperatures were maintained below 26.4°C on the warmest upper floor, 24.4°C on the ground floor, a depression of 7°C from the peak external with zero energy consumption in place of the substantial mechanical cooling of a 'business-as-usual' building.

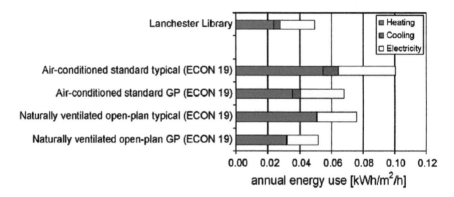

Figure 3.42
Relative energy performance against a rather outdated 1980s benchmark for air-conditioned buildings with small cooling loads. Even then, energy requirement is halved (Lomas, IESD, De Montfort University).

Figure 3.42 sets the results for the library into the context of UK Standard air-conditioned and naturally ventilated buildings. There is an order of magnitude saving in energy and cost achieved through the fundamental configuration of the two buildings, the library, despite its deep plan, improving considerably on the relatively very low Queens Building performance. It is at this point in the design, the order of magnitude reduction in energy requirement, that on-site renewable energy technologies become relevant, against these very low figures renewable sources could realistically make a major contribution, mopping up the residual load, not liberally applied to little effect to the business-as-usual buildings at the top of the bar chart.

Lanchester Library: reception history
Librarians like the building – both those who work in the building and others from elsewhere. It was, perhaps, a glimpse of the future of libraries in teaching and learning oriented higher education institutions. The building was co-winner of the SCONUL (Society of College, National and University Librarians) Library Design Award.[22] The citation gives some insight into the librarian profession's criteria for a 'good' library. It reads:

> This stunning 'castle' like building successfully provides an important new focal point on campus for attracting and supporting both students and academic staff. Users enjoy an open, varied and well-planned interior which is enhanced by natural light penetrating through the five glazed atria.[23]

The reception by architects and others in the construction supply side is also positive. Stephen Pidwell (2001) writing in *Architecture Today* remarks:

> Already it is attracting 6000 visitors a day against initial predictions of 2500 … the building seems to soak up this extra traffic effortlessly, with few complaints about comfort so far, and these have been easily resolved through fine tuning of systems.

He continues by discussing the problem of providing legibility and free movement in libraries without destroying the quietude, the 'atmospheres' of calm concentration gathering around the readers. Here he observes that:

> ... vertical circulation is relegated to staircases at the perimeter of the plan and there is little joy derived from the movement of people and no way of sensing the whole building from one point.

He absolves the designers by reporting, 'this was the librarian's preference'; but actually it was also the design team's preference. There is a late modern tradition of folding the service spaces out of the plan, not least in Louis Kahn's idea of 'servant spaces' developed at the Richards Memorial Laboratories and pursued subsequently by architects as diverse as Richard Rogers and Peter Salter, as a letter to a later issue of the same journal observed with some irony. This is, Pidwell (ibid.) argues, both a high- and low-tech architecture. He considers the details:

> ... the general treatment of the building can be disarmingly matter-of-fact. There is no fetishism here with special devices or elaborate castings, rather a leaning towards systems that can be adjusted and replaced easily and cheaply, appropriate to a busy facility maintained on thin budgets.

He baulks at the stacking of domestic scale windows on the northerly elevations, 'some views are cruel'.

Tony Monk of Hutchison Locke and Monk Architects is less guarded in his review of the building celebrating its Brick Development Association Building of the Year Award (Monk, 2001):

> It is rare for a new building to be truly significant and develop a new approach to design. The new Library and Resources Centre for Coventry University is just such a project, however. It contains a unique spark of originality and its concept is likely to have a profound influence on the design of buildings that provide large open-plan accommodation.

He reprises the design team's chain of stack-vented buildings and concludes:

> But in the Lanchester Library we see the culmination of these evolving ideas ... forms that at first sight might appear to be decorative exuberance are in fact derived from fundamental, functional elements and may be regarded as architecturally eloquent expressions of how the building works.

Monk's subsequent book (2004) on contemporary hospital design devotes its introduction to the Lanchester Library and asks if the design strategy could be transformed to produce an acute hospital, a proposition taken up by the Department of Health Chief Architect, the consequences of which are discussed in Chapter 6.

Notes

1. Leoni (1726) *The Architecture of L.B. Alberti*, original edition in the Middleton Library of the Faculty of Architecture and History of Art, University of Cambridge.
2. US Department of Energy requirement to remove a minimum of 99.97% of particles of 0.3 microns from air passing through (US Dept of Energy, The Office of Health, Safety and Security, DOE Technical Standard Specification for HEPA Filters Used by DOE Contractors. See www.hss.doe.gov/nuclearsafety/qa/hepa/index.html (accessed 6 March 2013).
3. The author is very grateful to Professor Cath Noakes and her colleagues at the Pathogen Control Research Institute based in the Department of Engineering at the University of Leeds, collaborators in a series of research projects for the UK Department of Health and the NIHR for their guidance and insights into this emerging science.
4. Also available at: https://archive.org/details/tallofficebuildi00sull (accessed 12 October 2016).
5. See Padovan, R., 'Building towards an ideal: progressive architecture in Holland', Chapter 5, p. 144, in Russel, F. (1979) *Art Nouveau Architecture*.
6. See 'Headquarters of the Algemeene Nederlandsche Diamantbewerkersbond', p. 152, in Polano, S. (1988) *Hendrick Petrus Berlage, Complete Works*.
7. See 'The Crypt – "A Tectonic Cave"', p. 366 in Jensen, T.B. (2009) *P.V. Jensen-Klimt, The Headstrong Masterbuilder*.
8. 'The City of Leicester: Social and administrative history since 1835', pp. 251–302 from *A History of the County of Leicester: Volume 4, the City of Leicester*, originally published by Victoria County History, London, 1958, available online at: www.british-history.ac.uk/vch/leics/vol4/pp251-302 (accessed 3 August 2015).
9. Cragside 1869–84, Saint explains that the plan straggles because of the incorporation of an original hunting lodge but Shaw's multi-gabled building at the approach demonstrates the point, see Saint, A. (1976) *Richard Norman Shaw*. New Haven: Yale University Press.
10. The programme of physical experiments to illustrate the complex coupling of floors and stacks in the real building was conducted in a small-scale Perspex model of the building immersed in a tank of water. The model reproduces the fundamental building geometry and the configuration and relative size of openings. The technique is mentioned as having been used in the 1940s in The Nuffield Report on hospital design as described in Chapter 6. It was intensively researched by Professor Paul Linden at Cambridge in the later 1980s and used to model the Queens Building. Saline solution was injected to add density to the fluid, inducing gravity driven flow so that the models were inverted. More recently Professor Woods at the BP Institute and Professor Hunt, the Dyson Professor in Engineering at Cambridge, pre-cool or pre-heat air to the building. The experiments are visualized using various coloured dyes. The complex geometry of the model enabled observation of the highly complex dynamic flows. Perhaps most usefully the technique reveals alternative modes of flow, perhaps unsuspected but equally probable under particular definable conditions.
11. *The Architects' Journal*, editorial, (1990) 'Into The 1990s, Project Review' 3 and 10 January.
12. The Editors (1999) *Architecture Today* 100, July, p. 62.
13. From the very interesting recent review of the genre by Debora Domingo Calabuig, Raúl Castellanos Gomez, Ana Abalos Ramos (2013) 'The Strategies of Mat-building', *The Architectural Review*, 13 August 2013. See www.architectural-review.com/essays/the-strategies-of-mat-building/8651102.article (accessed 7 August 2015).
14. Recent surveys cast doubts on the great majority of dampers available commercially as installed, and across secondary heating elements, which in the heating season raise the temperature further, slightly above the desired ambient level. Hence the need for the Lanchester Library to have dampers which were actually airtight.
15. $P = Cp * (1/2pUz^2)$ where p is the density of air (varies around $1.2kg/m^3$) and Uz is the site wind speed at height z in m/s.
16. LPC 1999, Section 3 'Risks requiring special consideration …b. concentration of value … bulk storage of goods and stock'. Fire Protection Association FPA (1999) *The Loss*

Prevention Council LPC Design Guide for the Fire Protection of Buildings – A Code of Practice for the Protection of Business. Association of British Insurers and the Fire Protection Association.

17. Dlugolecki, A. Thoughts about the impact of climate change on the insurance industry, Climatic Research Unit, University of East Anglia.

18. Dlugolecki was a Visiting Research Fellow at the Climatic Research Institute at the University of East Anglia but had previously been an insurance insider until 2001, Director of Insurance Development at Aviva, lead editor of the section on the effects of climate change on financial services in the 2001 IPCC assessment.

19. ODPM, 2004a, b Office of the Deputy Prime Minister (2004a) Modernisation Agenda for the Fire and Rescue Service (Home Office 1985), London: ODPM. Office of the Deputy Prime Minister (2004b) Regulatory Reform (Fire Safety) Order, London: ODPM.

20. An alpha t^2 fire is defined as $Q = a(t - t^1)^2$ where Q is the rate of heat released from the growth of the fire during the growth phase at time t (kW), a is the fire growth rate (kJ/s^3), t is the time (s) and t^1 is the time (s) of ignition (CIBSE, 1997, table 9.1). For a medium growth rate a = 0:012 kJ/s^3.

21. *Building Bulletin 101: Ventilation in School Buildings*, (2006) Department for Education and Skills.

22. McDonald, A. (2002) *Celebrating outstanding new library buildings*. Society of College, National and University Librarians. See www.sconul.ac.uk/pubs_stats/newsletter/27/ARTICL27.RTF (accessed August 2006).

23. Report of the Library Design Award Panel, SCONUL Library Design Award 2002, unpublished but available to members as hard copy.

4

Urban heat islands

Schools, hospitals, hotels, prisons, reformatories and the like cause us to be confronted with entirely new problems, utterly unheard of in the past ... in these circumstances, it is nothing less than grotesque to build a hospital or a school in the style of a palazzo.

(Bruno Taut, 1929, *Modern Architecture*, p. 136)

The additional complexities of the 'urban heat island effect'

Significantly higher temperatures were measured over London on an early August day in 1999, relative to the countryside beyond Heathrow airport. The findings confirmed a long experienced phenomenon which came to be christened the 'urban heat island effect' (UHI) but the extent of the uplift in city centre temperatures recorded was astonishing (Watkins *et al.*, 2002). The data revealed that although London may be several degrees warmer than its rural hinterland during the day, a very much greater variation may occur at night, in excess of 9°C at the somewhat unexpected epicentre: the British Museum in Bloomsbury (Figure 4.1).

The 2012 Climate Change Risk Assessment (CCRA) confirmed the existence of the phenomenon unambiguously (Figure 4.2).[1]

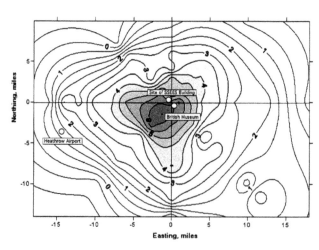

Figure 4.1
Variation in the urban heat island intensity across London on 2 August 1999 at 02.00 a.m. The temperatures (K) are relative to the rural reference. The location of the School of East European and Slavonic Studies is shown relative to the UHI epicentre, the Director's House in the British Museum. Reproduced from Short *et al.* (2004) based on original map by Watkins *et al.* (2002).

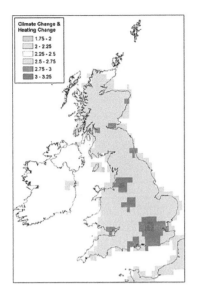

Figure 4.2

UK Met Office model predictions of increases in minimum temperatures plotted across the UK (°C) in the period 2041–50 compared with 1961–90 minima. The Climate Change Risk Assessment warns that the minimum temperatures predicted at their most intense will exceed the national Heatwave Action trigger temperature thresholds, heralding a heat emergency. Reproduced from Figure 5.2 in the CCRA, original source Met Office Hadley Centre, Crown copyright.

The first measurements had been corroborated by a series of colourfully titled research projects confirming regular temperature uplifts of 9°C in London, in Manchester and a little lower at 5–7°C in Birmingham, not so far from the Lanchester Library (Figure 4.2; Capon and Oakley, 2012).[2] Although the identification and measurement of heat islands above cities is random depending on local academic initiatives, UHI phenomena continue to be identified over a growing number of cities, some surprisingly modest in size.[3]

This revelation seriously compromised the interest in extrapolating the ventilation cooling strategies of the last chapter devised for the climate of the English Midlands to become exemplars for very low carbon buildings in major cities across the temperate zone. There is no longer a reliable possibility of passive night cooling during hot spells. The passive strategies are stretched beyond their capacity to deliver comfort to whichever of the current UK or US comfort standards are applied.

The UHI phenomenon scrambles reliable delineation of temperate climate regions. UHI footprints may be geographically small, insignificant perhaps in their total contribution to climate change as some scientists argue, but the proportion of the regions' population affected is significant.[4]

For the design professions, it is important to realize that due to the UHI effect urban sites may be shifting latitude markedly, several decades in advance of predicted future climate for the region as a whole. This suggests that air-conditioning installations within lightweight buildings may not be gratuitously oversized by a risk-averse industry as some commentators suspect. Instead, these installations may prove to be undersized in the urban heat islands, particularly in glass towers, requiring a major retrofit exercise within the amortized commercial life of such an office building. Institutional investors may wish to consider this vulnerability and cost when rejecting more passive design strategies on the basis of risk and future-proofing.

No doubt the heat island effect had long been familiar to suburban and rural commuters from the Home Counties, wondering what to wear for the daily journey into London. Novelist and writer J.B. Priestley described the oppressive City microclimate of the early 1930s in *Angel Pavement*, his novel of working life in a veneer importer's office: 'Angel Pavement and its kind, too hot and airless in summer, too raw in winter, too wet in spring, and too smoky and foggy in autumn'.[5]

Over a century earlier, the chemist and meteorologist Luke Howard (1772–1864) had identified anomalies between temperatures he had measured at three rural sites and those taken by the Royal Society in the city at Somerset House. He published his analysis in 1818 and 1820, and later as a re-edited edition (Howard, 1833).[6] He advises in his summary of findings in Volume II (ibid. p. 289), with temperatures in Fahrenheit:[7]

> The *Mean Temperature* of the *Climate,* under these circumstances, is strictly by the effect of the population and fires, to 50.50°; and it must be proportionately affected in the suburban parts. The excess of the Temperature of the city varies through the year, being least in spring; and greatest in winter; about 48.50° Fahr: but in the denser parts of the metropolis, the heat is raised, and it belongs, in strictness, to the *nights;* which average three degrees and seven tenths warmer than in the country; while the heat of the day, owing without doubt to the interception of & portion of the solar rays by a constant veil of smoke, falls, on a mean of years, about a third of a degree short of that in the open plain.

Howard identifies a night-time urban heat island over what was a very much smaller central London of about 2°C by the second decade of the 19th century. This is surprising perhaps, but London's atmosphere at that time was rich in anthropogenic particulate matter. His records of monthly mean temperatures in London between 1797 and 1806 (his Table A, Vol. II) show a highest August mean of 67.56°F in 1802, a highest June mean of 64.00°F in 1798, and the highest July mean of 66.28°F in 1803. The monthly means in the country are clearly lower by about 2°F. He cites a highest August mean of 65.27°F recorded in 1807, and in July 1811 of 61.84°F and in the same year, 61.58°F for June. Yet more fascinating diurnal variations are given in the extended data of Table IX for 1807 in Vol. I. On 10 July temperatures ranged from 54°F (11.65°C) to 81°F (27.2°C), a depression of 27°F (equivalent to 15°C) offering excellent prospects for a successful passively cooled building and one which could generate the prodigious ventilation flows required to exhaust the various products of combustion within, not least from oil lamps. A heat wave clearly extended from 8 July until 10 August 1807, maintaining peak temperatures throughout above 70°F with 82°F achieved on 25 July. Howard also suggests that the climate varies in the long term by about 4.5°F but to an identifiable cycle, probably Lunar, as devised by 'The Divine Intelligence', who or which, he complains, is not given due credit in the many publications on meteorology. There is nothing 'natural' about the variability of the climate.

The editor of the modern e-version of Howard, Professor Mills (2008),[8] explains that Howard had correctly deduced what are now offered as the principal causes of the urban heat island: man-made heat sources which warm the atmosphere contributing to the winter uplift in temperatures; dense urban configurations which entrap heat, restrict re-radiation to the sky and suppress cooling airflows through the city and the absence of moisture so that there is no evaporative cooling.

A changing climate

It was self-evident that the climate databases hitherto used to simulate building performance within these cities, e.g. the Kew 1967 data, had become compromised. The existing databases contained summer temperature predictions that were outdated and too cool. Reliance on these models meant the potential overheating risk may not be picked up in modelling. The Kew 1967 data contains just three hours of dry bulb temperature in excess of 27°C, whereas the then new 2002 CIBSE London data predicted 107 hours over 27°C, and this not accounting for UHI effects (Holmes and Hitchin, 1978).

In fact, the 2009 iteration of the UK Met Office climate change exercise, UKCP09, did not include for UHI effects across the whole of the UK. The grain of its matrix of predictions is too large to capture city centre anomalies but it indicated not insignificant rises in mean average summer night-time temperatures, 2–3°C by the 2050s under the Medium Emissions scenario and 3–4°C by the 2080s. Onto this one then has to factor the UHI effect.

Heatwave mortality

The London School of Tropical Hygiene and Medicine predicted the likely uplift in mortality during an unusually hot spell. Researchers deduced that the heatwave between 4 and 13 August 2003 was responsible for 14 802 excess deaths in France and 2045 in England and Wales (Johnson *et al.*, 2005).[9] Citizens aged 75 or older were at particular risk, their mortality rates increased by 59%, as the effectiveness of heat-shedding physiological responses reduces with age, particularly after 70 years of age. Papers contributed to the 2015 NHS Heatwave Plan meeting advised that mortality starts to rise when temperatures reach as little as 17–20°C. The stakes are very high and mis- or underestimation of city temperatures in the design of buildings will have serious consequences.

School of Slavonic and East European Studies (SSEES)

The Watkins *et al.* (2002) paper was published as the staged competition to design the new School of Slavonic and East European Studies (SSEES) was concluding.[10] The SSEES site is close to the British Museum in Central London. The design team's opening gambit was destroyed when viewed against the example year of 1989, calculated using the Holmes and Hitchin method (1978), and the Design Summer Year (DSY) calculated to best represent likely warm summers in Bloomsbury.[11] The DSY recorded an absolute peak temperature of 33.68°C and 267 hours with a dry bulb temperature in excess of 25°C – this is very, very different to the Kew data maxima. Even the crude CIBSE rule that internal temperatures should not exceed 25°C for more than 5% of occupied hours or 150 hours in practice was exceeded by the number of hours recorded by the Heathrow DSY at the outer edge of the London UHI (CIBSE, 2002). Satellite imagery shows the intensity of raised temperatures over Central London at a particular late daylight moment, varying from 290K at the periphery to 301K (Figure 4.3).

Even if the internal temperatures in a building within London could be maintained passively throughout at ambient temperature during the peak summer period (quite an achievement) the CIBSE criterion could not be met even before including the likely uplift in the city's summer night temperatures.

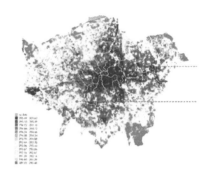

Figure 4.3
London, land surface temperatures at 9.00 p.m. on 12 July 2006, prepared by the LUCID project from ASTER satellite land surface temperature data 'Surface Kinetic Temperature Product Level 2 AST 08'. The literature indicates validation of these images is complex. From image reproduced in Nickson (2011), acknowledging NASA and the EPSRC LUCID project available at http://unhabitat.org/wpcontent/uploads/2012/06/GRHS2011 CaseStudyChapter06London.pdf (accessed 9 August 2015).

The 2012 Climate Change Risk Assessment (CCRA) (Cope *et al.*, 2012, para vii) warned: 'there is a very real danger that the UHI could be exacerbated in the future by autonomous maladaptation in the form of widespread installation of air-conditioning for comfort cooling'.[12]

This raises a serious challenge of whether and how a 'comfortable' low energy, as-passive-as-possible building could be made at the centre of a major UHI. It can and a solution is presented to the theoretical approach taken in devising the SSEES building. It is 'theoretical' because the strategy had never before been attempted at this scale in a city centre. This academic building achieves very low energy cooling, the radical design requirements emerging as an upset to contemporary custom and practice in the construction industry and the regulatory authorities.

SSEES was conceived by Robert Seton-Watson and Tomas Masaryk to study the languages, literature, history, culture and social conditions of Eastern Europe. It opened in October 1915 whilst Masaryk, soon to become the first President of Czechoslovakia from 1918, was successfully lobbying France and Britain to support the liberation of the Austro-Hungarian Empire into democratic states including a unified Czech–Slovak nation. The new school in London was a useful base for the campaign but, ultimately, the intervention of the United States was critical. The US Secretary of State Robert Lansing's Declaration of 1918 removed the possibility of a resurrection of the Austro-Hungarian Empire sanctioned by Europe's surviving monarchies.[13] In 2005, SSEES was 90 years old as the project for a new building commenced, the institution was already drenched in its own history (Figure 4.4)

However, despite this highly impactful provenance, its well-being as an autonomous institution stuttered in the 1996 iteration of the UK Higher Education Funding Council's Research Assessment Exercise, the national audit of research quality of enormous significance to UK academic institutions, against which five year blocks of funding are awarded. With a substantial reduction in income, SSEES contemplated merger with a number of other institutions but in 1999 the School became a component of University College London (UCL).[14] The prospective arrangement with UCL included an incentive: an undertaking by the UCL Provost to fund a purpose-built headquarters for SSEES in the heart of Bloomsbury in Central London (Figure 4.5).

The only vacant site available on the UCL campus was the loading bay of the highly regarded Chemistry Department in a re-entrant of their own cruciform building on Taviton Street. The site had been earmarked for the expansion of

Figure 4.4
The entry hall to the new SSEES building incorporating Frantisek Bilek's sculpture. The piece had been known as 'The Spirit of the Slavs' for decades, very appropriate for its installation here, but it is now thought Bilek originally named it 'Passion Unchained', yet more appropriate (photograph by Peter Cook).

Figure 4.5
The SSEES building: Façade to Taviton Street (photograph by Peter Cook).

the Chemistry Department, and the piles had already been sunk for their new building (Figure 4.6).

The Vice-Provost admitted to the President of the Czech Republic at the opening ceremony that it had not been at all clear that it would be possible to squeeze 3500m² onto this unpromising site, a hole in a vestigial Georgian terrace in an important conservation area, whilst maintaining deliveries to the Chemistry Department behind and complying with the health and safety implications of a large laboratory building. A rectangular plan for SSEES in this context would have been a grim prospect. The D-shaped plan develops a half-cylinder form at the rear, admitting daylight to the chemistry building and the rear elevation of the SSEES building. The day-lighting efficiency of this form, compared with other options, emerged through very revealing iterative simulations using Radiance software at De Montfort University (Ward-Larson and Shakespeare, 1998). Team reflection on the speculative daylight and solar penetration modelling at this very early stage had a profound influence on the invention of the SSEES building. Figure 4.7 shows the sketch massing resulting from the diagnostic exercise.

Whereas the UCL Estates Department officers were interested in building generic open floor space, the school was anxious to replace its members' spacious offices in Senate House and a pair of important Georgian houses in Tavistock Square with something characterized to the same intensity, even if differently expressed. The sketch of a building climbing up through the landlocked site with symmetrical staircases on its one public elevation, a central lightwell and stacks breaking the Georgian cornice line persuaded the SSEES members of the interview panel and the Director of Estates, but probably dismayed the project managers briefed to encourage a more universal proposition (Figure 4.8).

The brief is bi-partite: a research library with a unique collection along with accommodation for a school of 47 established academics and their dependent research students, post-docs and visiting fellows. The building would also have

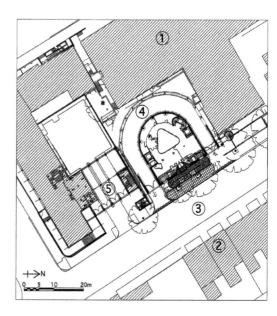

Figure 4.6
The SSEES building: Site plan.
1. Existing chemistry building.
2. Georgian terrace.
3. Taviton Street.
4. Vehicle access.
5. Examination hall.

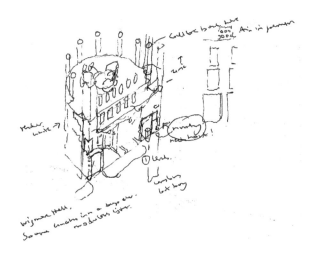

Figure 4.7
Short's final massing sketch for the SSEES building showing
the D-plan, the hint of a free-running buffer zone, the use of
the residual space for stairs.

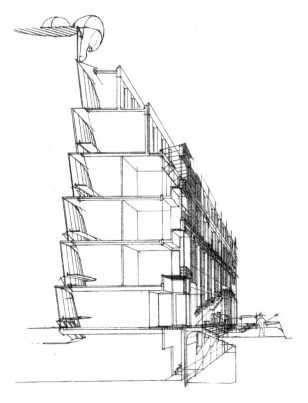

Figure 4.8
Short's sketch of a putative street front, the compositional
challenge of the symmetrically rising stairs implicit, as
explained by Blunt in his study of Neapolitan façades.

ceremonial functions as it would be regularly visited by ambassadors and
occasionally by heads of state from the region, a number of them being alumni
(Figure 4.9).

It was quite clear that the renewed research success of the school was
being driven by its academics' publications, written around demanding teaching
duties. Academics had built up substantial private libraries in their own rooms,

Figure 4.9
Interior of the SSEES
Library (Peter Cook).

expediting the work and so a great deal of attention was given to the design of individual offices. The school floors are radially planned so that Principal Investigators' offices encircle open research space (Figure 4.10).

This has proved to be a very effective arrangement, much more potent in developing integrated research endeavours than the experience of researchers placed within lines of offices along double-loaded corridors. The school won a major Arts and Humanities Research Council framework grant as it opened, the research environment is significant in these bidding exercises. The semi-circular D form is, of course, not unprecedented in the early modern era, architect William Burges' New Speech Room (1877) at Harrow School is a particularly muscular example.

The building enters into the rhythm and organizational principles of a Class 3 Georgian street, following architectural historian John Summerson's (1945) explanation of the classificatory system, but does not replicate the former five Georgian houses occupying the site. They were not felled by enemy bombing but demolished by the University of London in 1963. SSEES reads as an institutional building: a small urban palace with a symmetrical elevation within which red-carpeted staircases, domestic in width but palatial in height, take the 450 students and staff, bourgeois visitors, grander diplomats and the occasional head of state up to the *piano nobile* (Figure 4.11).

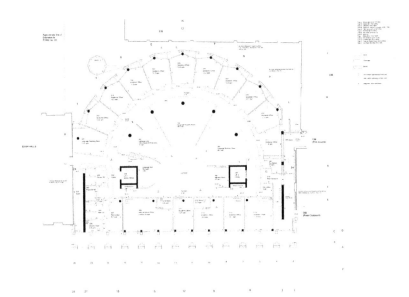

Figure 4.10
Plan of the fourth floor showing academics' offices encircling research space for funded projects, part subdivided here but open on other floors.

Figure 4.11
A cipher for an ambassadorial stair occupying the buffer zone (Peter Cook).

The explanatory piece in *Architecture Today* (Harbison, 2006) referred to Kruschev's notion of accessible chandeliered grandeur which he delivered to the citizens of Moscow in the stations of the underground railway. The stair hall and its elevation are realized in continuous load-bearing brickwork, engineered to eradicate movement joints, buttressed by cross walls pierced to take the stairs rising through them (Figure 4.12), all intended to evoke the brick stepped gable churches and town houses of the Backsteingotik architecture of Lower Pomerania and the Baltic ports beyond (Figure 4.13).

The openings in the elevation do not fall neatly into repetitive bays but are intended to float across the seamless brickwork (Figure 4.14).

Architectural antecedents

Although the buffer space strategy is likely to be effective in stabilizing internal conditions in any climate it is expensive in space and therefore cost, and requires a function, here as necessary circulation. Imprinting the progress of a staircase onto an elevation is not unprecedented, but the device is rarely used now, not least because of fire and smoke regulation restrictions. It is difficult for interior spaces to look through the stair chamber and be compliant, but the SSEES building achieves this.

Whilst stair turrets abound with openings spiralling up around them, the principle unfolded example of a stepping stair elevation in the modern era may be architect H.H. Richardson's magnificent but spectacularly overspent Boston Trinity Church (1872–77) inspired by the churches of the Auvergne in France. Its precision makes it a modern invention and he repeats the theme in the

Figure 4.12
The buttresses bracing the buffer zone behind the façade (Short and Associates, Michael Ritchie).

Figure 4.13
Early 20th century *Backsteinform Lagerhallen* on the quayside at Stralsund (author).

Figure 4.14
Detail of façade (Peter Cook).

corner stair turret to the Salamanca Tower above, so called because it closely follows the medieval tower of Salamanca cathedral, by projecting the line of the stair treads behind, through onto the massive stone elevation (Street, 1914, p. 99). It crops up periodically throughout his work (such as the Woburn Public Library, 1876–79). Frank Furness also employed the device on the stair tower on the eastern elevation of the long demolished Baltimore and Ohio Railroad Passenger Station at 24th and Chestnut Street in Philadelphia, Pennsylvania of 1886 (in effect for Johns Hopkins, whose new hospital will be described in Chapter 6), three super crisp, asymmetric arched openings in the sheer, flush-pointed brick face sweep up the stair (O'Gorman, 1973, plate 26-5). Symmetrical opposing staircase elevations seem to be a rarer sub-species.

The art historian Anthony Blunt (1968) includes a sequence of diagrams of symmetrical stair arrangements in the 'Notes to the Plates' of his book *Sicilian Baroque*, including the Villa Palagonia and the Villa Valguarnera in Bagheria. He includes a historic photograph (Plate 117), taken before the 1943 bombing of Palermo, of the staircase of the Palazzo Castel di Mirto, accredited to Andrea Giganti in the 1750s. It is part external but climbs through and up an arcaded elevation. Blunt records this arrangement as a Neapolitan innovation dating from the building boom when the city was established as an independent kingdom in 1734 after a period of culturally benign governance by Austrian viceroys. The fully fledged stair arrangement across the façade of Jakob Prandtauer's Benedictine monastery of St Florian near Linz, is somewhat earlier, 1709, perhaps an example of architectural contagion from north to south.

In his book on Neapolitan architecture, Blunt (1975, p. 138) identifies the compositional challenge:

> The basic problem in designing a staircase which is visible on the façade of a building is that, if the flights run parallel to the façade, the windows or arches by which they are lit form a diagonal line which cuts across the horizontals of the floors … Neapolitan architects showed the greatest ingenuity in either concealing or exploiting these irregularities.

He credits the rather grand, well-connected architect Fernando Sanfelice as a particular master of this device: 'Sanfelice took advantage of the experiments of his predecessors, but he went far beyond them in the invention of new forms and new dispositions of staircases'.

Although ultimately derived from Spanish and Genoese prototypes, Sanfelice's innovation is that: 'his staircases normally start with two flights, running left and right parallel with the façade, leaving an arched opening in the middle leading through to the garden'.

This is the SSEES arrangement. Blunt explains that the symmetrical parting of staircases also served, usefully and diplomatically, to separate the core family from less closely related dependents. At SSEES the right hand stair climbs up to the library, the left more privately to the school (i.e. the family). Blunt describes the implications of distorting openings to take up the inclination of the stair (1975, Figures 233–236):

Sanfelice was faced in an acute form with the problem of the sloping line formed by the openings on the flights, but he turns it to good account and, particularly in the staircase in the right hand court of his own palace, makes the slopes of the flights the principle theme of his design, which therefore takes on the form of a series of V-shapes.

Blunt observes that the lines of both the top and bottom of the sequence of openings follow the slopes of the flights. He quotes Sanfelice's biographer De Dominici, describing the effect of the symmetrical stair building as, 'like a bird stretching out its wings to rest on the buildings along the sides of the court', something to aspire to in building for Higher Education today (Figure 4.15).

The elevation set behind the outer street elevation is more repetitive. Opening windows are set into a white facetted plastered matrix, an oblique reminder of the 1912–13 Cubist Villas of Chochol and his contemporaries in Prague (Figure 4.16).

Figure 4.15
Sanfelice's stair façade at Sanfelice's Palazzo dello Spagnolo, Naples, 1738. With very grateful thanks to Andrea Pane, Assistant Professor of Conservation at the Department of Architecture at the University of Naples Federico II, for sharing her images of Sanfelice's work.

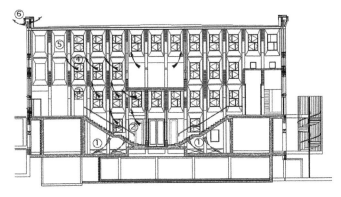

Figure 4.16
Drawing of the facetted white 'inner' elevation behind the masonry street façade. Longitudinal section parallel to the 'double facade' indicating use of the stair hall as an air exhaust plenum:
1. intake to the plenum.
2. exhaust from the upper ground floor.
3. exhaust from the first floor.
4. exhaust from the scond floor.
5. third floor is not connected into the double facade.
6. 'breathing parapet' exhaust void.

The Czech ambassador and his entourage recognized these small clues at the opening ceremony for SSEES. This disparate community has, of course, a deep sense of its complex regional identities.

Configuring passive downdraught cooling at scale
Having established a working form with which the school empathized and judged likely to fall within the terms of Camden Borough Council's 1994 'Supplementary Planning Guidance for Bloomsbury' and those of the many conservation groups and agencies which protect the area, the question remained how to animate this object to keep it cool through the overheating season at the centre of the UHI? The solution is a 'centre-in/edge-out' scheme through all its seasonal modes. The innovation was to reverse it seasonally so that the building operates in low-level supply stack-ventilated mode during winter and the cooler mid-season moments. As temperatures rise, the strategy is reversed. The building becomes sealed at its base and pre-cooled air is admitted at the head of the central lightwell (Figure 4.17) allowing cool air to fill

Figure 4.17
The central lightwell, looking down into the SSEES library (Peter Cook).

the glazed void and wash across the surrounding floors. In turn, this develops some stack pressure difference to draw warmed air up out of the building through exhausts of various forms at the perimeter. Airtightness is absolutely key to success, hence the analogy drawn with the attempts to construct a perfect experimental vacuum pump in the late 17th century.

In this arrangement, cellularization is possible without interfering with the ventilation strategy because the exhaust stacks are distributed around the rear elevation, thus serving each module of potential office. At the street front, facing west-south-west, the double façade forms a buffer space performing as an exhaust plenum drained by a continuous ventilating cornice (Figure 4.18).

Figure 4.19a shows the essential arrangement as a cross-section broadly east–west. Figure 4.19b depicts the building as a piece of apparatus showing the valves, plena and conduits available to direct airflows through the constructed spaces. Each valve, be it a tight sealing damper or window, is independently controllable in meaningful increments. This is extremely important. There are over 300 controllable elements in the 'apparatus', all critical. A 3% failure rate in operation, as experienced in commissioning, is enough to destabilize the whole intended ventilation regime. Risk can be mitigated through attentive commissioning. This building enjoys very particular seasonal modes, a major development of the Queens/Lanchester strategies. Optimizing the potential performance will periodically require quite rapid cycling through modes in abrupt shifts in weather. How will the system cope with that?

The following diagrams describe the four basic seasonal modes. Figures 4.20a and b show the winter heating and fresh air ventilation configuration. Fresh air is drawn into a concrete lined plenum between ground and lower

Figure 4.18
The ventilating cornice in construction (author).

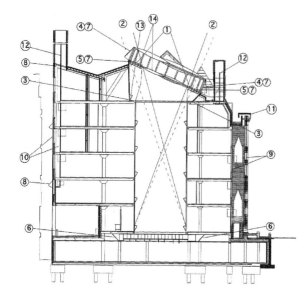

Figure 4.19a

SSEES building: cross-section southwest to northeast.
1. Cooling coils introduced below the air inlet.
2. Cooling coils set back from the lightwell perimeter so that daylight penetration is not unduly compromised.
3. Design anticipates condensation on the cooling coils and enables its capture and safe removal.
4. Dampers controlled to prevent winds driving ambient air down into the lightwell, disrupting the cool air reservoir.
5. Louvres exercise fine control of the airflow to contend with periods of low cooling demand.
6. Preheating coils set back from the base of the lightwell.
7. Dampers and louvres can open without cooling when in operation to enhance the supply of ambient air for night ventilation and mid-season ventilation cooling.
8. To exhaust air, which is below ambient temperature – options include: (i) low-level openings in the stacks; or (ii) heating coils in the exhaust stacks.
9. Double facade with transparent elements to the inner and outer surfaces.
10. Linked stacks produce a double-envelope facade with glazing to the inner and outer surfaces.
11. Linear exhaust termination: the 'breathing parapet'.
12. Internal partition preventing backflow to the top floor.
13. Cool air is delivered directly to the top floor.
14. Transparent ethylene-tetra-fluoro-ethylene cushion creating a ventilated void.

Figure 4.19b

Mechanical controls driven by the Building Management System.
1. Actuated opening windows.
2. Dampers.
3. Cooling coils at the atrium head and waste heat batteries in the exhaust stacks.
4. Acoustic transfer ducts at partitions.

Key
⊘ Opening vent
＼ Chilled plates
ɪ Heating elements
⌂ Acoustically treated transfer ducts (passive)

ground floors through a series of openings facing as many orientations as possible. The lightwell temperature is permitted to rise to 16°C by the start of the day's business. As air enters each floor from the lightwell through a ring of independently controlled bottom hung windows it is heated to 21°C as it passes through slim, low resistance 'hospital' pillar type hot water radiators. The temperature is then maintained at 21°C throughout the occupied day. Although each of the 300 or so opening elements is independently operable, a much more refined control system would be required to enable individual

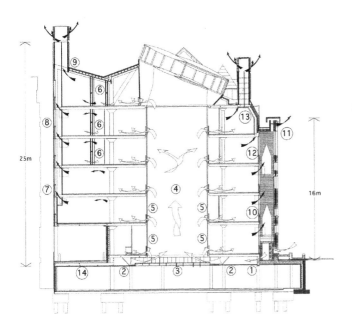

Figure 4.20a
Winter mode heating and fresh air ventilation.
1. Dampers at the intake.
2. Dampers and heating elements at the perimeter of the lightwell base.
3. Fully glazed lightwell base.
4. Lightwell fills with warmed air.
5. Air enters each floor through a bottom-hung opening window across a secondary heat source.
6. Acoustically treated transfer ducts in deep partitions at high and low levels.
7. First and second floor exits coupled.
8. Third and fourth floor exits coupled.
9. Fifth floor dedicated vent stack.
10. Street-side spaces exhaust into the stair hall, double façade.
11. Stair hall exhaust via the 'breathing parapet'.
12. Third floor dedicated exhaust.
13. Fourth floor dedicated exhaust.
14. Lower ground floor is exhausted by a dedicated full-height stack.

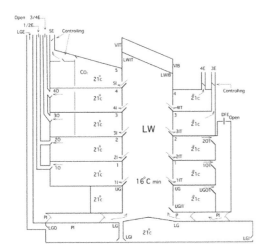

Figure 4.20b
Winter heating and fresh air ventilation mode, set points: $T_{air} = 21°C$ and CO_2 less than 1000 ppm. Key to mode of control:
VIT, Ventilation Inlet Top of lightwell.
VIB, Ventilation Inlet Bottom of lightwell.
LWIT, Lightwell Inlet Top, the inlet dampers around the upper edge of the inclined lightwell top ring.
LWIB, Lightwell Inlet Bottom, the inlet dampers around the bottom edge of lightwell ring.
1I, 2I , 3I, Low level inlets from the lightwell to the various floors of the building on the northeast (inboard) side.
1IT, 2IT, 3IT, Low level inlets between the lightwell on the Taviton Street side of the building.
1O, 2O, 3O, Outlets from each floor at high level into the stacks on the Chemistry side of the building.
1OT, 2OT, UGOT, High level outlets from each of the floors to the double façade and its high level continuous cornice exhaust DFE on the Taviton Street side of the building.
LGE 1/2E, 3/4E and DFE, Exhaust outlets from the Lower Ground, First & Second, Third & Fourth, and Double Façade.
PI, Supply air inlets to plenum at base of lightwell.
LGI, Inlets to lower ground from outside, from the ventilating corridors and from the lightwell and to remain closed until the first call for fresh air ventilation is received.

occupants to fine tune their immediate environments without destabilizing the whole system. Achieving sufficient understanding to deliver capability routinely would be a very important development in delivering a national programme of very low carbon buildings which achieve their performance passively.

Warming air is driven through the building by natural buoyancy and allowed to enter each floor through a ring of low level actuated windows (steel Crittal windows were chosen to reinforce the associational connection to the 1920s, decision-making which may well baffle non-architects given the superiority of double-sealed frames of aluminium extrusions, fatter in elevation), from which it is drawn across the floors to the exhaust plena formed by the double façades connected to stacks above. In the heating season, batteries of heaters within the supply plenum and low level heating elements at the point of entry to each floor heat the supply air in two stages, a tactic that had worked well in the Lanchester Library building.

Figures 4.21a and b depict the 'Ventilation cooling mode', triggered if the average of the slab and air temperatures monitored by the BMS exceeds 24°C. The free area for the supply air is maximized by opening all intakes at low level and the top of the lightwell. The mode is entered incrementally. Air is supplied at low level first, until the lightwell air temperature falls to 16°C, or, until the total free area of all the supply opening areas on all floors exceeds the maximum free area of the inlets supplying the plenum, 13m², at which point the windows in the lightwell ring at high level open. In this scenario the designers thought that exchange between the occupied floors and the lightwell could correct the warming temperatures through exchange flow, without the need to stack ventilate the building to the exterior. This fully open mode would be maintained until the assisted cooling (passive downdraught cooling, PDC) set point temperature was reached and PDC mode was triggered. The set point assigned at design stage was 25°C and when this was reached then the PDC mode commenced. This mode failed entirely in the commissioning exercises, upward airflow merely bypassed all of the occupied interior. This mode was deleted from the controls protocols but is worth reinvestigating by including high-level openings in the lightwell to each floor to induce a kind of internalized single-sided ventilation, a suggestion from the BP Professor Woods.

Ventilation cooling is enhanced by the use of night ventilation, so-called 'night purging'. Night settings for inlets and outlets were designed to be as for day-time ventilation. Night ventilation would commence at 11.00 p.m. and

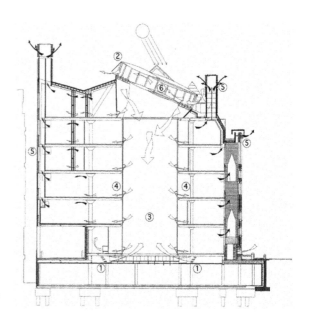

Figure 4.21a
Ventilation cooling mode, the originally intended mid-season ventilation strategy.
1. Intake dampers open at the lightwell base.
2. Inlet dampers at lightwell head open.
3. Lightwell, air at ambient temperature.
4. Air enters each floor at low level.
5. Exhaust stacks.
6. Void between inner and outer ethylene-tetra-fluoro-ethylene layers vented as necessary to reduce solar heat gain.

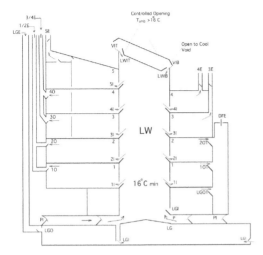

Figure 4.21b
Ventilation cooling mode, AvT$_{slab}$ and T$_{air}$ greater than 22°C and T$_{ambient}$, 22°C (see key for Figure 4.20b).

cease at 6.00 a.m., or whenever ambient temperatures started to rise at dawn, mindful that the building might over-cool, triggering the heating system.

If temperatures continue to rise, the PDC mode is adopted, reversing the bottom up airflow, admitting ambient air at the top of the central lightwell and pre-cooling it so that it flows downwards, filling the space with a static reservoir of dense, cooler air (Woods *et al.*, 2003a and b). Figures 4.22a and b show the reversal of airflow within the cross-section and the controls settings which enable it.

Cooled water at about 16°C is fed into the pipes below the inlet dampers to cool the incoming air, mindful of the likely dewpoint temperatures to avoid condensation. The volume flow of pre-cooled air to the occupied floors is controlled, as in all modes, at the inlets to each floor from the lightwell.

The base of the lightwell was to be closed, the cooling coils switched on, the lightwell top opened and the inlets at each floor from the lightwell to be closed for a period to enable a head of pre-cooled air to develop, at say 20°C. It was thought this might take two to three minutes, at which point the lower inlet dampers to the inclined lightwell head should be closed. This was to avoid the development of an asymmetric supply of pre-cooled air to the southeast side of the building.

Would it work?
Research was needed to examine and validate the proposed strategies (described above), as well as ascertain what changes (if any) would be necessary before the design was actually executed. Three modelling techniques were employed: computational fluid dynamics (CFD) modelling of airflow; the dynamic thermal modelling of space temperatures through a full year; and a physical water bath model, a more refined version of that used to test likely airflows in the Queens Building. There was a high level of confidence in the behaviour of the building in its 'normal' winter heating mode, the bottom-up centre-in/edge-out ventilation of the Lanchester Library. The interest was in the likely effectiveness of the downdraught cooling. The CFD model, constructed

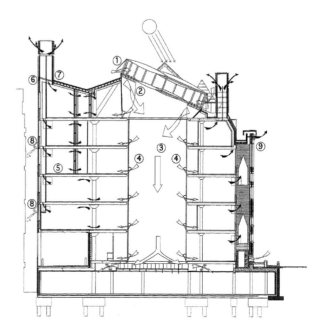

Figure 4.22a

Summer Ventilation Mode Passive Downdraft Cooling.
1. Intake dampers at the light well head open.
2. Louvres open above the cooling coils.
3. Lightwell acts as a reservoir of cooled air.
4. Bottom-hung windows opening out.
5. Acoustically treated transfer ducts in office partitions.
6. Air exhausted via compartmentalized stacks.
7. Waste heat from a chiller is dumped into stacks to promote buoyancy driven flow.
8. Stacks can open at the base of each substack to allow cooled exhaust air to escape.
9. Street-side exhaust is below the level of (cool) air intake to the lightwell.

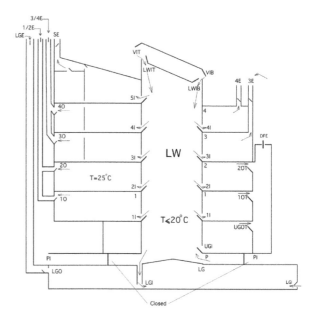

Figure 4.22b

Control settings for the dampers in Passive Downdraught Cooling (PDC) mode (see key for Figure 4.20b).

by a leading scientist in this field, predicted a healthy flow with warm air merrily rising up, across the floors and out of the building through the stacks, drawing in fresher air from the lightwell at low-level in PDC mode in which an ambient air temperature of 30°C was modelled (Figure 4.23).

In the Design Summer Year weather data, 33°C was exceeded for only two hours of the year (Short *et al.* 2007). In both PDC and passive modes, the air picks up 4°C en route through the floors and a further 2–3°C through the perimeter spaces. It might rise towards 27–28°C at ceiling level as it departs the last

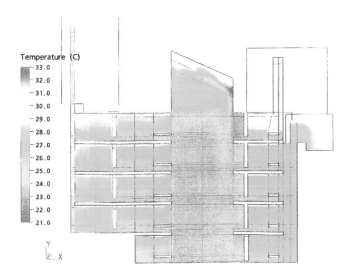

Temperature (C)
33.0
32.0
31.0
30.0
29.0
28.0
27.0
26.0
25.0
24.0
23.0
22.0
21.0

Y
Z X

Figure 4.23
Computational fluid dynamics analysis, cross-section through the SSEES building in passive mode (IESD, De Montfort University). Predicted temperature distribution in the PDC mode. The ambient temperature is set at 30°C, stack internal temperatures rise to 32°C, a warning sign.

occupied space. The model suppresses the all-important radiant cooling effect created by the building structure following night-time cooling, increasing confidence. Radiant effects were included in the dynamic thermal simulations. The model ESP-r (ESRU 1998) was used to predict temperatures, a combined airflow and thermal model which very usefully generated the likely passive airflows internally from calculating relative temperature differences in and around the building. These often have to be programmed by the operator into other thermal models on the basis of prejudice, a hunch and/or work using CFD models in parallel. The first iteration of results revealed a risk of exhaust air from floors one, two and three flowing back into the fourth floor, the familiar shared stack problem. The stacks venting the fourth floor were then separated from those exhausting the other floors. In this the SSEES building started to become rather more complex than the Lanchester Library. The digital models suggested that if the lightwell temperature was kept at or below 23°C, it would be possible to keep these most vulnerable spaces below 28°C under DSY hot summer conditions. The winter simulations predicted adequate fresh air ventilation. In fact the output suggested that perimeter heating of the building might not be necessary but it was decided to provide it to mitigate the risk of the significant cost of a retrofit. And so the computational models confidently predicted the building would deliver conditions within the guidance available throughout the year.

However, the Director of the BP Institute thought that buoyant outflow may not readily develop when internal and ambient temperatures were close or identical, quite possible in a hotter summer. He predicted the potential coupling of flows from different floors, so that a lower floor might drive flow in an upper floor and vice versa. Multiple ventilation regimes could develop, raising challenges for the control system and they clearly do. Different possible flow regimes were determined by identifying different hydrostatic pressure profiles within the lightwell (Woods *et al.* 2003a). The first laboratory experiment corroborated the anxiety. The technique is controversial, there are scaling challenges, air is replaced by a different fluid, water, its density varied by injecting saline solution or by physically heating or cooling it with localized pads. Simplified J- and U- tube geometries produced stalling flows in high summer temperatures and recirculation on higher floors, induced by heat fluxes on a lower floor. A Perspex model of the whole building included all the stacks and their connections on all floors and the central lightwell supplying pre-cooled or pre-heated air into the floors. The experiments were visualized using various coloured dyes: red for warming fluid and blue for pre-cooled fluid in PDC mode. The dynamic flow regimes were clearly very complex, the stacks fired up in unexpected sequences (Figure 4.24).

Figure 4.24
Water bath modelling of SSEES in warming conditions. The top three photographs show the development of the steady ventilation regime was non-uniform, with different stacks building up flow at very different rates, although eventually a well-ventilated space became established. Here the heat was supplied to the ground floor and the ventilation flow down the atrium, across the ground floor and up the stacks is revealed (BP Institute, University of Cambridge).

Ventilation flow. The lower three photographs show that in the case when pre-cooled (darker) fluid was supplied in the lightwell, the dense blue (darker) fluid filled the lower two floors reducing but not arresting the flow. As the flux of blue (darker) fluid increased relative to the buoyant red (lighter) fluid, the ground floor fluid became dense relative to the exterior, leading to stalling. As may be seen in the side photograph (centre) the different stacks were convecting either light red (buoyant) or dark (mixed red and blue) fluid, so that the density of the fluid rising in different stacks from the same floor may be different, leading to temperature gradients across the floor space. This was related to the location of the source of the heating on each floor. In the last photograph of the series, the pre-cooling had by now really suppressed the flow and cooled the two lower floors substantially. They are shown to be, after some time, dark (blue) and quite well mixed. The introduction of heater batteries in the stacks was intended to overcome this loss of buoyancy (BP Institute, University of Cambridge).

In winter and mid-season, whenever internal temperatures were higher than external, flows were healthy and more vigorous as the stack height was increased, but under high exterior temperatures (the peak mid-summer condition) the flow decreased as the height of the outflow vent in the outflow stack was lowered, i.e. as the building became a U-tube, the recipe for stalling. This could be countered by introducing heat into the top of the stacks, which was effected by capturing waste heat from the chiller or allowing the base of the stacks to open. Both precautions were included in the tender documentation in the nick of time. To counter a reverse flow developing on upper floors connected to the same stack as a lower floor, the stacks were further de-coupled, serving two floors only at a time and the front of the building was de-coupled from the rear by introducing full height glazed screens, to counter the effect of differential stack heights between front and rear. These different stack heights were imposed by Camden Borough Council's planning and conservation officers – thus revealing the baffling complexity of the competing forces determining our built environment. The asymmetry of the top of the building, the street frontage stacks are several metres lower than those behind, is determined by town planning and conservation anxieties about conserving the spirit of the important Georgian context (Figure 4.25). However, this asymmetry is unhelpful in terms of the physics of airflows. The Great Redeemer will not relax the natural physics of this world to retain the line of a Georgian street front.

Ironically, there are contradictory public policies, even within the same local authority. There is absolutely no intersection with, or acknowledgement of, the physical principles driving the low energy environmental strategy within the planning department. The intent for low energy buildings was strongly encouraged by another parallel borough council policy (Short *et al.*, 2004; Short *et al.*, 2009). Further alignment of policies, particularly at the local level, would assist clients and their design teams to realize low energy buildings.

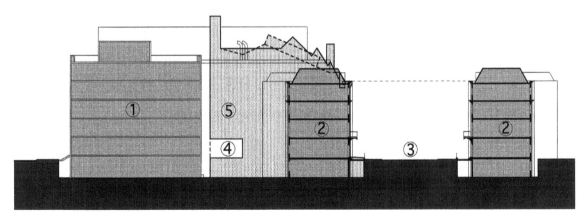

Figure 4.25
Prising SSEES into the Georgian void, asymmetrizing the U-tube into a J-tube. Cross-section through the SSEES building overlaid on the Georgian Terrace of Taviton Street.
1. Chemistry building.
2. Georgian terrace.
3. Taviton Street.
4. Vehicle access.
5. Proposed SSEES building.

Regulatory obstacles

In terms of UK fire and smoke legislation and the specific guidance for London, the situation of the building is even more intense than that of the Lanchester Library, smaller but on a landlocked site in Central London, with a basement. The principal amendments relevant to the design of naturally ventilated, non-domestic buildings since the approval of the Coventry building concerned the redefinition of compartment size, relating it to floor area, rather than a volumetric measure. The strategy would evolve as follows. Due to the close proximity of neighbouring buildings, quite typical in an urban context, the current guidance could be interpreted to recommend that the floors be compartmented from each other. Figure 4.26 shows the requirements of a compliant scheme which denies any opportunity for viable passive cooling and ventilation.

The air supply lightwell, of course, becomes a breach of the compartmentation and is, in effect, an atrium, despite being unoccupied in this scheme, technically requiring one-hour fire resistance. The additional cost of a one-hour rated atrium/lightwell glazing assembly, £250 000, was logged in the risk register by the project manager. All floors would be required to be sprinklered throughout, requiring a tank and additional plant space. Effective smoke extract is required from the lightwell, it would have to be mechanically driven, the compartmentation

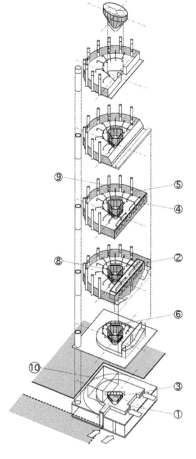

Figure 4.26
The code compliant scheme within 2004 codes and regulations:
1. Ambient air admitted to the plenum between the lower ground and the upper ground floors.
2. Lightwell/atrium to achieve 60-min fire resistance in insulation and integrity.
3. Lightwell/atrium door to be 60-min fire rated.
4. All other walls/floors between compartment floors to achieve a 60-min fire resistance.
5. Elevations to adjacent buildings made fire-resistant to comply with requirements of the 'enclosing rectangle' calculation in Approved Documents.
6. All floors to receive sprinklers throughout.
7. Lightwell/atrium to be mechanically smoke vented (not indicated on diagram).
8. All floors to be mechanically smoke vented, potentially via slot exhausts around the lightwell.
9. All dampers to be 60-min fire rated and all sealed shut on the detection of a fire.
10. Additional plant room space for sprinkler pumps, mechanical smoke ventilation equipment and back-up power'.

of the floors and the lightwell would make passive ventilation completely unviable in its cost and complexity given the number of one-hour rated punctures required through compartment floors and walls. An engineered fire and smoke control strategy was brokered with the London Fire Brigade, as shown in Figure 4.27, a single, naturally smoke-ventilated compartment to give an equivalent standard of fire safety to the prescriptive formulae of Approved Document B (2002) for a purpose Group 3 non-domestic office type building.

It was tested for the worst-case scenario, a fire growing exponentially in the lower ground floor. The strategy is based on the timely failure of the glazed lens at the base of the lightwell releasing the smoke and inducing a prodigious throughflow of fresh air across the lower ground floor and up through and out of the lightwell (Short *et al.*, 2006).

Does it work?

Shapin and Schaffer's (2011, p. 29) observation on the sensitivity of the early vacuum pumps discussed in Chapter 1 could not be more pertinent:

> Among the chief difficulties was the problem of leakage. Great care had to be taken to ensure that external air did not insinuate itself back into the

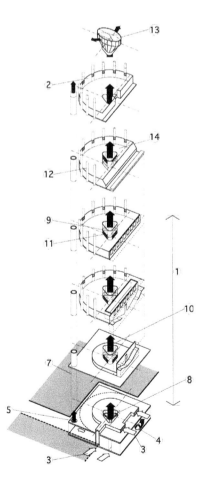

Figure 4.27

SSEES configured as a single, naturally smoke-vented compartment, scenario of fire in lower ground floor (LGF) below glazed lightwell base.
1. Building less than 18m to highest floor level.
2. Rear elevation sealed to give 60 minutes' resistance.
3. Air intakes to plenum open to admit air into LGF.
4. Theoretical location of a fire in the LGF.
5. Damper to dedicated exhaust stack to LGF open.
6. LGF supply dampers open.
7. Stack to LGF exhausts smoke.
8. Lightwell base designed to fail at critical temperature achieved by the growing fire.
9. Lightwell glazing to be 'smoke retardant'.
10. Smoke breaks through lightwell base, cooling as it rises.
11. All dampers to unaffected floors close.
12. All dampers to exhaust stacks close.
13. Vents at top of lightwell open to clear smoke.
14. Intermediate floors retained at positive pressure to resist smoke entry.

pump or receiver through a number of possible avenues. This is not at all a trivial and merely technical point. The capacity of this machine to produce matters of fact crucially depended on its physical integrity, or, more precisely, upon collective agreement that it was air-tight for all practical purposes.

This exactly records the fragility of the SSEES building ventilation and cooling during its painfully slow commissioning stage. Figure 4.28 summarizes the serial defects and their effect on airflows. The most revealing monitoring exercise was to release smoke into the top of the lightwell as the downdraught cooling was engaged in early August. PDC had been operational in the building for a four-hour period on Thursday 9 August and then from 8.00 a.m. on 10 August. Most of the building felt cool, pleasant and fresh to the researchers, despite warm outside air temperature in excess of 28°C and bright sunshine hitting the vertical windows at the upper face of the lightwell lantern. The next version, discussed later in the chapter, will disassociate natural lighting from the PDC mechanism. Figure 4.29 shows the results of the test, temperatures are maintained at or below 23.5°C when external temperatures reached 28°C.[15]

This peak temperature may not satisfy a property agent marketing a new City office building, for reasons explained with candour by Guy and Shove (2000, Chapter 7)[16] but the adaptive comfort model suggests that through a period of hot days many occupants are likely to be quite comfortable at 23.5°C. These results were achieved despite a number of failed windows at the lower end of the lightwell lantern wheel inhibiting the cross-ventilation of the lantern to remove heat generated by solar gain. At least two window mechanisms around the lightwell wheel were malfunctioning, misreporting their status to the BEMS. In one case the chain had sheared, in another the actuator motor was inoperable. Short *et al.* (2009) catalogue the pattern of construction and product failures.

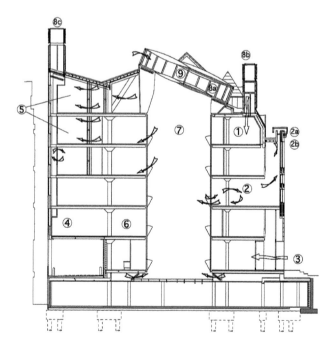

Figure 4.28
Commissioning diagnostics.
1. Director's office, fourth floor, suffered downdraughting in winter, stagnation in summer.
2. Library administrators' office, poor airflow rates.
2a. Parapet to office, damper control adjusted.
2b. Parapet to stair, similar control intervention.
3. Entrance doors actuated for easier access for wheelchair users add substantial free area of supply (and exhaust) at random.
4. Controls adjustments to first and second floors of the Library.
5. Increased flow to fourth and fifth floor offices.
6. Increase flows to side meeting rooms.
7. Increase flow rates in atrium throughout seasons.
8a, b and c. Increase all exhaust flow rates.
9. Increase air supply at inlets to plenum, i.e. open dampers further and sooner.

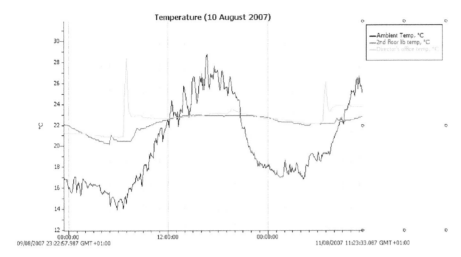

Figure 4.29
Temperatures in second floor library and in fourth floor director's office both maintained below a maximum of 23.5° C by PDC as afternoon ambient external temperatures exceeded 28° C.

Figure 4.30
Lightwell at fifth floor – smoke injected at high level is descending with the cooled air.

Smoke was injected into the lightwell lantern wheel to test airflow. This showed air movement from the lower to the upper end of the wheel. On reaching the top there was some recirculation, much as predicted in the CFD analysis: smoke was observed flowing back along the top layer of the ETFE roof, the gift of the conservation officers who required the front portion of the building to genuflect towards the adjoining Georgian buildings' cornice line, some was observed leaving through the windows at the top of the ring, thought to be due to wind, but was also observed moving down through the coils. Large quantities of smoke could clearly be seen from the fifth floor walkway flowing downwards from the coils. The smoke was estimated to take about 30 seconds to fall from the cooling coils to the base of the lightwell; Figure 4.30 records the moment.

This equates to a ventilation flow rate of approximately 18m^3/s which significantly exceeded the design flow rate of 12.76m^3/s. PDC was much faster than predicted and effective, an extraordinarily low energy cooling technique.

Reception history

> The first time I saw UCL's new School of Slavonic and East European Studies (SSEES) I nearly fell off my bike. It was the strangest, most intriguing building I'd seen for a long time. At twilight the interior was glowing through a double-layered façade – it was just possible to glimpse a staircase rising up between the two layers beneath stacked brick arches, which I later found out were 'borrowed' from Antoni Gaudi. It occurred to me there is more shock value now in a building that has such eccentricities than in one that is sensibly, fashionably 'modern'.

The above quote (Richardson, 2006) was written by the editor of *Blueprint* in an issue part devoted to a new 'category' of contemporary architecture, 'Eccentric Architecture': 'Blueprint celebrates three offbeat new projects demonstrating the courage to rebel against the strictures of the architectural establishment.' Richardson and lead staff writer Peter Kelly advance their argument (ibid.):

> The architects featured here (FAT, Featherstone Associates, Short and Associates) are a diverse bunch and will no doubt feel uncomfortable about being bracketed together [Not at all]. You're most unlikely to find Professor Alan Short … hanging out in a bar with the Clerkenwell dudes of FAT [Happy to] … though they have many ideas in common and have both been labeled postmodern. [My responses are interspersed and shown in brackets.]

Richardson and Kelly speculate on contemporary taste:

> One thing the three buildings featured here have in common is their rejection of the notion of good taste in favour of rigorously pursuing an idea … There's a kind of accepted good taste that is too prominent for comfort.

They credit this sentiment to the Director of the Architecture Foundation Rowan Moore. There may be an interesting resonance with the theory of 'Social Practices', practices of contemporaries swept up in an irresistible 'will to form', implanted in designers as a 'social practice' transmitted by 'carriers' of 'polite modernism' (Shove *et al.*, 2012). These ideas will be explored further to reflect on the prospects for practice-oriented public policy.

Charles Jencks (24 November 2011) wrote a heart-warming account of the SSEES building in the *Architectural Review*. He summarized the survival of what he inaugurated as the 'post-modern' tendency in later 20th century architecture:

> It went underground during the 1990s, the decade of Default Modernism, the reigning approach. But since the Millennium, it has roared back, happily without its moniker, and now it is very much alive with complexity architecture, digitised ornament, contextual counterpoint, and a thousand metaphorical buildings with enigmatic shapes, intent on communicating with a global culture, sans religion, sans belief, sans a public iconography. The results are decidedly mixed.

But, he continues, venturing a list of notable works in the idiom:

You cannot create a very deep iconic architecture without clients and a society that knows what they want, culturally. But a few, good postmodern buildings since 2000 show it is possible. What are the dozen recent, canonic PoMo buildings that reset the paradigm? A list is only a list, but it shows the direction. Herzog and de Meuron's CaixaForum (Time City) in Madrid and the Bird's Nest in Beijing (designed with Ai Wei Wei); Foreign Office Architects' Ravensbourne College in London, and Toyo Ito's and Cecil Balmond's Serpentine Pavilion … Norman Foster's 'Skypricker' … and obviously Frank Gehry's Guggenheim Bilbao … Peter Eisenman's Critical architecture … Memorial to the Murdered Jews of Europe in Berlin. Every movement of architecture has its green buildings and Edouard François has produced some postmodern gems, but a modest and brilliant essay in contextual counterpoint is Alan Short's UCL School of Slavonic and East European Studies, hiding away in Bloomsbury, London.

In 2013 Jencks elaborated on the musical analogy, citing Walter Pater's contribution that, 'All art constantly aspires towards the condition of music'. In fact Pater, the ultimate University of Oxford aesthete, made this observation in 'The School of Giorgione' within *The Renaissance: Studies in Art and Poetry*, the 1877 hygienicized republication of an earlier 1873 text omitting a chapter deemed anti-religious in its inquiry into homoerotic content. For Pater, Eastham (2010) explains, the 'Giorgionesque' refers as much to the 'quality of air as an acoustic medium', a sublime acoustic atmosphere formed within and around the landscape depicted'.

Eastham concludes that Pater's aesthetic aspiration was to become 'an end in itself' and music the most formal medium. Jencks identifies a 'creative background art', unique to Architecture and in this mode, it becomes urbanism. The SSEES building exhibits this phenomenon (Jencks, 2013):

> … several designers … have built a background contextualism: Alison Brooks … And above all Alan Short, who has worked at impossibly constrained sites with low budgets. For instance, his building for Slavonic studies, in a traditional London brick context, produces a low-key music.

Powell and Strongman (2007, pp. 130–131) concluded their entry on the SSEES building with a declaration of the appeal of the 'eccentric':

> The street façade is actually an outer screen: the real façade lies behind it, with a dramatic entrance space between the two. The metal and glass internal structure provides a memorable contrast to the solidity of the street elevation. But the latter is a delight. Bucking most current trends to create a highly individual, even eccentric, addition to the variegated public domain of London's central academic quarter.

The editors of *Architecture Today* (Issue 167, April 2006, Editorial) were less lyrical, more angry in leader writer mode, pointing to the profession's failures:

> The SSEES building is effectively a major research project in its own right. Given the huge global investment in urban buildings, it is morally untenable

that so few projects are seen as innovative opportunities to combat climate change.

This chapter has offered a little insight into the reasons for that detachment, it's difficult and risky. Critic Robert Harbison (2006) commented that the building:

> ... is the closest thing near at hand to a building about air. A great deal of ingenious thought has gone into the movement of air throughout the building, to the point where special organs of respiration appear on the façade.

He bemoans the fact that the brickwork of the front layer of the building does not penetrate within to form all of the spaces. So do we, SSEES is a rehearsal for a more fully realized masonry building.

Postscript: further thoughts on the use of PDC in the urban heat island

The SSEES building shows it is viable to create a building in an urban heat island without the need to resort to traditional air-conditioning. The avoidance of air-conditioning not only improves resilience and reduces the CO_2 emissions, it re-connects architecture to the provision of air, light and thermal comfort. However, the experience of creating this building shows it takes determination and effort to overcome the impediments and reduce risks. Planning, building regulation, fire regulation and building insurance, etc. are often designed to facilitate the status quo which places a large burden of proof for alternative designs. Clients, the design and engineering professions and the research community have a responsibility to not only create low energy building solutions, but to assist with the creation of more appropriate public policies for doing so.

The SSEES building shows it is viable to create low-energy, naturally ventilated buildings that can work in an urban heat island. These buildings can be designed to be future-proofed and resilient – by anticipating future climate scenarios such as heatwaves.

Subsequent opportunities have arisen enabling further exploration of the PDC reversing flow technique within the London urban heat island. The first is for another academic building, a postgraduate research centre. The innovation here is to attempt to achieve a more uniform distribution of pre-cooled air through the cellular arrangement of a broad base diaphragm wall. The scheme has a double attic plenum, the upper layer for gathering exhaust and the lower for distributing pre-cooled air across the top of the multi-cellular spine wall below. A labyrinth below services the mid-season, winter heating stack driven buoyancy flow regime. The stacks at roof level are combined intakes and exhausts owing something to the early stack designs for an earlier Mediterranean example (see Chapter 7). The exhausts are collected into an attic plenum to enable more effective heat recovery in winter, the heat being re-directed into the labyrinth supplying tempered air. Figure 4.31 depicts the passive stack mode of winter and mid-season, air is supplied at low level and rises through the central diaphragm wall and across the floorplates to exhausts at the perimeter, centre-in/edge-out.

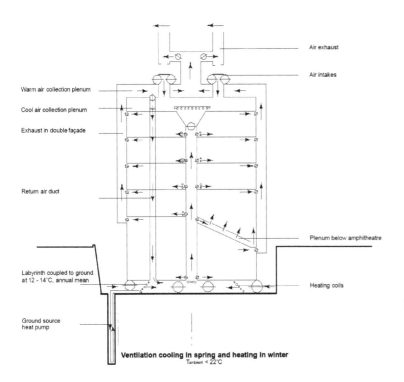

Warm air collection plenum

Cool air collection plenum

Exhaust in double façade

Return air duct

Labyrinth coupled to ground at 12 - 14°C, annual mean

Ground source heat pump

Air exhaust

Air intakes

Plenum below amphitheatre

Heating coils

Ventilation cooling in spring and heating in winter
Tambient < 22°C

Figure 4.31
Ventilation and cooling infrastructure, Postgraduate Research Centre, Marylebone, Central London. In the summer PDC mode, the labyrinth is sealed off, supply air is pre-cooled across cool water pipes at the attic supply plenum level and drops through the cells of the diaphragm wall exiting at low level into the floors below, two per cell. As the air picks up heat it rises towards the perimeter double façade exhausts. In exceptionally hot periods it can be dropped out of the base of the stacks comprising the double façades.

The building is in masonry and masonry vault and concrete soffits are exposed to the air volume inside. Air supply to an auditorium is similar to that of the Queens Building amphitheatre. Figure 4.32 shows plan levels at the top of the building, the two layers of the roof plenum and the labyrinth below ground level, the combination of which permits more precise engagement of more modes of operation.

The radiused elevation is depicted in Figure 4.33 crowned by H-pot stack terminations. Figure 4.34 shows the southeast elevation comprising a deep diaphragm wall organizing airflows and combined intake/exhaust terminations above connected into the two level roof plenum.

The technique is also applicable to major covered internal spaces, atria formed between adjacent buildings as in this example, also in Central London. Such immense internal volumes with lightweight translucent roofs cannot be heated and cooled conventionally within the latest building regulations unless they are relatively small and the surrounding buildings are of exemplary performance so that the two may be averaged to achieve the new target energy figures. The challenge remains how to conserve heat in winter and shed it in summer. This scheme for a 1500m^2 quad proposes very efficient thermal storage with a seasonal cycle of draw down, exploiting passive solar water collectors placed above exhaust/intake wheels of a similar design to that on the SSEES building. Figure 4.35 shows the basic principles in exploiting thermal storage across the seasons.

Triple layer ETFE inflated cushions make the translucent canopy with a reduced thermal transmission over that of glass and a capability to self-shade by manipulating interference patterns printed on their surfaces.

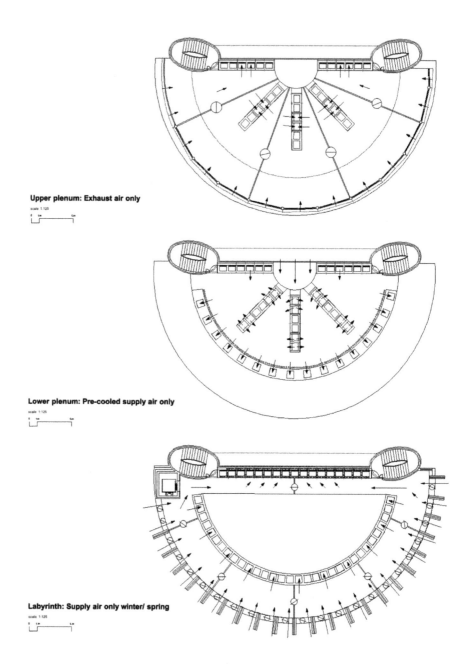

Upper plenum: Exhaust air only

scale 1:125

Lower plenum: Pre-cooled supply air only

scale 1:125

Labyrinth: Supply air only winter/ spring

scale 1:125

Figure 4.32
Plan levels: Top, the exhaust air plenum gathering exhaust from the perimeter stacks through elementary heat exchangers; the summer PDC intake plenum below, drawing air down cells within the central diaphragm wall; the labyrinth plan below the lower ground floor supplying the diaphragm walls in the centre and northern perimeter.

Figure 4.33
North elevation accommodating exhaust stacks.

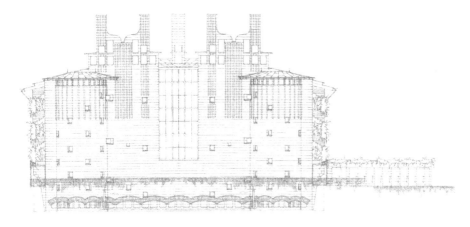

Figure 4.34
Southeast elevation incorporating a supply and exhaust diaphragm wall and combined intake/exhaust terminations.

The warmed water is circulated through a tank farm of water vessels at low level. The farm is subdivided to store water at a variety of temperatures to respond to the different seasonal needs. At night, water is cooled and pumped through coils in cold water tanks which will supply the PDC coils the following day or in the next heatwave or potentially the following season. The system is reinforced by geothermal heat exchange beyond the building to limit the extent of the tank farm, the main cost of which is in its weight and the impact of that

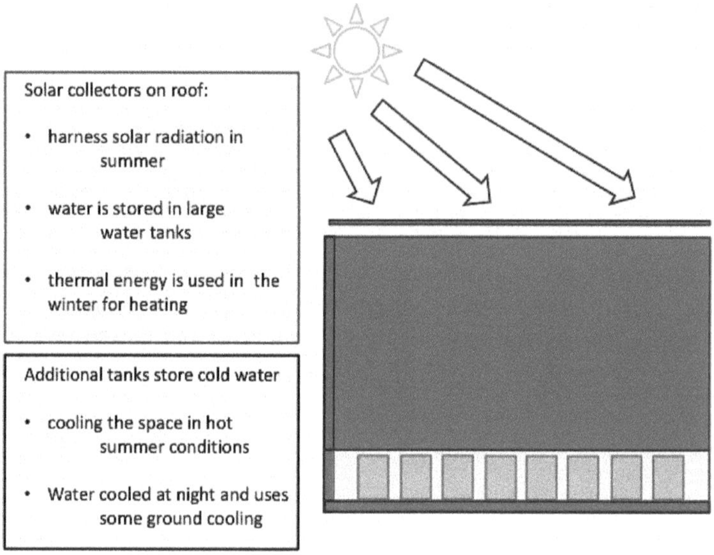

Solar collectors on roof:

- harness solar radiation in summer

- water is stored in large water tanks

- thermal energy is used in the winter for heating

Additional tanks store cold water

- cooling the space in hot summer conditions

- Water cooled at night and uses some ground cooling

Figure 4.35
The basic principle of thermal storage applied to a large atrium space (BP Institute, University of Cambridge).

Winter System

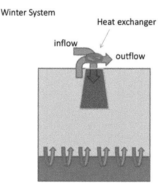

Heat exchanger

inflow

outflow

Winter air supply system – incoming air passed through heat exchanger with outgoing air – Heat recovery minimises the heat loss associated With ventilation

Air inflow is regulated by a large fan system which Brings in air and drives it to low level in the space

Low level water tanks, with hot water from the summer heat storage provides heating of the space through air-water heat exchanger, using small tubes to carry water, and low wattage fans driving low level air over the air.

Figure 4.36
Winter mode: Heating a very large space through exchange flow.

Low level heating from stored thermal energy supply in the Water tanks

on the foundations. The winter mode exploits exchange flow at the head of each array such that incoming air dropping into the space is pre-warmed by the exhaust air rising through simple heat exchangers (Figure 4.36).

A large diameter slow revolution fan drives incoming air down. The warmed water in storage below delivers heating at the trafficked surface directly at pedestrian level through low temperature trench heaters driven by low wattage fans. Figure 4.37 shows the principles driving the PDC mode.

The PDC mode will be familiar now but the innovation here is to 'pump' warmed exhaust up around the cooled downflow up the perimeter of the fabric skirts and out through the same termination arrangement, a two-storey wheel of actuated openings and cooling coils.

Figures 4.38 and 4.39 show the principles translated into a more detailed scheme operating in Winter and Summer modes.

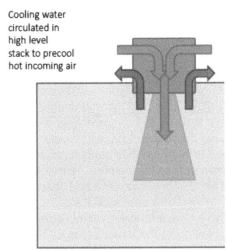

Cooling water circulated in high level stack to precool hot incoming air

Summer inflow and Outflow system

Incoming air at high level is passed through cooling pipes in which cold water circulates

Large fan draws down external air through fabric structure to provide fresh air in interior

Outflow of hot air around side of the structure allows for escape of heat

Prevailing wind determines opening orientations

Figure 4.37
Summer mode:
Exploiting Passive Downdraught Cooling.

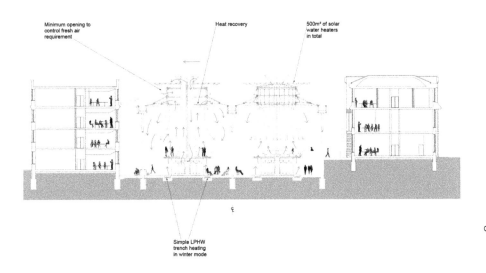

Minimum opening to control fresh air requirement

Heat recovery

500m² of solar water heaters in total

Simple LPHW trench heating in winter mode

Figure 4.38
Buoyancy-driven convective heating with heat recovery in winter and mid-season.

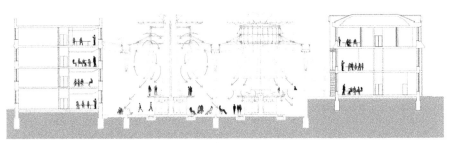

Figure 4.39
PDC in operation guided by skirts of sound absorbent material hanging from the PDC wheels at roof level.

To produce the low grade heat required here in winter the passive solar collectors are broadly viable, receiving some 1.2MW hrs of incident daily light energy, perhaps realizing some 300–600MW hrs of energy to put into the system, 300–600MJ/m^2. The water is then heated from 40°C to 60°C. Some 20MJ can be captured within 1m^3 of water so that 15–30 m^3 of water will store the season's passive solar energy. The 160m^2 of solar collectors shown on the drawings at 50% efficiency will deliver 2400–4800MJ of thermal energy. The winter heating load beneath a roof with a U-value of 0.2W/m^2/°C at a 10°C temperature difference is about 3000MJ so that some 1200m^3 of water may deliver the heating requirement, whereas in summer hot periods to deliver 3–4°C of cooling only 300m^3 of water storage may be required. Nonetheless 1200m^3 of water is a prodigious quantity to store in a safe and thermally organized way, it will be at least part substituted by a geothermal array.

This is an arrangement of known technologies capable of considerable refinement, not least in the progressive heating of the water storage. This is very necessary to preserve the possibility of delivering a relatively comfortable civic scale atrium throughout the seasons in a warming temperate climate without a huge energy penalty. In hot climates, the glazed atrium is misconceived, there are quite different ways of enclosing large and ambiguous indoor/outdoor spaces.

Notes

1. Now firmly labelled as having been published under the Coalition government, the CCRA 2012 report is available at: www.gov.uk/government/uploads/system/uploads/attachment_data/file/69487/pb13698-climate-risk-assessment.pdf (accessed 12 October 2016).
2. The EPSRC funded LUCID project. See www.homepages.ucl.ac.uk/~ucftiha/ and SCORCHIO. See www.sed.manchester.ac.uk/research/cure/research/scorchio/ (both accessed 29 April 2013).
3. See for example, Decheng Zhou, Shuqing Zhao, Liangxia Zhang, Ge Sun & Yongqiang Liu (2015) *Scientific Reports* **5**, Article number: 11160, Nature.com, doi: 10.1038/srep11160 identifies UHI over 32 cities in China but that the UHI footprint decays rapidly to suburbs and varies dramatically day to night whilst at a quite different scale UHI effects also occur over Mediterranean island capitals. Also M.K. Theophilou, D. Serghides (2014) Heat island effect for Nicosia, Cyprus. *Advances in Building Energy Research* **8**(1) www.tandfonline.com/doi/abs/10.1080/17512549.2014.890538#.VcZnZOup3dk (accessed 12 October 2016).
4. See for example, Stanford Report, 19 October 2011, '"Urban heat island" effect is only a small contributor to global warming, and white roofs don't help to solve the problem, reporting on research by M. Z. Jacobson, J. E. Ten Hoeve: Heat emanating from cities – called the "urban heat island" effect – is not a significant contributor to global warming, Stanford researchers have found,' Louis Bergeron.
5. Priestley, J.B. (1930) *Angel Pavement*. Harmondsworth, Penguin, 1977 reprint, p. 36. The fictional City of London lane he calls Angel Pavement was probably at the southern end of Wilson Street, about the only point from which one could just see both Moorgate and Broad Street Station, as Priestley describes.
6. Volumes 1 and 2 published 1818 and 1820. The volumes are rare; the Cambridge University Library has both 1818 and 1833 editions and has scanned them, but the International Association for Urban Climate has reproduced the text online with a commentary at: http://urban-climate.org/documents/LukeHoward_Climate-of-London-V1.pdf (accessed 12 October 2016).
7. A fascinating exercise awaits researchers to reconstruct the history of the London urban heat island.

8. http://onlinelibrary.wiley.com/doi/10.1002/wea.195/pdf (accessed 12 October 2016).

9. Available online: www.eurosurveillance.org/ViewArticle.aspx?ArticleId=558 (accessed 12 October 2016).

10. It is believed that the finalists included architectural practices: John Miller & Partners, Feilden and Clegg, and Stanton Williams.

11. The Design Summer Year consists of an actual one year sequence of hourly data, selected from the 20 year data sets to represent a year with a hot summer, See www.cibse.org/Knowledge/CIBSE-other-publications/CIBSE-Weather-Data-Current,-Future,-Combined-DSYs (accessed 12 October 2016).

12. CCRA (2012), para viii, Risk descriptions, BE1-Urban Heat Island.

13. See Masaryk's manifesto in Masaryk, T.G. (1918) *The New Europe: The Slav Standpoint*. London, reprinted by Bucknell University Press 1972, University of Michigan, and Perman, D. (1962) *The Shaping of the Czechoslovak State: Diplomatic History of the Boundaries Volume 7*, Leiden: Brill.

14. *Times Higher Education Supplement* 16 November 1988, 'Four London colleges say "Yes" to SSEES'.

15. Short and Associates *et al.*, (2007) unpublished record.

16. Their interpretation, written some two recessions ago, based on the analysis of semi-structured interviews with representative professionals in the office building world belongs within the general notion of social practice, that 'decision-makers' are not wholly unconstrained autonomous actors with an entirely free choice as their course of action but persons 'embedded in a network of social relations that limits and controls the technological choices that she or he is capable of making' (Cowan, R.S. (1987) 'How the refrigerator got its hum' in McKenzie, D. and Wacman, J. [eds] *The Social Shaping of Technology*. Milton Keynes: Open University Press, p. 262). They identify the property agent–developer–institutional investor communities operating in the highest value city centres and their design and construction industry dependents as a particularly intransigent and influential network. This is discussed further in Chapter 10.

5
Theatres

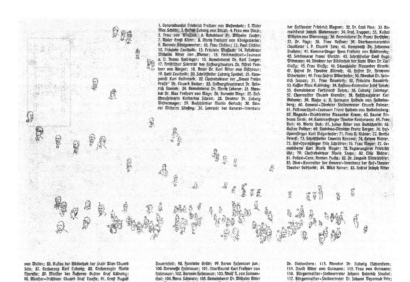

Figure 5.1
Gustav Klimt, Etching, Das Alte Burgtheater im Vienna, No. 45 (24 × 35cm). A rare depiction of a habitual audience, many of whom are identified on the accompanying overlay, including the Habsburg Erzherzog Karl Ludwig and his third wife. Every audience member will have had their own expectations of tolerable conditions in full fashionable or ceremonial dress. The auditorium ceiling is tight to the upper balcony, relying on the exhaust infrastructure above the central gasolier to maintain some airflow, but it must have been hot (author's collection).

I have had many abortive discussions with architects building new theatres – trying vainly to find words with which to communicate my own conviction that it is not a question of good buildings and bad: a beautiful place may never bring about explosion of life, while a haphazard hall may be a tremendous meeting place; this is the mystery of the theatre, but in the understanding of this mystery lies the only possibility of ordering it into a science … It is not a matter of saying analytically what are the requirements … this will usually bring into existence a tame, conventional, often cold hall.

(Peter Brook, 1972, *The Empty Space*, p. 71)

Relating his structure to precursor buildings, while yet attempting to make an original contribution, the architect adds a link to what is a continuing chain.

(Karsten Harries, 1983, pp. 9–20)

Why would one even consider investing time in attempting to build a 'free-running' theatre environment when the overwhelming imperative would have to be to make a 'good' space for performance? At first glance, the problems appear insuperable. Would not the density of people and lights generate so much heat that the design would require an absurd increase in the number and size of punctures for air inlets and outlets, much more than for other non-domestic building types. The form would probably disintegrate. What hope would there be of containing sound? And would not the audience – volunteers after all – paying to be in the space, find it uncomfortable, unpalatable even: they would encounter layers of hot and hotter air. They would be listening involuntarily to parallel performances in neighbouring auditoria, all of which will be jostling to take their entitlement of fresh air from a common source and cast off the foul.

In reality, theatre auditoria tend to be occupied very intermittently, full at most for two performances a day, with intervals, alternating with long periods of low occupancy during rehearsals and technical set-up, presenting a real environmental design conundrum. Theatres came to be regarded as notorious fire-traps in London, harshly dealt with in the evolving web of regulation, unyielding in its insistence on compartmentation.[1] Theoretically, no collapse of notional compartments into one contiguous space is permitted, as in the Lanchester Library. Ventilation airflow paths to and from every compartment need to be entirely separate to prevent sound transmission and to have a chance to control the distribution of fresh air and its eventual exhaust. Moreover, fire and smoke legislation has propelled the design of theatre environments towards the artificial mode, eschewing natural ventilation: an unintended and unhappy consequence. This chapter presents some recently built examples of naturally conditioned theatres in city centres, theatres in which it was successfully argued that auditorium and stage could safely comprise a single compartment, in which actors and audience populate the same surface, so that the presence of a proscenium is entirely at the behest of the director.[2] In fact the voluntary introduction of a proscenium in contemporary staging has become rather novel and provocative. Pre-training for younger directors is available.

Some theatre companies very much prefer to work in natural environments. Why? For Arts organizations made capital rich and revenue poor by patterns of subsidy, fierce containment of life-cycle costs is critical to survival. The National Audit Office review of Arts Council Lottery funding showed organizations incapacitated by their running costs.[3] For some the burden would prove fatal. But there is another reason to recover natural environments in which to perform: air-conditioning washes out 'atmosphere' in theatre. It contributes to what theatre director Peter Brook has called the 'Deadly Theatre'. It reduces the immediacy of the theatre for both the audience and the actors. The actor Timothy West, patron of the Theatres Trust, told us, 'theatres are all about Geometry and Atmosphere'. The actor David Suchet wrote forcibly to the director Stephen Daldry about the future form of the rebuilt Royal Court Theatre in Sloane Square, London. The director forwarded it to the engineer Max Fordham as briefing material. Max showed us the letter as we all commenced the design of the Contact Theatre. Suchet wrote:

> I maintain that air-conditioning has a negative effect on theatre performances. My own experience has proved this statement to be true. In 1981 at Stratford I managed to get the air-conditioning switched off for the trial scene in 'The Merchant of Venice' and the difference was very noticeable. In 'Separation' at the Comedy Theatre I managed to persuade Front of House to turn it off and only have it on during the interval. The difference was extraordinary. The other two most notable occasions were 'Othello' in 1986 at Stratford and the Barbican (the air-conditioning in this theatre requires a whole article as it is definitely the most destructive to both comedy and tragedy and wins the prize for the noisiest and 'draughtiest' of any I have experienced) ... and 'OLEANNA' at the Royal Court and The Duke of York's Theatre. In both theatres air-conditioning was only used before the audience came in and the interval. At the Duke of York's when the air-conditioning was left on inadvertently the difference was very noticeable indeed. I would go so far as to say that the play and the performances, of us the actors, were noticeably worse when the air-conditioning was on ... I have no doubts about this whatsoever. For every theatre play I am in I will fight for no air-conditioning during the actual performance. I know that this is a fight that I will not always win but I will fight none the less!!! I believe it's that important!!!

And so an important but elusive variable is introduced into the assessment of 'success' in the forming of internal environments, the production of 'atmospheres'. Can 'atmospheres' be usefully discussed other than anecdotally? How can the designer 'engineer' geometry and atmosphere? Scholarly attention to 'atmosphere' has enjoyed a recent revival.

'What is Atmosphere? Is it air and weather? Or is it the in-between – effect, matter, immaterial, space, ephemera?' asks tutor and landscape architect Silvia Benedito of her students at Harvard University.[4] Bohme (2013) attempts an explanation in the context of theatre:

> What matters is that, in speaking of atmospheres, we refer to their character ... atmospheres are totalities: atmospheres imbue everything ... they bathe everything in a certain light ... It is the art of the stage set which rids atmospheres of the odour of the irrational: here it is a question of producing atmospheres ... to provide the atmospheric background to the action.[5]

Bohme concludes that this would imply some common sensation, some 'intersubjectivity', so that the character of atmospheres should then be communicable, the audience might be expected to have some broadly comparable experience of the set and the actions on and within it. But this emerging discipline is still entirely based on anecdote. Need it be? Bohme cites the perhaps obvious and extreme example of the staging of Wagner's operas, in which atmosphere was, by contemporary accounts, very energetically contrived within the Bayreuth Festspielhaus. Audiences were intoxicated, certainly the German constituency, if not all of the French. True believers were anaesthetized by the atmosphere against acute discomfort. Frederic Spotts describes the complex mechanism by which interconnected physical and psychological states were induced in a calculated way to captivate the Bayreuth audiences. There seems to be very little work on this folding of psychological responses through contemporaneous physiological response, that is the natural exchange between the 'felt-body' and the enclosure, the flux of radiative, conductive and convective heat and coolth, the sound pressures, the relative humidity, brightness and darkness, the latter exploited to great effect at Bayreuth.

How does one actually engineer an intense intentional 'atmosphere', communally experienced. Is this always, as Peter Brooke suggests, a happy accident? Bayreuth suggests not. Here lies a rich and rewarding piece of work. There is, unfortunately, an ideological divide between wholly complementary fields of research endeavour.

Although borrowed from atmospheric science and meteorology, the term 'atmosphere' and the perception of atmospheres is now discussed very pointedly out of any scientific context. This is regrettable and there is a precedent for a more holistic exploration. The 'ecological approach' to understanding perception, termed 'visual perception' in Professor James Gibson's revelatory work, is firmly rooted in science (Gibson, 1986). He denounces the concept of 'space' as the basis of human perception of the world, 'Space is a myth, a ghost, a fiction for geometers', a shocking dismantling of mid 20th century mainstream modernist architectural belief. Gibson introduces the idea of the terrestrial environment as comprising, 'a medium, substances and the surfaces that separate them', the recognition of the useful in the environment, but that does not quite encapsulate the phenomenon of 'atmospheres' descending upon navigable landscapes (Gibson, 1986, p. 16). It is a utilitarian analysis. It does not then capture the inviting atmosphere, the warm glowing intensity of the buzzing auditorium caught by Adolph Von Menzel in his 1856 painting of the Theatre du Gymnase in Paris, all the more evocative because it was painted from memory (Figure 5.2). Audience members are turning towards each other to swap opinions on the performance, the musical accompaniment builds in volume as the actors unveil a wholly unanticipated development in the unfolding narrative.

Its catalogue entry for the travelling exhibition of German 19th century art at the Metropolitan Museum of Art includes the following passage:

Like Degas, Menzel seems to have been fascinated by the spectacle, the artifice, the light effects, and the phenomena of the theater. However, unlike Degas … Menzel was primarily interested in a specific event and a specific moment in time … a Realist statement.[6]

Figure 5.2
Adolph von Menzel,
'Theatre du Gymnase',
1856. Oil on canvas, 46 ×
62cm. Courtesy of bpk/
Nationalgalerie,
Staatliche Museen zu
Berlin/Joerg P. Anders.

Griffero (2014), in a recent work, muses on the meteorological origins of the concept of atmosphere: 'a sort of nomad feeling that is, so to speak, in the air … that we feel *in* the felt-body.'

And (ibid. p. 55):

Think of the emblematic case of wind, whose enigmatic dynamicity – one *'cannot see it, nor does one know where it comes from and where it goes'* [quoting Volz 1910: 59, German only] significantly influenced even the theological notion of spirit.

This recalls the tainted Willy Hellpach and his interesting and persuasive concept of 'geopsyche'. In 1911, Hellpach (1977, p. 56) concluded 'climate' was a fabrication, 'weather' is simply domesticated as 'climate': 'climate [is] an excellent example of atmosphere as an unwilling feeling present in the space'.

Griffero (2014, p. 105) muses, following Hellpach's train of thought, that:

… we perceive air as being cumbersome, unnerving, feeble, heavy, rigid, harsh, biting, stingy, suffocating, sneaky, and by these words we attribute to its mildness, warmth, heat, coldness, windiness or freshness a particular 'note', which cannot be simply identified with the heat, coldness, calmness and the like … An example of a specific climatic atmosphere is the oppressive heat of a summer afternoon, influencing everything and almost casting a monochromatic aura over all things.

It is not difficult to imagine a part physical explanation for the perception of hot atmospheres, the effect of humidity or aridity, particulates in suspension, high radiant temperatures torching through the air, convective air movement rising off luminous surfaces and the dissolution of sightlines and horizons, the phenomenon of atmospheric perspective. Predictably perhaps, but no less disappointing, Griffero (2014, p. 123) 'denounces' scientific definitions of 'atmospheres' by physicists who, he explains, simply do not understand: 'unlike other phenomena considered invariable by naïve physics, atmospheres are relatively (perceptually) amendable'.

They may be in flux, not steady state. Revulsed by 'modern science', sceptical that quantification could contribute to an understanding of the 'experience' of architecture (Perez-Gomez, 1983), compounding a distrust of geometry as a form-generating tool with all the consequences of all other potential scientific insight, the post-post-modern, 'anti-modern' commentator, anxious to organize and classify his/her collection of experiences is, in the end, to paraphrase architectural historian John Summerson (1957, p. 309), left with a disconnected 'rag bag' of anecdotal observations.

However, taking an 'unmodern' stance, Giffero (2014, p. 143) raises the possibilities of the 'de-subjectification' of 'feelings'. This might yield some more reliable taxonomy of atmospheres to help designers, relying solely on a personal archive of atmospheric experiences to create authentic new places, to avoid the production of unsympathetic atmospheres.

Elie Haddad (2010) is sceptical that strings of such anecdotal observations can be employed to deliver resolved designs:

> … [the] weakest point in Norberg-Schulz's theoretical proposition: his desire to translate phenomenological discourse into a tool for the generation of architectural forms that recreate a semblance of meaningful environments.

But, could a more systematic approach help to provide the hoped-for insight? How interesting it would be to attempt to relate, in a systematic way, the physical exchange between human bodies and their environments, with emotional states: contentedness, intellectual efficacy and enquiry, the self-reported environmental 'sensation', interpreted through a developing understanding of 'adaptive' response, all discernibly climate related. Such a body of work has been undertaken by Professor Gail Brager and Professor Ed Arens at the University of California at Berkeley, and Professor Richard de Dear at the University of Sydney. They had systematically collected 86 000 responses across the principle climates by April 2012, by which their theory of adaptive comfort is validated. The work continues. This understanding of the human experiences of inhabitation has not yet become mainstream practice. As a result, engineers have insisted the occupants of their buildings should be comfortable when quite clearly they are not. The theatre environment has been very particularly sacrificed, the victims being both the actors and their audiences.

The sensation of radiant temperature has for so long been considered to be important that it has acquired mytho-poetic status. Examples from literature are particularly instructive in conveying both the 'atmosphere' of a place and its effect on a character. For example, in *The Count of Monte Cristo* by Alexandre Dumas (1844, 1966 edition, p. 79) character Edmond Dantes, on arrest, was thrown into a:

> ... room that was nearly underground, its bare dripping walls seemingly impregnated with a vapour of tears ... he found himself alone in the silence and the darkness, as black and noiseless as the icy cold of the vaults which he could feel pressing down on his feverish brow.

Dantes was experiencing much more than the dry-bulb air temperature, the appallingness of his experience was compounded by the very high relative humidity of the chilly motionless air, the low surface temperatures of the stone enclosure, characteristically tracking the annual mean temperature of the interior of the rocky promontory, his relatively elevated skin temperature on arrival, the physical shock of sudden adaptation, a sense of betrayal but still not utter hopelessness. Isabella in flight from Manfred, her lascivious prospective father-in-law, the Prince of Otranto, in Walpole's (1811) *The Castle of Otranto*, running through 'intricate cloisters' was compelled to flee into the feared caverns beneath:

> An awful silence reigned through those subterranean regions, except, now and then, some blasts of wind that shook the doors she had passed, and which grating on the rusty hinges, were re-echoed through that long labyrinth of darkness.

In *The Stones of Venice*, John Ruskin (1893, Book 2, Chapter III), on entering St Mark's Basilica in Venice noted the sudden, marked drop in temperature as he felt it (a symptom of the cool radiant surfaces encircling him), the sharp increase in relative humidity, the transitory presentation of the particles in suspension in the narrow 'phosphorescent streams' of sunlight admitted through 'a remote casement'. The atmosphere builds from physical phenomena: the extraordinarily long reverberation time, the chiaroscuro created by the severe rationing of openings to daylight, the structural temperatures, the dust and the sudden demands they make on the sensory organs to adapt and their delay in achieving accommodation.

An infamously 'bad atmosphere' pervaded the Chamber of the House of Commons in the Palace of Westminster by the third decade of the 19th century. This and the correctives proposed are interesting because the Chamber was, even as a temporary space, an auditorium, of the 'runway' type. The 1836 Inman Report, commissioned by a Select Committee appointed in 1835 to 'consider the best mode of ventilating and warming the new Houses of Parliament', heard evidence that the air in the House was often, 'foul, rancid, or pestiferous'. The Committee, determined to correct this, trawled the 'ventilation community' and subpoenaed Dr Reid (discussed at length in Chapter 1) to consult on the problem. The scheme which arose was authorized to 'be submitted to the test of actual experiment', an exciting prospect at a prodigious scale (Reid, 1844).

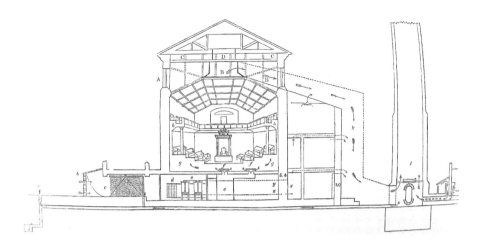

Figure 5.3
Dr Reid's economical visualization of his ventilation scheme for the House of Commons, bottom-up with extract at the highest level but connected to the J-tube of an aspirating chimney, the buoyancy effect accelerated by hot air injected at its base (Reid, 1844).

The progress of this experiment has been subjected to scrutiny by a series of commentators, most recently by Dr Henrik Schoenefeldt. It all ended badly for Dr Reid in London. The scheme was not implemented correctly, he was dismissed, he objected vociferously. As we discovered, the succeeding administration was pleased to exonerate him at its predecessor's expense, so that, liberally compensated, he resettled, dignity restored but still vengeful, in America. He published his well-resolved unbuilt scheme. The essential principles are recoverable (Figures 5.4, 5.5a and b). The air quality problem in the Chamber persisted into the 20th century. The atmosphere was still reported as being intolerable in 1908, prompting the calling of another Select Committee and a Royal Commission on Ventilation led by Dr Armstrong. The architect A. Saxon Snell was a key witness, he of the circular hospital ward controversy which we will encounter in Chapter 6. The pertinent issue became the desirability or otherwise of 'plenum ventilation'. Saxon Snell was unambiguous in his evidence, plena led to an enervating atmosphere. Saxon Snell as the House of Commons Architect evidence quoted in the proceedings of the 1904 Select Committee on the Ventilation of the House of Commons (*Building News*, 1904, Vol IX.5.100, item no. 9):

As to the monotony that you speak of, the effects of it may be seen in the plenum you know that the full Plenum System where the supply is exceedingly uniform [sic]; but those who use that system a great deal – those who, for instance, live in a room that is absolutely ventilated by that system – only complain of the air being oppressive and giving rise to lassitude.

Reid's House of Commons scheme is credible, bottom-up ventilation driven by a stack arranged as a J-tube to give additional stack height, the airflow forced

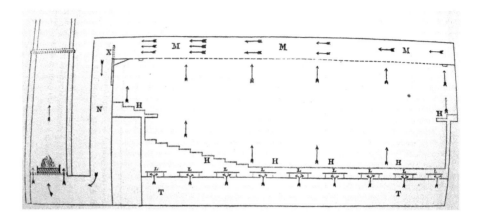

Figure 5.4
A surprisingly diagrammatic depiction of Dr Reid's ventilation arrangements for the House of Commons (Reid, 1844). Reid apologizes that he couldn't afford more elaborate diagrams but this is fortunate in preserving clarity. Air is drawn through floor registers from a full plan plenum below, rising through the seating rake, supplied independently to the gallery, and exhausted through a perforated ceiling, Reid being an enthusiast for perforated wall, floor and roof planes to ensure equilibriated distribution of air. He doubles the effective height of the stack by folding it into a J-tube, fuelled by an open fire at the base of the aspirating chimney. A more sophisticated device injecting hot air appears in subsequent drawings.

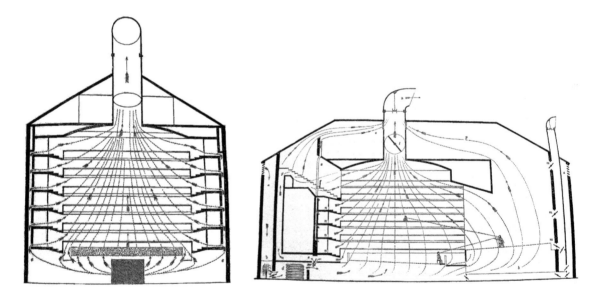

Figure 5.5a and b
Reid's ambitious but optimistic scheme for an opera house, the backstage stack is lower with multiple entry points poised to scramble the directions of the neat arrows (Reid, 1844).

by an open flame. It is still highly diagrammatic but Reid subsequently applied the principles to the problem of ventilating a vast opera house. Here (Figures 5.5a and b), air for the audience is brought down behind the main façade from cornice level to a chamber of heater batteries beneath the foyer before being discharged through the tilted platform of stalls seating, some diverted behind the foyer up to the highest level balconies or 'gods'.

The stage is supplied from the auditorium and all exhaust rises through an immense gas-fired chandelier, a gasolier, to a revolving termination which is estimated to be some 16ft in diameter, controlled by a rotating damper of similar size. An additional supply drops from cornice level to a horizontal supply route below the lower stage void, connecting to the main distribution plenum below the stalls, whilst an exhaust rises alongside to almost the height of the central exhaust, but not quite, at the rear of the stage, with multiple damper controlled entry points, all of which may be problematic. It is a scaleless proposition, a 'big picture' concept. One can appreciate all the more the refinements introduced to make a strategy like this work 20 years later in Vienna and Paris and thereafter by their imitators, the exception to this growing concern for the audience, being, most notably, Wagner's opera house at Bayreuth.

Surgeon-Major Billings, encountered in Chapter 1, records the rapid advances in environmental understanding and its proponents' achievements in making large, apparently successful, public assembly spaces by the late 1860s, less than 20 years later (Billings, 1884). Helpfully, he reviews the then-current state of the Chamber in Westminster in 1876 as an overture to his Chapter 9 on theatres, with an account of Dr John Percy's experiments there, subsequent to Reid's attempts. Billings disagrees forcefully with Reid's prescription for air supply, 'wholly inadequate'. He explains that the high level supply had tapped into the most polluted air and the low level intakes in two adjacent courtyards, drawing air in summer over ice blocks on wooden racks, depressed temperatures by only 1°F. Should air be supplied from above or below in these densely occupied buildings? He describes the lunchtime experiment at the Smithsonian Institute in Washington, DC on 27 March 1865 as part of the curative programme for the US House of Representatives at the Capitol, Washington: all air currents, that is the plume off the warm body and the exhaled air, were observed to rise. The upwards method was vindicated. The subject then smoked a pipe, super-heating his exhalations and emitting liberal particulates, and the same phenomenon was observed, only more so.

The paradigm for Billings was what he called 'The Grand Opera House', actually the Wiener Hofoper, in Vienna, built between 1861–69, the architect being August Sicard von Sicardsburg, accommodating an audience of 2700 including, periodically, various generations of Habsburgs, a fearful responsibility. With confusion reigning still over the mechanism of cross-infection, Sicard subscribed to the general belief in harmful miasmas and carbonic acids given off by the body, particularly in crowded and poorly ventilated environments. One imagines the designers would take no chances with the Imperial Court, and they didn't (Billings, 1884):

> Probably no theatre in the world excels the Grand Opera House at Vienna in the extent and completeness of the special arrangements for securing ventilation, and in no theatre of the same size, and under similar climatic conditions, have better results been obtained.

The ventilation strategy was the work of a distinguished doctor, surgeon and hygienist Dr Karl Böhm, later to become medical director of the Hospital

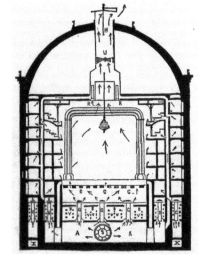
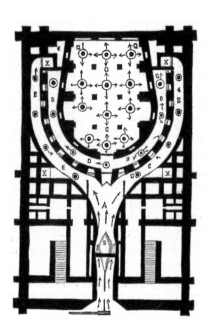

Figure 5.6a and b
Wiener Hofoper, 1861–69
(Billings 1893).
A. Fresh-Air Chamber.
B, C, D, E. Heating Chambers.
G. Tubes for Fresh Cold Air.
H. Foul Air Shaft.
S. Fresh-Air Fan.
U. Foul-Air or Aspirating Fan.

Rudolfstiftung in Vienna. How did Böhm crystallize his thoughts into such a complex configuration and then cajole the architect and constructors to implement the scheme sufficiently accurately to work and continue to work under close control? Figures 5.6a and b show the sheer scale of the building and thereby the ventilation routes and plena. It was fan assisted, one at the intake S, one at the exhaust U. Every gallery, compartment and box has its own dedicated supply, heated by steam pipes as required, and exhaust routes and has some measure of independent control. The helical supply fan delivered up to 3.5 million cubic feet per hour through two 19.5 × 13ft supply ducts supplying a basement chamber equivalent in volume to the auditorium above, 388 000 cubic feet, organized in three horizontal layers. The central layer was dampered to admit or deny air through immense heating coils and was enabled to bypass all heating in spring and summer, presumably connecting through both in mixing mode. 807ft^2 of inlet supplied an auditorium with a plan area of 14 608ft^2, so that Dr Böhm sensibly persuaded the authorities to provide a prodigious free area to enable air supply, equivalent to 5.5% of the floor area, twice that of the more recent Contact Theatre reviewed later in this chapter. The 90-burner chandelier assists with impulsion at the summit. Billings was amazed by the centralized control room. Electrically-driven controls and indicators made this concentration of controls in one spot possible. We do not quite know if Billings is reporting on personal experience. He certainly stayed in Vienna on 13 November 1876 in the Hotel Metropole. His letter home to his wife suggests a worthy day looking at hospitals and art galleries but he omits to mention a highly agreeable night at the opera. The Johns Hopkins Hospital archivists suggest that relations in the Billings home were brittle. Dr Böhm was inventing an intensely occupied environment, 143.7ft^3 (4.07m^3) per audience member, 'producing an atmosphere', in which the Habsburgs and all their extended family, diplomats and visiting dignitaries should, feeling safe and well, enjoy temporary transportation from Imperial tensions through the full length of the performance. The volume per person is less than half that recommended today but, of course, the same volume lay beneath the audience's feet. Fricke *et al.* (2006) reprise the custom and practice of contemporary concert hall design, their Figure 1 records 10m^3/person as the optimal volume. The authors conclude that the custom and practice formula of volume/audience number of between 6 and 8m^3/person is actually unreliable.

Billings drew much of his understanding of the principles and strategies for good ventilation from General Morin's substantial 1863 manual *Etudes sur la Ventilation* with its set of intricately detailed plates. Morin (1795–1880) was a high-ranking military officer whilst serving as the first Professor of Mechanics at Metz, an impeccable military and academic provenance which must have greatly assured Surgeon-Major Billings (Morin, 1863).

Morin's scheme for a generic Gaity Theatre (Figure 5.7) is yet more sophisticated than the Hofoper, incorporating multiple supply routes across the basements from opposite directions, threaded beneath the stage house and the foyers from the notional street front. Their width is consistent as more ducts connect into the system, suggesting options for switching supply directions as wind direction and levels of pollution outside vary. They converge on two heating chambers as at the Hofoper. Passing through the chambers, air is fed into a complex web of ducts leading to discharge at the leading edge of apparently all balconies and boxes and, at the 'fourth wall', from the edges of the proscenium itself. The occupied volume of the auditorium is used as a mixing chamber. Air is exhausted back under all seats in boxes and balconies and taken up and over into a very large chamber above the immense gasolier, comprising stacked rings of burners, intended to deliver a tremendous accelerating force to the draught.

Morin provides an alternative scheme for a Lyric Theatre (Figure 5.8). The differences are subtle, perhaps performances were envisaged to be longer and quieter with less churn in the audience. Air is supplied to the balcony and box edges and the proscenium sides as in the Lyric Theatre scheme but exhausted at ceiling level at the back of each box and through a close grid of exhaust registers across the stalls floor. Perhaps more intense deployment of cigars was anticipated. The stalls exhaust is driven up large stacks either side of the theatre envelope accelerated by the exhaust flues from the boilers. All exhausts discharge into a volume across the top of the entire auditorium ceiling above banks of vast rings of gas jets and up through a more thoroughly developed roof stack. One might speculate that a smarter audience paying higher ticket prices might demand greater comfort in the form of more forceful and apparent ventilation.

In sharp contradistinction was Wagner's Festspielhaus completed in 1875 (Spotts, 1994). Determined to depart completely from the developed model of the grand opera house and progressing through a series of architects of decreasing prominence (Gottfried Semper, Wilhelm Neumann and Otto Bruckwald), it is perhaps not surprising that the published drawings, particularly cross-sections (Spotts reproduces a perspectivized reconstruction of the Festspielhaus as of 1882), contain no indication of the inclusion of any voids or paraphernalia to help with the conditioning of the interior, except for an octagonal lantern with small louvred panels above the fabric ceiling plane to the

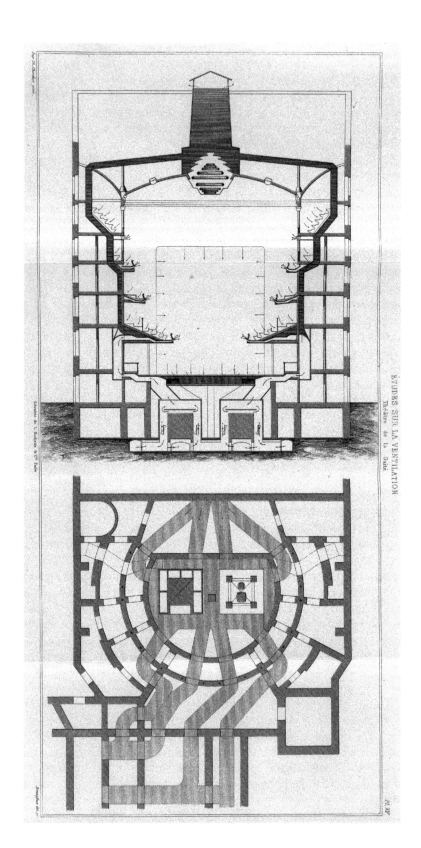

Figure 5.7
Morin, *Etudes sur la Ventilation*, Plate XV 'Theatre de la Gaite', plan and section. Reproduced by courtesy of Cambridge University Library.

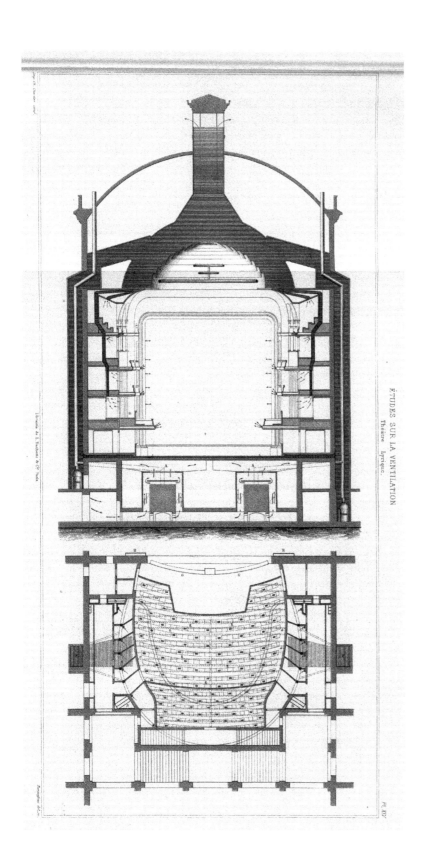

Figure 5.8
Morin, *Etudes sur la Ventilation*, Plate XIV 'Theatre Lyrique', plan and section. Reproduced by courtesy of Cambridge University Library.

auditorium. Contemporary accounts suggest that, dressed formally to an unforgiving code, members of the audience were both stressed and mesmerized. Intimidated on taking their seats, indeed frozen to them because they squeaked audibly if the occupant moved, the audience was plunged into darkness, apparently for the first time in a Western theatre, stunned by what was revealed behind the curtain, layers of proscenia framing a fantastical scene, the conductor and orchestra mysteriously concealed. In this hallucinatory state (as the *Figaro* correspondent observed), in this sacred atmosphere, audience complaints about excessive cold or heat or stuffiness remained unrecorded. A triumph in the production of an atmosphere.

Spotts (1994) observes:

> Bayreuth demands seriousness. Who would go to the inconvenience of travelling to an out-of-the-way place to endure the discomfort of sitting on a hard seat for hour after hour in an auditorium without air-conditioning at the height of the summer except out of deep devotion to Wagner's works. The composer deliberately planned it that way.

A reporter in 1956 caught a notorious Diva commenting, counterintuitively: 'In Vienna, I'm fed up with a rehearsal before it starts. Here I enjoy it right down to the end. The place is inspiring. Even the air is inspiring.'

However, a peripheral insight into the environment emerges from the orchestra's complaints that the pit was draughty, complaints met unsympathetically by Wagner. A recent reminiscence by Christian Thielemann (2015) records that Karl Böhm stood in two bowls of cold water while he conducted to counter heat developing in the crowded pit.[7] The fly tower has very large opposing Diocletian windows which look to have been openable. A river of cooled evening air may have descended between the hanging backdrops onto the raked stage before spilling over the edge into the orchestra pit, from which only the heat produced by the musicians themselves could induce the plug of cool air to move. Conversely, the height of the fly tower would drive a stack-induced flow, drawing up air from everywhere below, straining through cracks and fissures at the rear of the pit, both modes quite possibly happening simultaneously and through each other, a phenomenon which will be observed below, in the large stacks of the Contact Theatre.

Contemporary guidance for tourists proposing to attend a performance warns of the lack of ventilation, excessive heat and the hardness of the furniture. Izenour's idiosyncratic but compelling argument of 1992 (Izenour, 1992), originally put in a paper in the Yale architectural journal *Perspecta*, suggests that Wagner's model was the open air Greek amphitheatre, 'the Odeum' of Agrippa, superimposed on the Agora of Athens in approximately 14BC, described specifically as an 'odeion' rather than a theatre by Pausanias (Nielsen, 2002; Dinsmoor, 1950). In this interpretation, the Bayreuth Festspielhaus is a late 19th century return to the 'steeply raked, long-radii-generated, truncated auditorium' (Izenour, 1992, p. 177). Izenour asserts that the size, shape and scale of the auditoria of these two buildings, built 1700 years apart, 'is almost identical'. Bizarrely, it is almost impossible for Wagner or Bruckwald to have known the physical site, it was not excavated until 1931 by the American School of Classical Studies (SCSA, 1990). Nonetheless, in generic terms, Izenour observes a historical tendency to 'interiorize' performance, perhaps in order to reduce or remove visual and acoustic interference and provide resilience to unpredictable weather further north from the Mediterranean. Not so for Wagner, for whom audiences should be at the mercy of the elements.

Wagner's deliberately primitive but hypnotic 1875 recreation can be compared with increasingly sophisticated contemporary equivalents across the Atlantic. The design for the Metropolitan Opera House in New York City was published in 1883, with five tiers of boxes, the work of Frederic Tudor. The whole building is positively and continuously pressurized, requiring vast, 'almost unlimited' quantities of air. It is a very mechanical solution incorporating a colossal 'blowing machine'. The airflow is controlled at the point of exhaust through a great valve in the centre of the dome, a rising winch-operated 16ft diameter dish. The strategy in winter was to heat the auditorium all day with air at 80°F to warm the surfaces so that the radiant temperatures would be comfortable during performances operated at 70°F, cosy but fresh, balancing air quality against energy costs. But this really is the moment of migration into artificial environments at the public scale. The delivery of air below the seats prefigures the much-feted innovations in Denys Lasdun's late 1970s National Theatre in London by a century.[8] The near contemporary 'wind-driven' Madison Square Theatre in New York City would appear to have been more interesting with its selective wind-catching tower and cheesecloth filters, a 360 seater, much smaller, cooled with four tons of ice nightly.

Controls were clearly becoming more and more sophisticated as the cost of sustaining an army of janitors was becoming less feasible. Billings gives warm praise to Dr Reid's work at St George's Hall in Liverpool, reproducing his Memoranda for the operation of the building, the controls manual, on p. 152. The full controls protocols were eventually published in *The Heating and Ventilating Magazine* in 1907 (Honiball, 1907, p. 15).

The retreat from naturally-driven environments

Innovative Edwardian environmental designers in the UK were captivated by North American practice. Architectural historian Nikolaus Pevsner (1970, p. 443) observed that Europeans had been dazzled by the mechanical equipment on display for enhancing comfort in buildings in the 1876 World's Exposition at Philadelphia. However, innovative and ambitious British ventilation and heating engineers have subsequently emerged from anonymity. Ashwell and Nesbit are credited with the 'design' and installation of the heating and ventilation systems in Lanchester, Stuart and Rickards' Methodist Central Hall in Westminster, vast by any contemporary standards, designed and built between 1898 and 1912, employing the American patented 'Nuvacuumette' vacuum steam system secured under licence in 1900.[9] The voids and passages for the transport of air through the building are so substantial that they form an essential part of the structure, so that the architects either delivered the design to precise instructions from the heating engineers, uncharacteristic behaviour given the contemporary debate about professionalization of the art, or were effectively co-authors, grappling with American air-handling machinery (Figure 5.9).

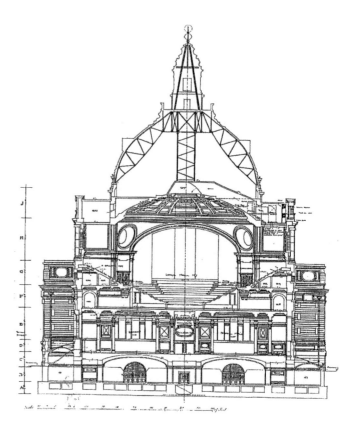

Figure 5.9
Lanchester, Stuart and Rickards' Methodist
Central Hall in Westminster 1898–1912. Cross-
section taken from the Conservation Management
Plan for the Methodist Central Hall Westminster.
Courtesy of Richard Griffiths Architects, authors
of the Conservation Management Plan.

Would not the entrenched social hierarchy of the trades and professions argue very much for the latter? The environmental vision in the Central Hall was the inverse of European practice as sanctioned, ironically, by the American Billings, an attempt to implement the 'downward method' within the second largest concrete dome in the world. Air is supplied through the crown of the inner dome by a large paddlewheel fan, still in place, if emasculated and exhausted beneath the seating rakes below. More warmed supply air is generated in the substantial basement plenum and sent upwards to be introduced into the hall at high level via the huge mixing plenum of the outer dome. It is a counter-Newtonian mechanical system, fighting the natural stack effect, relying on pressure differences generated by three double inlet centrifugal fans. The heating and ventilating design was acquired from the North American 'Atmospheric Steam Heating Company', the fans are suspected of being Sturtevant products, fitted with paddle blades. Rather later than in North America, designers in Britain were constructing profligate artificial environments, on licence from an emerging North American corporation, delivering 70 major buildings by 1914. The selling of this new mechanical system induced the calculated loss of almost a century of accumulated understanding of the configuring and control of naturally driven environments at the public scale. What empirically derived knowledge and understanding was being discarded?

Fin de Siecle understanding of the passive conditioning of theatres

i. A general weariness of vitiated air or miasmas. A consensus understanding of what was dangerous in 'bad air', slowly displacing fears of miasmas and favouring germ theory. Ironically, the design of effective natural ventilation would deal with both threats. Professor Elizabeth Shove (2003, p. 86) observes that scientifically, germ and miasma-based explanations of disease compete, parasitic micro-organisms may well not smell, filth may be harmless, but nevertheless contemporary behaviours and practices are still informed by these beliefs associated with miasma theory.

ii. A belief in the 'upward method'. In which air is introduced at low level and allowed to rise towards a high level exhaust, natural stack ventilation, as opposed to the 'downward method', the top down J-tube geometry employed by Reid's generation. The March 1865 Smithsonian Institute experiments on the behaviour of plumes driven off a human subject convinced an already persuaded Billings. The J-tube is unstable unless mechanically assisted, as observed in the SSEES building in Chapter 4.

iii. The productive use of plena, intermediate mixing chambers. A useful device, 'in large public buildings comprising many apartments, which require to be sustained at uniform temperatures, at particular periods, and are subject to great fluctuations at anothers [sic] so as to demand variable temperatures and velocities in the state of air supplied, a central chamber is found to be very desirable' (Reid, 1844, p. 332, para 73).

However, plena were condemned in Saxon Snell's evidence to a 1908 Royal Commission on Ventilation. The poor experience may have derived from insufficient flow or a cold plug of air lodged in a plenum, stagnating flow. Perhaps motionless air in interior spaces was just too unfamiliar, compounding anxieties about personal odour and the build up of carbonic acid. Plena have proved to be helpful.

iv. An appreciation of the sheer scale of the interventions required to make natural ventilation work in public buildings. Clients, architects, engineers and contractors appear to have designed their buildings holistically. Billings (1884, p. 332, para 73) adds that 'buildings planned in utter ignorance can only be made endurable', and strongly advises against bogus value-engineering, as relevant today as in 1884 (ibid. p. 107):

> So far as construction is concerned, the difference in cost between providing for an air supply of ten and one of sixty cubic feet per minute will often be not so great as to be a serious objection, provided the plans be made before the construction of the building is commenced. [Billings' emphasis]

v. A prescient suspicion that multiple stacks from a single volume may induce volatility in airflows. This was understood from observation, an insight into a more sophisticated model but an unanticipated hazard in contemporary buildings. Reid proposed a very promising 'push-me, pull-you' wind-driven type system which will revert to simple stack flow in still air conditions (Figure 5.10).

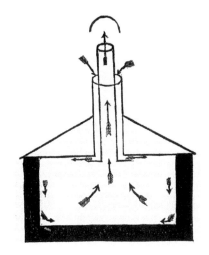

Figure 5.10
Billings' profoundly insightful diagram of a 'push-me-pull-you' mixing ventilation scheme attributed to M'Kinnell, his double tube ventilator (Billings, 1884, p. 95 Figure 20).

The Hon. Mr Ruttan reported on a very interesting example. Billings condemned Ruttan as a charlatan, in some way finding him wholly responsible for a notorious public hall in Detroit designed by the architect Lloyd in which the heating was not overly effective with hot air building up in the galleries (Billings, 1884, p. 130). Billings quotes Ruttan (1862):

> Four ventilation shafts were used, one in each corner, which, as Mr. Lloyd remarks, is an error, 'it being impossible to make several flues draw with equal power, and more often than not cold air will come down one or two of the shafts unless heat is employed to force the draught.'

This is an extraordinary observation. It speculates on exactly the experience at the Contact Studio as measured by the BP Institute and the predicted failure mode of the SSEES building. But reversing, mixing flow can be manipulated to positive effect, as will be seen below. Reid illustrates the principles of a wind-driven reversing flow scheme supplied at eaves or ridge level (Figure 5.11).

vi. An understanding of the fine grain of comfort across a large interior.
Direct and discrete treatment of every seat, box and gallery provides flexibility of control, the provision of several control points at various stages of delivery, attempted in Europe, certainly at the Hofoper, but was effectively systematized in North America. The need was great. Corbin (1996, p. 52) quotes Mercier on the public perception of the airborne menace in theatres, somewhat earlier than the Hofoper era: 'Other congested places spread infection ... theatres. Here, complaints were concentrated on the boxes, which were said to poison women of delicate constitutions',[10] and Senancour's character Obermann (2.48), who:

> felt the keenest repulsion for the Opera, 'where the breath of two thousand bodies, whose cleanliness and wholesomeness are to a greater or lesser degree suspect, makes you sweat all over'.

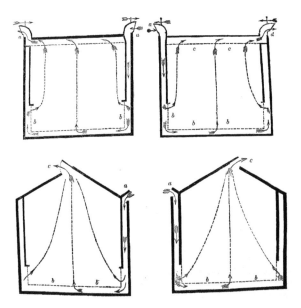

Figure 5.11
Reid's diagram of a space cross-ventilated symmetrically
from alternating wind directions.

Corbin (1986, p. 191) reports that the odour sometimes forced the worried spectator to leave the theatre.

vii. An interest in rapid centralized control. The installation of electricity made dispersed, centrally operated control mechanisms possible. This was as important to naturally driven as mechanically driven schemes, more so as the controlled devices are more distributed by definition, replacing human beings, such as nurses in Billings' hospital.[11]

viii. An understanding of the extreme necessity for airtightness. An unknown quantity, but this was little valued in the cheap fuel economy of post-war North America until relatively recently. Conversely, the close attention to detail in the Johns Hopkins Hospital of 1875–89 (explored in Chapter 6) created conditions that were too airtight according to monitoring records, so that basement windows had to be opened to achieve airflow rates. Le Corbusier knew that the quality of North American manufactured doors and window frames was greatly superior to European equivalents, plaintively illustrating a pressed folded metal section available in North America in *Vers une Architecture*. Much of the technology he illustrated of European origin was old and ineffective.

ix. Belief in building radiant temperatures, 'warm' encasement in plaster or timber panelling. Frank Matcham and numerous lesser theatre designers understood the value of thermal admittance, responsive dense plaster surfaces able to retain both high and low radiant temperatures.

x. Understanding that stage and flytower fight intended airflows within auditoria. A phenomenon discovered during postmortems into serious theatre fires. Frederic Tudor's Metropolitan Opera House in New York City, published in 1883, with five tiers of boxes is equipped with footlight enclosures to supply ventilation into the volume of the stage to counteract the 'onshore breeze' which removes 'atmospheres' as fast as they develop around the actions of the actors. The engineer Max Fordham drew the destructive effect for the author of air-conditioning in the Glasgow Empire theatre (Figure 5.12).

xi. The delivery of complex building configurations. Billings' (1893, p. 158) excellent advice on procurement, to issue proper and complete specifications, was to be specific about the radiating area to be provided, and if not possible, pursue a negotiated contract: 'the best course is to call for no bids, but employ a respectable and honest party to do the work as it should be done, and pay the price.'

By the beginning of the 20th century the ideas pursued by Reid, Böhm and others, as recorded and sanctioned by Billings for an American audience, add up to a viable formula for conditioning the environment in a contemporary theatre.

There is no discernible discussion of the determinants of acoustic quality within a theatre space. Wallace Clement Sabine's work leading to the quantification of the effect of material absorbent of sound energy on audibility of speech in the lecture room below the Fogg Institute at Harvard did not commence until 1895. It did not yield the equation for determining the reverberation time until a year or so later. He discussed 'The effects of air currents and temperature' in a paper published in the posthumous collection of his writings in the context of a suggestion that the ventilation in the Boston Symphony Hall, being designed by architect Charles McKim and on which he advised, worked from supply at the stage end and exhaust at the top rear to 'hasten' the sound energy down the auditorium. Sabine (1922, p. 117) patiently calculates through the prospective advantage, an equivalent moving of the orchestra one centimetre closer to the audience.[12] Exactly.

However, Sabine adds (perhaps unsuspecting of its import in a passive theatre ventilation system) that: 'Any opening into the outside space, provided the outside space is itself unconfined, may be regarded as being totally absorbing.'

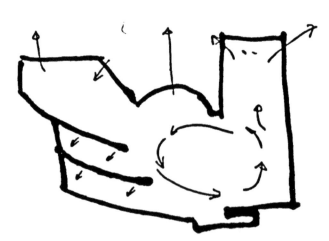

Figure 5.12
The onshore breeze effect, sketch of the Glasgow Empire given to the author by Max Fordham as a dire warning at the commencement of the Contact Theatre design.

Is this still a practical proposition? More is now known about the natural physics of the universe, but theatre design has not changed appreciably in the intervening period. In the context of a more egalitarian environment than that of the Habsburg court the 'theatre problem' remains: large, high, single volumes in which occupancy is very intermittent but periodically very dense with a full audience emitting 100 watts per audience member, each of whom is there voluntarily, paying for the experience, potentially stacked up through the stratified atmosphere. Additional prodigious intermittent lighting generates heat loads at high level. An acoustic imperative exists to contain and exclude noise between neighbouring auditoria, the foyers and the city beyond. All these volumes are enveloped in perhaps the most ferocious fire regulation of all building types, the historic legacy of serial dressing room fires. Nonetheless Figure 5.13 shows an entirely feasible generic approach which solves the 'theatre problem', delivering comfort, refreshment and atmosphere, silently, as we now demonstrate.

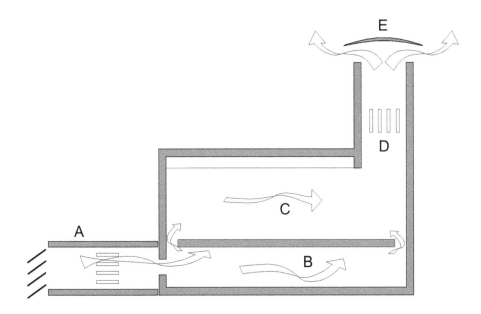

Figure 5.13
A generic diagram for a naturally ventilated passively cooled theatre (from Short *et al.*, 1998).
A. Air inlet: Locate to minimize pollution and noise in a location unsusceptible to negative wind pressure, configured to take static acoustic attenuation.
B. Plenum: Geometry to evenly distribute air intake into auditorium, incorporating thermal mass to pre-condition supply air, warming or cooling depending on the season.
C. Theatre space: Environmental design strategy to dispel prodigious heat inputs from stage lighting and audience but airflows limited by speed and temperature. Exposed thermal mass beneficially affects radiant temperatures but hard surfaces impair acoustics so that careful brokering of hard and soft surfaces is required.
D. Exhaust: Free area to equate with supply, all supply controls and acoustic attenuation in reverse.
E. Terminations to stacks: Aim for design resilient to wind direction and turbulence to always enjoy positive pressure gradient from intake level, defend against wind-induced downdraughting which would neutralize the stack effect. The H-pot/Cross-pot is indubitably superior to all other terminations in its resilience.

A viable alternative solution

The two lecture amphitheatres in the Queens Building for De Montfort University, discussed in Chapter 3, were originally expected to be air-conditioned. It was inconceivable that conference standard auditoria with 180–190 seats would not be mechanically conditioned. Max Fordham suggested that they need not be. The tight programme demanded the proof of concept, or disproof, be quick. A Department of the Environment grant funded the research in Cambridge and Leicester.

Figure 5.14 shows the amphitheatres sitting above flat-floored classrooms on part conical concrete slabs with a serpentine design studio above, gently raked to enable tutors to address a whole studio, reprising the experience of inhabiting the 'trays' at Harvard University's Graduate School of Design and the form of William Burges' Harrow School New Speech Room (1877), as modified by Sir Herbert Baker.

With hindsight, it can be seen that the design inadvertently reinvented the wheel, recovering some of the lost understanding of naturally driven ventilation schemes. The following figures show air introduced through belfry louvres to protect against weather into a plenum packed with large static acoustic attenuators and heating elements before being discharged below each seat (Figure 5.14).

The enclosure is heavy brick masonry and *in situ* concrete but much is masked by lightweight acoustic absorbent behind timber panelling to reduce the reverberation time but every square metre denies the effect of the admittance of the structure in stabilizing radiant temperatures. The emerging design was tested using a physical model in which water was used in lieu of air, the first such experiment for a building design (Figure 5.15).

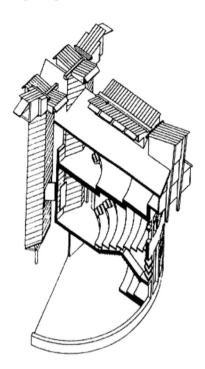

Figure 5.14
The Queens Building, Leicester, stacking of a design studio above a 135° fan, 187-seat amphitheatre, above a flat floored classroom, all naturally ventilated and passively cooled, an arrangement Billings strongly advised against. The twin stacks are structural and therefore essential, and drain warmed air from the amphitheatre. Air supply is via sound attenuating plena beneath the seating rake.

Figure 5.15
The water bath model of the Queens Building auditoria constructed by Professor Paul Linden and Dr Greg Lane-Serff in their Fluid Dynamics laboratory in the foetid Silver Street cellars of the Applied Maths and Theoretical Physics Department at Cambridge University. The tests are gravity-driven and so the model is inverted as saline solution drains through.

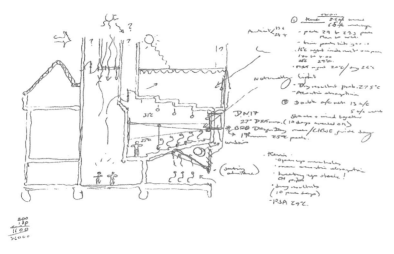

Figure 5.16
Author's sketch of the agreed design for the Central Building made at the final design research meeting in Cambridge, brokering the environmental design strategy for the Queens Building. The sketch incorporates the final recipe of thermal mass, structural form, free areas, internal heat gains, comfort criteria, heating, passive cooling, air flow, acoustic isolation, absorption, natural light, solar gains and sight lines. Exercising this holistic judgement and leadership is the architect's contribution. The architect needs to be sufficiently literate in the various disparate disciplines to be able to arbitrate a 'best' outcome. These judgements are rather different to those concerning expediency required of project managers.

The model works on the basis that the stack effect is inverse to the gravity effect such that warmed air tending to rise can be modelled with denser water tending to fall, the density being achieved by introducing saline solution. The whole model is inverted. Scaling effects intervene; there is no thermal capacitance in the acrylic enclosure. Nonetheless the model gives a good indication of the strength and direction of flow and the extreme complexity of 'reality'. In the built building the stack effect is dramatic and the difficulty is in achieving airtightness, the smallest leak admitting cold draughts. The design brokers many variables championed by many interests, acoustics against thermal comfort, costs against size and so on. The final sketches agreed across 31 attendees at the concluding design research meeting are reproduced below (Figure 5.16).

Figure 5.17 shows the interior of the amphitheatre facing back onto the street front, the less successful orientation. The built theatres are very robust to high external temperatures as the heat load test demonstrates (Figure 5.18). It is entirely possible to drop the temperature inside the building well below the peak external temperature with no expenditure of energy whatsoever. Could this strategy be elevated to the scale of a 'real' theatre for drama?

Figure 5.17
The amphitheatre fronting onto the Queens Building façade, less successful than its reversed twin but still delivering a creditable performance. Acoustically absorbent panelling encroaches on the exposed thermal mass, rendering in physical form the fractious atmosphere which developed between the acoustician, energy physicists and structural engineers. There are no 'correct' answers, such integrated designs have to brokered through a series of judgements over the actual requirements of the space, here to accommodate discourse leading to excellent engineering (photograph by Peter Cook).

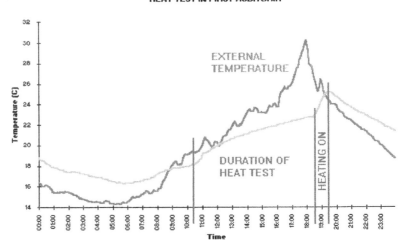

Figure 5.18
Heat load tests in the Queens Building in June demonstrate that it is possible to reduce the internal temperatures in the building below the peak ambient temperature by several degrees. External temperatures have risen to 30°C but internal temperatures peak at 22.5°C, achieving 7.5°C of zero carbon cooling.

Contact Theatre, Manchester

> The actor's work is never for an audience, yet always is for one. The onlooker is a partner who must be forgotten and still constantly kept in mind: a gesture is statement, expression, communication and a private manifestation of loneliness … yet this implies a sharing of experience, once **contact** is made.
>
> (Brook, 1972, p. 57)

In 1965 Manchester University rewarded its highly regarded Drama Department with a new 300-seat theatre, designed by Grenfell Baines' interdisciplinary Building Design Partnership (BDP). Despite this provenance the theatre was formulaic, one of many rectangular multi-purpose halls built at the time. We will see another at Lichfield. The formula delivered, consciously or otherwise, an unusually wide stage at 16 metres (52 feet 6 inches). An apocryphal tale at the time was that Peter Brook chose Contact to premier 'The Man Who' in the UK, but the author can find no record of this (Brook and Estienne, 2002). Its 'broken back' seating rake, a leitmotiv of 1960s multi-purpose halls, flat at the front and steep at the back, its unaltered and by then unfashionable (but now highly fashionable) 1960s interiors and the unfeasibly noisy mechanical ventilation almost ground down the most loyal audience members. The noise problem was solved by not switching on the ventilation machine under any circumstances, generating Bayreuth-like physical atmospheres. Temperatures in excess of 40°C were recorded by the author and others during the 1993 December pantomime season.[13] Theatrical smoke, introduced briefly on stage early on in the performance, hung motionless at shoulder level amongst the first seven or eight rows (ibid.). There was virtually no air movement. The university department was overwhelmed by the responsibilities arising from owning a public theatre and handed it to the newly constituted Contact Theatre Company. Its remit was to produce drama for, and more importantly, with young people. It became known as the 'Young Vic of the North'.

The first award to the Northwest by the UK Arts Council Lottery enabled radical rebuilding and extension of the Contact Theatre to provide an end-on Space 1 for 384 people, a studio-type Space 2 for an audience of 120, a rehearsal room for 100 and a foyer for comedy/drinking/eating, in the very noisy city centre (Figure 5.19). The acoustic challenge was profound with or without a passive intent but the intent, established at the interview and wholly supported by the Company led by the Chief Executive Vikki Heywood, later to run the Royal Court and the Royal Shakespeare Company, and Artistic Director Brigid Larmour, was indeed to make a wholly naturally ventilated passive theatre. Extraordinarily and quite unknowingly, the design as finally built adopted or attempted to adopt all of the late 19th century and Edwardian principles identified here for making a naturally conditioned theatre, with the infrastructure to deliver it at the necessary scale (Figure 5.20).

Figure 5.21 is a worm's eye view up into the theatre showing how air is introduced through brick and terracotta intakes thrown out around the perimeter, supplying plena and labyrinths below Space 1, the main auditorium, the foyers and, hovering two floors above, Space 2, the Studio (Figure 5.22). The driving force is generated by stacks above Space 1 and growing up out of the corners of Space 2. The stacks climb to beyond the height of surrounding obstructions and terminate in complex terminations designed to overcome city centre wind

Figure 5.19
Contact Theatre: The original 1963 theatre on the day of the Lottery Award, and the rebuilding, which retains the original theatre, with its massive brick walls, within.

Figure 5.20
The rebuilt Contact at nightfall. The night is very pertinent to theatre. Griffero (2014, p. 57): 'But we must mention, above all, the primitive and poetic charm of the night: its obscurity … leads to the disappearance of distances … and a paralyzing isolation from things … perhaps twilight is even more atmospheric than the night'.

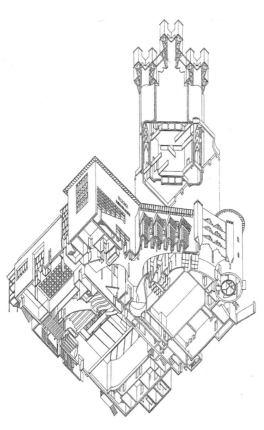

Figure 5.21
Contact Theatre, upwards view in axonometric showing air intakes, plena and labyrinths below the main auditorium, supplying the foyers and, hovering above, the Studio.

Figure 5.22a and b
Contact Theatre air intakes across the elevations, conceived as projecting objects presenting as much free area as possible to the exterior and as perforated screens to enable the stack effect to draw the required quantities of fresh air.

turbulence. The reconfigured main auditorium within the existing wide box now has a 'waisted' single parabolic rake with no centre aisle, continentally seated but with every row accessed from within the auditorium (Figure 5.23). Plush bench-seated boxes line the outer walls, set out to twist in towards the centre of the deep stage. The illusion of rotation is intended to optically re-centre the periphery of the seating rake as the stage is either widened or choked down, so that there is some sense of the audience positively oriented to the stage, a species of virtual flexibility with no moving parts (Figure 5.24).

Fresh air is introduced into the main auditorium from one side only, by default, from the enclosed Humanities courtyard to the west, passing through a bank of acoustic splitters, 2.4m in length, directly into a plenum below the raked seating. Static acoustic splitters to deal with the lower frequency noise expected here are

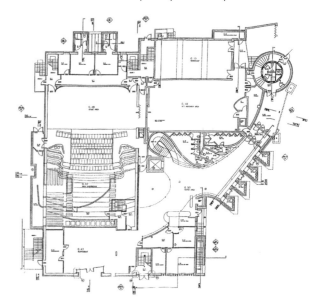

Figure 5.23
Contact Theatre ground level plan.

Figure 5.24
Interior of Contact Space 1, the parabolic rake seen across a box (Ian Lawson). In the words of the then Artistic Director (quoted in Short *et al.* 2011): 'It was a terrible, terrible, terrible theatre and I wanted that cure to give a more human line and I don't know why we went for those sort of boxes but I know that we thought they might be quite fun for a young audience … you could squeeze lots of people in and that they would feel a little less regulated than (in) the seats … I think, those boxes felt a bit special, also because I think they were a level higher so you were quite aware of them from where you sat in the middle. So it was more a feeling of that horseshoe and yet quite good sight lines. I think maybe it was because we wanted it to be possible to close in the proscenium and I think that would allow us … to shift those seats a bit or at least you'd be able to shift your seat so you could focus in more.'

large. This is the only orientation available for locating inlets because the new foyers are arranged across the east side of the auditorium, blocking potential intake routes. General Morin would be unimpressed. Incoming air enters the plenum and is then distributed between four compartments below the raked seating, each compartment individually controlled by BMS-controlled dampers to balance the distribution of air entering the auditorium. The thermal mass inside the concrete-lined labyrinth pre-cools the air during peak summer conditions. Heating elements are hung below each seating platform in the airflow paths. Air then enters into the inhabited volume of the auditorium beneath each seating row through continuous openings in the risers, an old trick as we saw in Frederic Tudor's opera house of early 1880s New York.

To avoid the possibility of warm air stagnating below the cantilevered control room, potentially overheating the audience below, extra air intake grilles were added in the floor of the rear gangway providing $1.25m^2$ of free area fed from a dedicated plenum, as witnessed in the Wiener Hofoper. In addition, the control room is detached from the rear wall to allow warm buoyant air to flow upwards into the upper zone of the space, just the kind of gambit that was played in the Wiener Hofoper.

Air is exhausted through a 5m high chamber cut into the existing roof profile. Five stacks, each with a free area of $4m^2$, sit above the void, with dampers and low-speed fans to dispel warm summer stagnation, located at the junction between the void and the auditorium. The stacks contain arrays of vertically mounted acoustic splitters, the same size and proportion as at entry. One needs to mimic in reverse every precaution introduced at entry. The stack terminations comprise two integrated orthogonal H-pots, each the size of a small house, minimizing the risk of wind-induced flow reversal for all wind directions.

During commissioning, a heat load test was carried out in Space 1 in which 60kW of theatre lighting and 40kW of simulated occupants were deployed. Figure 5.25 records the outcomes. The test operated for longer and with higher

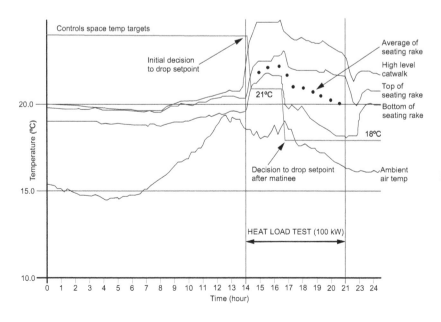

Figure 5.25
Results of a heat load test during commissioning: temperatures rise rapidly at the start of the afternoon performance but the ventilation system abruptly arrests the temperature build up and can reduce temperatures in a full auditorium as external temperatures rise (Richard Quincey of Max Fordham and Partners).

heat gains than were expected in reality. The design temperature differential of about 38°C was achieved in most of the seating locations except in the top seating rows, which were slightly warmer. Lowering the set point after the matinee simulation caused the fans to operate in the exhaust stacks but prior to this the fans were not needed to maintain environmental conditions. As Figure 5.25 shows, a good environment is delivered under heavy heat gains with almost no expenditure of energy. Space 2 is yet more intriguing, all new and with an uncompromised geometry, bi-axially symmetrical with four large corner stacks, a full sized laboratory model.

Space 2, the Studio Theatre

> I think it's important that a space has character and that it is specific. And again I think there are pluses and minuses with the design of Contact but it has character and people respond to it. It has personality and people respond to that. And young people respond to that. You don't want a generic, multi-use thing that has that institutional arts centre feel to it.
>
> John E. McGrath, Artistic Director

The Studio is flat-floored space propped above the double height scene dock of Space 1. Here air is introduced from the north side only, 6m above street level, into a 600mm high plenum below the entire floor, comprising a high thermal mass labyrinth. Incoming air is split symmetrically into two paths via banks of acoustic splitters before dividing again into chambers beneath the studio floor (Figure 5.21). It passes through finned heating elements suspended below grilles in the plane of the floor on all four sides of the Studio. Waste air is exhausted through four vertical openings up at soffit level connecting to an exhaust plenum above the perimeter corridor on all four sides. Four tapering acoustic splitter chambers connect into the brick masonry stacks above, terminating in cross H-pots at the same level as those venting the main auditorium, some 11 storeys above ground level. Three small, half-bladed, short-cased axial fans are placed at the base of each stack above the attenuators. This is a more desirable location, acoustically, than the position of the main auditorium fans which are located below the acoustic splitters for ease of maintenance.

A simple wall-mounted dial in the Studio enables staff to input the expected occupancy level. Their prognosis informs the building management system (BMS) which makes a decision regarding the opening extent of the dampers in the inlet and outlet stacks. With a heavy programme of use each day, the Studio is indeed reported as being comfortable. But it works in a way that was wholly unanticipated by the author although Surgeon-Major Billings and the Hon. Mr Ruttan would have predicted with confidence the bizarre observed behaviours.

The first concern emerged in the wind tunnel. Earlier versions showed stalling of through airflow from the dominant wind direction. An 11-storey slab building occluded the theatre. Figure 5.26 shows the stacks raised in height and given the cross H-pot form shown to be most resilient to turbulence. Figure 5.27 shows the principle transformed into lightweight zinc clad stacks above Space 1. Figure 5.28 shows the wind tunnel model hacked open and rebuilt into its final iteration of higher stacks and interlocking cross-top terminations to give the greatest pressure differentials between top and bottom. Computational fluid dynamics analysis represented perfectly symmetrical airflows through the Studio plenum and up into the space above (Figure 5.29). Figure 5.30 shows the more substantial insulated brick masonry stacks draining the Studio, Figure 5.31 the original drawing, and Figure 5.32 a detail of the steel structure propping the masonry H-pots, unashamedly plagiarized from the figure in Viollet le Duc's 1863 *Entretiens sur L'Architecture* depicting immense shallow vaults propped above a major public space.

Figure 5.26
Contact wind tunnel testing of the final design for the Studio Space 2 cross-pot stack terminations. The first run of tests demonstrated that the obstruction of high adjacent buildings would neutralize any pressure differentials, thereby stagnating all airflow until heat had built up within the interior (author).

Figure 5.27
Lightweight stacks and cross-pot terminations sitting on the existing Space 1 lightweight roof. Ideally stacks should have thermal mass to retain warmth to reduce the prospect of cold downdraughting at their perimeter. In this the idea of glazed stacks is wholly baffling (author).

Figure 5.28
The wind tunnel model in its final iteration (author).

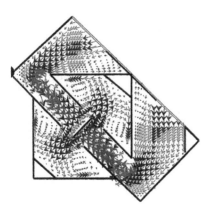

Figure 5.29
Computational fluid dynamics simulation of
the symmetrical airflows generated within
the new Studio theatre plenum (Professor
Phil Jones, University of Wales, Cardiff).

Figure 5.30
Cross H-pot stack terminations to the Studio theatre.

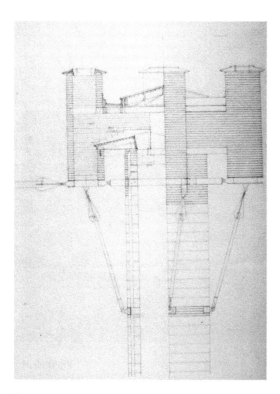

Figure 5.31
Author's drawing of the stack terminations.

Figure 5.32
Structure of the Studio stack terminations with a clear debt to
Viollet-le-Duc.

Figure 5.33 shows the conventional model of the airflows developing within the Contact studio ventilation scheme, 'Advanced Natural Stack Ventilation', defended by H-pot terminations, draining warmed air away and maintaining a displacement layer of warm air well above everyone's heads. In fact, the four-stack Studio appears to perform generally in mixing ventilation mode. BP Institute measurements within the stacks reproduced in Figure 5.34 revealed that air is flowing downwards as well as up, simultaneously sometimes, and/or down one stack whilst exhausting through another and that these modes are adopted at perhaps the slightest provocation. Although accidental, this mixing mode is an attractive prospect. Billings (1884, p.158) (Figure 5.10) reported that 'Many attempts have been made to combine the inlet and outlet in the same ventilator … the best of them is probably the double tube of McKinnel'. He promotes their use for larger single volume spaces. The findings at Contact suggest that Billings is entirely correct. Here is potentially a very low energy approach to warming and ventilating a largish single space from the top only. Incoming air is warmed through outgoing air as it descends, mixing, into the space. Figure 5.35 proposes a very different account of actual airflows through the Studio as deduced by the BP Institute. Figure 5.36 shows the BP Institute's measurement of temperatures within the four Studio stacks at Contact over five September days. It is clear that temperatures fall suddenly, representing reversal of flow. Figure 5.37 shows this phenomenon exploited as an environmental design strategy, incoming air is warmed by escaping warmed air. It is a winter heating strategy, cooling in summer would be delivered by a different mechanism but a passive downdraught cooling configuration could be incorporated within or around the intake/exhausts, not unlike that proposed for the atrium scheme in Chapter 4.

Reflections
The design team's later Garrick Theatre at Lichfield operates in this way, quite unintentionally but well. The theatre demonstrates a high level of resilience to hot summer spells. Its original dimensions were identical to those of the original Contact. The theatre opened during a heatwave. The situation is more intense than at Contact, the audience is larger and stacked in a full raked balcony above

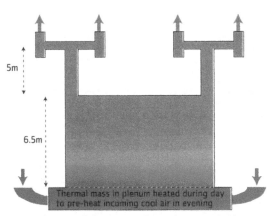

5m

6.5m

Thermal mass in plenum heated during day to pre-heat incoming cool air in evening

Concept - displacement ventilation when warmer inside

Figure 5.33
The theoretical model of the Contact studio ventilation scheme, 'advanced natural stack ventilation' defended by H-pot terminations draining warmed air away, maintaining a displacement layer of uncomfortably warm air well above everyone's heads (BP Institute, University of Cambridge and Breathing Buildings Ltd).

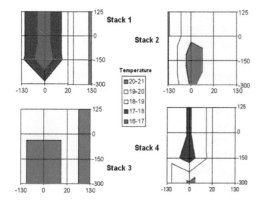

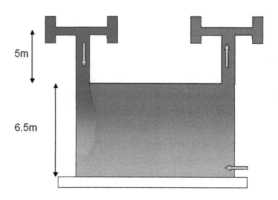

Figure 5.34
Temperature distributions measured across Studio stacks 1–4; 1 and 4 appear to be 5–6°C cooler than 2 and 3, suggesting reversal of flow whilst accommodating some upward flow simultaneously.

Figure 5.35
A depiction of what is very likely to be happening within the Studio with various volatile flow regimes equally probable, triggered by gusts and perhaps other phenomena (Shaun D. Fitzgerald and Andrew W. Woods, November 2006).

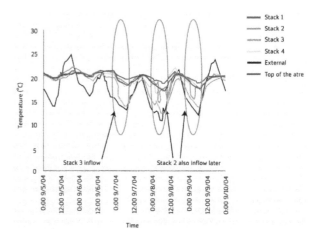

Figure 5.36
BP Institute's measurement of temperatures within the four Studio stacks at Contact over five September days, the external temperatures and the temperatures 3m above the flat floor of the Studio (BP Institute, University of Cambridge).

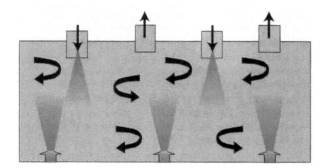

Figure 5.37
The BP Institute's theoretical model of a balanced mixing ventilation scheme, the outflow of warmed air balances the inflow of cool air at the top, the cool air descending but diluting and mixing before it reaches the occupants represented by the plumes of heat rising at the bottom of the space. It represents a flat-floored hall. Of course summer cooling benefit has to be delivered in a different way but adding passive downdraught cooling in hot conditions could engender exhaust flows by pumping air back out through the top of the space, as in the atrium scheme shown in Chapter 4 (BP Institute, University of Cambridge).

the stalls (Figures 5.38 and 5.39a and b). Air is deliberately supplied across the width and rake of the balcony exactly as in the Vienna and New York theatres and as General Morin recommended. It is exhausted through acoustic silencers above (Figure 5.40). The intake area is prodigious and effected through banks of terracotta cavity ducts mounted within a skin of terracotta mathematical tiles to avoid any reference to the visual paraphernalia of air-conditioning and all that is associated with it (Figures 5.41 and 5.42). Monitored data is very encouraging, internal temperatures peak at 28°C as external temperatures reached 37°C. Far more air is exchanged than theoretically possible through the intakes only, hence the presumption that air is moving freely in and out through the exhaust terminations (Figure 5.43).

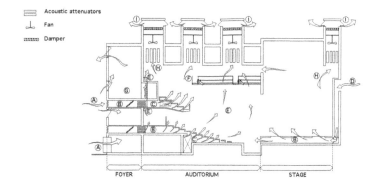

Figure 5.38
The Garrick Theatre at Lichfield, a larger example with a full balcony. Long section showing air supply directly to plena below each seating level exactly as General Morin recommended, and exhaust through acoustic silencing boxes below ridge vents. The ventilation system has to maintain the displacement layer well above the highest placed seats at the back of the balcony rake.

Figure 5.39a and b
The main auditorium at the Garrick, 'A new and rosy coloured theatre … has to be a cause for celebration' (*The Mail on Sunday* September 21, 2003. Culture Section 'a bright welcoming building' *The Sunday Times Culture Section*, September 21, 2003) showing an audience perched high up in the volume of the auditorium. (Peter Cook)

Figure 5.40
The Garrick, air exhaust containing
banks of acoustic splitters hung below
a stack (Peter Cook).

Figure 5.41
The new front to the Lichfield Garrick showing liberal punctures connecting into the plena of the
stalls, boxes and balcony (Peter Cook).

Figure 5.42
Air intakes formed in terracotta pots
held in a stainless steel frame in a
matrix of mathematical tiles. The
paraphernalia of air-conditioning
installations is studiously avoided
(author).

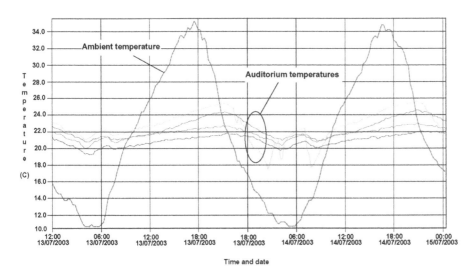

Figure 5.43
Internal temperatures recorded in the Garrick during the second hot spell in the summer of 2006 peak
at 28°C as external temperatures reached 37°C. More vigorous night-time ventilation would further
suppress internal highs, the pre-cooling plena (in blue) were maintained at 24°C, absurd in retrospect.

What might an urban block of free-running buildings such as Contact be like? How would each set of spaces and all their environmental machinery affect each other? Incredibly perhaps, with all the funding that has been put into researching 'sustainable urban design', it is not at all clear. Excessive modelling will not design built propositions. There is some very interesting work to do to explore the actuality of the low to zero carbon block in all its huge complexity, but Figure 5.44 shows an early synthesis, the scheme for the Coventry Arts and Media Centre, a close collection of passive and very low carbon performance spaces and studios, passed over by Arts Council England in favour of the wholly disastrous Jubilee Arts Centre which ate its client company.

One of the roles of a practitioner should be to systematically gather evidence from field and practice. Eleven free-running auditoria have now been built and monitored by this design team. With this knowledge, it is becoming possible to develop a taxonomy of approaches related to theatre types derived from styles of drama from end-on proscenia to fully in-the-round studios. Figure 5.45, unashamedly derived from J.-N.-L. Durand's systematic catalogue of theatres, relates seating plan forms to sections distinguishing plena, supply and exhaust configurations. One could build on the gathering experiences and data and animate the matrix of options quantitatively binding science to artistic intent.

Reception history of the Contact Theatre
The design of the Contact Theatre was declared by some commentators to be idiosyncratic, green and deliberately (which implied 'unnecessarily') provocative. However, the 2000 RIBA Awards panel took a positive view and decided (Chapman, 2000):

> This is brick-built Gaudi come to Manchester. What was previously understated and uninspiring has been transformed into an elaborate and

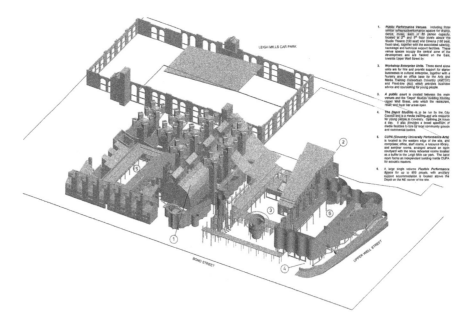

Figure 5.44
A concatenation of Contact theatres for the Coventry Arts and Media Centre, a naturally conditioned urban block.

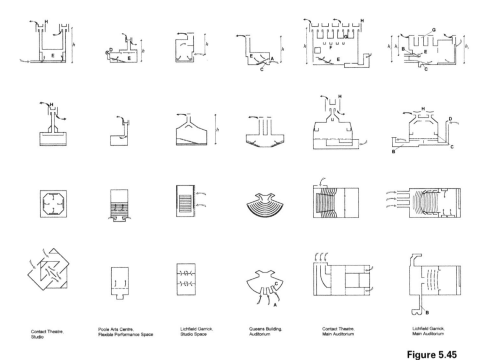

Figure 5.45

Contact Theatre,
Studio

Poole Arts Centre,
Flexible Performance Space

Lichfield Garrick,
Studio Space

Queens Building,
Auditorium

Contact Theatre,
Main Auditorium

Lichfield Garrick,
Main Auditorium

Figure 5.46

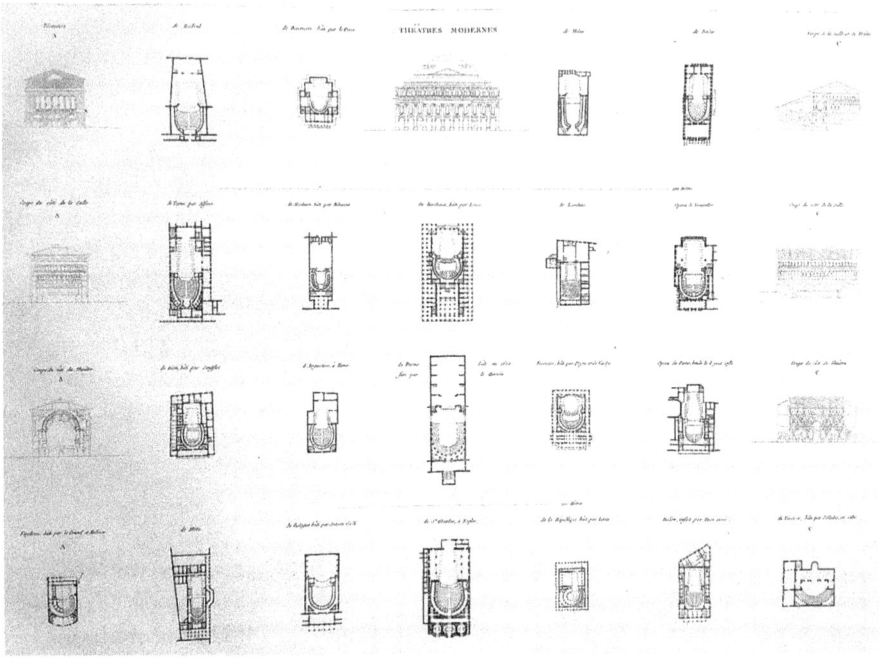

Figure 5.45 and Figure 5.46
The principles behind the naturally ventilated theatre types described, showing, for each type, from top to bottom: long section; short cross-section; seating level plan; plenum level plan [A locates fresh air intake; B air supply path; C air supply plenum; D high level intake; E occupied volume; F lighting gallery; G exhaust stack; H stack termination]. First published in Short and Cook (2005), after Durand (c. 1801, Figure 5.46).

engaging monument with a strong independent image, perversely a highly appropriate response to its context. It is all done with absolute conviction and it is now certainly highly visible in the street scene. The project has a strong environmental agenda made manifest in the strong root forms of the natural ventilation system ... with references to Gaudi and the castles of southern Germany.

Others, such as the critic Sutherland Lyall was 'bemused' by Contact:

You're driving up Manchester's Oxford Road and you've almost got through the university quarter when you skid to a halt because over there on the right ... is this extraordinary sight. A pale brick, three- or four-storey ramble of a building ... balancing precariously 40m up and above, this amazing jumble of dislocated brick chimney things. You think Expressionism, maybe Steiner, maybe Ton Albers, perhaps Michael Hopkins and his vast chimneys opposite Big Ben and Nottingham ... Never before have you seen anything like those gigantic, crazy, chimney structures tottering up there in the sky. This is England. This is England under the planning system yoke and you find yourself astonished that the planners didn't go into terminal, total paroxysmal mode before letting such an extraordinary show to [sic] go ahead. But yes they did and here it is and what can anyone possibly make of it?

In the *Sunday Times,* Hugh Pearman (1999, p. 12) claimed to have identified a 'will to form':

There's something in the air – chimneys are making a comeback. This is not postmodern whimsy, however, but a rediscovery of form following function ... The Contact Theatre ... has been transformed into something genuinely fantastical ... this is heart-on-sleeve, manifesto architecture. It is saying that ecological thinking can and should result in buildings that not only are different, but look different, with new shapes, new roofscapes. For the aesthetic chosen by its young architects, Short and Associates, is a world away from today's prevailing style of Polite Modern or its godfather, high-tech ... Obviously Short is something of an extreme case. His earlier Queen's building at De Montfort in Leicester ... got the modernists spluttering in anger. Having a theatre to do suits him perfectly, since he can indulge his taste for drama.

The artistic director who inherited the design, Ben Twist, was never wholly convinced by the design strategy:

And I think rightly there was a desire by the University, by the theatre company to create a building that was environmentally and financially more sustainable, which on the surface I think is great. And therefore probably to appoint Alan Short and Co at a time before the physics had been done and the modelling had been done, and that happened, you know, over quite a period, and I think that probably nobody including Alan and Co knew quite what the implications of that would be ... I really wonder, and I don't know how much extra these towers cost but I really wonder for half a million quid or whatever probably it was, whether you couldn't have got a really good insulation, really good boiler system, other environmental measure, solar panels, a windmill on a hill in Oldham somewhere, that wouldn't provide the same or better savings – greater environmental and financial savings ... I think that one of the difficulties of the project was that I think the ventilation system ended up driving the architecture and the engineering which ended up driving the project quite a lot. I think that Alan might hate me for saying that. And I think that had quite a detrimental effect on the building project, because we were constantly cutting, because the towers got higher, the cost got greater and so we had a limited budget. So we kept on having to cut down on the studio and so on to accommodate what is basically a ventilation system so there's a bit of a danger that you've got a ventilation system with a theatre attached rather than a theatre with a ventilation system. I think that's quite an issue about that building.[14]

Ben Twist refers to a structural conundrum in the funding mechanism adopted by the Arts Council, it cash-starved projects until they were well under way. Contact was funded right at the beginning by the Lottery, procedures were still being formulated but far too much credence was given to the apparent resolution achieved at the feasibility stage design and its estimated cost. The implications of this 'sense' of completion at a very early stage reverberated through the entire process. The essential specialist inputs – acoustician, structural and environmental engineers, building physicists, fire engineers – each carried a cost for which there was already no provision. Innovative design is of necessity iterative but the standard plan of work adopted by this and many other funders heavily penalize the revisiting of each successive design stage. This was compounded by the near absence of design contingency funds, a consequence of the then prevailing Arts Council of England policy. Managing this creeping overspend during the later stages of design and construction was very harrowing for those involved. But the client team hung on to their vision, which was delivered almost in its entirety, no spaces were lost. However, the quality of finishes was heavily diluted and in some cases finishes were omitted in their entirety. The 'vision' which propelled the project of a fully fledged 68 person strong production company became outmoded and redundant during the delivery of the building, theatre had moved on. Contact was violently restructured into a largely receiving platform but for emerging artists, the new vision inhabiting a shell from an earlier era as a 'found space'. McGrath enjoyed the irony and challenge in this.

Notes

1. The Theatres Trust reminds us that the London theatre, His Majesty's, in the Haymarket burned down in both 1790 and 1868. In 1794 Drury Lane introduced the first iron safety curtain separating the auditorium from the stage house, possibly in response to the Haymarket disaster, but it was the loss of 200 lives in 1887 in the Royal Theatre in Exeter that galvanized the governance of theatre building construction in England. See www.theatrestrust.org.uk/resources/exploring-theatres/history-of-theatres/nineteenth-century-theatre (accessed 13 December 2015).
2. Ove Arup and Partners report on various simulated scenarios in which a fire broke out on the stage or above it. Results showed that smoke would spread out to the seating

areas quickly, that the plume would lean toward the back of the stage as air is drawn into the stage volume through the proscenium and that conventionally located fire protection devices would not trigger until the heat release rate reached 22MW. They do not model specific air supply to the stage. (Ove Arup & Partners PC (2009) *Fire Safety in Theaters – A New Design Approach*. National Fire Protection Association, Fire Protection Research Foundation report). But modelling of the naturally ventilated Garrick at Lichfield shows that judicially located and plentiful air supply at low level into the stage house and the orchestra will lift smoke away cleanly to high level exhaust terminations at the rate at which it is generated.

3. The AHRC funded (£353 000) research project 'Designing Dynamic Environments for the Performing Arts' confirmed this view, its findings were recently published in the book of the project, *Geometry and Atmosphere*; see Short, C.A., Barrett, P. and Fair, A. (2011) *Geometry and Atmosphere*. Gower Ashgate. The film of the project narrated by Timothy West is available at: http://sms.cam.ac.uk/media/1095045 (accessed 12 October 2016).

4. Silvia Benedito, Harvard University, Graduate School of Design raised this question in the preambles to her seminar course ADV 09128 'On Atmospheres and Design', Spring 2013.

5. Gernot Bohme credits the scholarly use of the term 'atmosphere' to Hubert Tellenbach (1968), *'Geschmack und Atmosphare' (Taste and Atmosphare)*. See also Bohme's paper, 'The art of the stage set as a paradigm for an aesthetics of atmospheres' in *Understanding Atmospheres Culture, materiality and the texture of the in-between*, University of Aarhus, 16–17 March 2012 and, which is similar in content, Bohme, G. (2013), 'The art of the stage set as a paradigm for an aesthetics of atmospheres', *Ambiences* online, proceedings of the 1st international congress on ambiences 'Faire une ambience, Grenoble, 2012'. See http://ambiances.revues.org/315 (accessed 27 April 2014).

6. *German Masters of the Nineteenth Century: Paintings and Drawings from the Federal Republic of Germany*, New York: Metropolitan Museum of Art, Harry N. Abrams, Inc. (1981). Unattributed between catalogue entries by Alison de Lima Green, Lucius Grisebach, Verena Haas, Denise McColgan and Charles S. Moffett.

7. As reviewed by Neil Fisher in *The Times,* 15 August 2015.

8. *The Architects' Journal* (12 January 1977) **165**(2).

9. See Methodist Central Hall Westminster London 2003 Homepage and CIBSE Heritage Group archives.

10. Mercier *Tableau de Paris* 7.309, as quoted in Corbin (1996) *Foul and the Fragrant*, p. 52.

11. A comprehensive history of building environment controls, entrepreneurs and their companies, what drove the innovations and what they changed in design responses, is yet to be written. This would be a fascinating doctoral project.

12. Published three years after Sabine's death; scan available online at: https://archive.org/stream/collectedpaperso00sabi#page/128/mode/2up (accessed 12 October 2016).

13. Unpublished records reprised in Short *et al.*, 2011, Chapter 3.

14. Ben Twist to Zeynep Toker, researcher for the Arts and Humanities Research Council Major Project Design and Delivery of Dynamic Environments for the Performing Arts (DeDEPA).

6

Hospitals

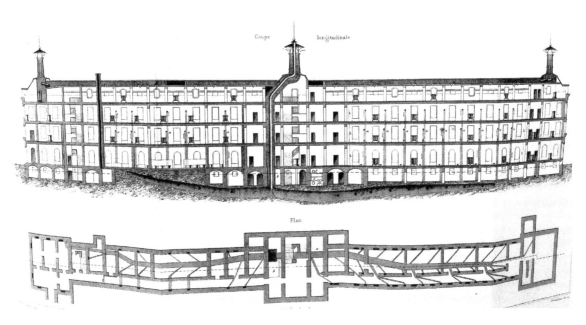

Figure 6.1
'Hopital Militaire Vincennes – Ventilation par appel par en bas', from Morin (1863). Courtesy of Cambridge University Library.

The interesting point is that what seems to be a predominantly masochistic people (the Nacirema) have developed sadistic specialists … The medicine men have an imposing temple, or *latipso*, in every community of any size. The more elaborate ceremonies required to treat very sick patients can only be performed at this temple. These ceremonies involve not only the thaumaturge but a permanent group of vestal maidens who move sedately about the temple chambers in distinctive costume and headdress.

(H. Miner, 1956, Body ritual among the Nacirema, pp. 503–507)

If the Anglo-Saxon no-taste is so inveterate that even three thousand miles of tumbling water and a whole century of embargo with independence have failed to diminish it, what amongst all the perfumes of Arabia shall we apply to its disinfection?

(Editorial, 1874, *The Architect: A Weekly Illustrated Journal of Art, Civil Engineering and Building*, 17 November, p. 251)

Notwithstanding Miner's satirical account of organized healthcare in America and *The Architect*'s deliberate confounding of taste, style and infection, the National Health Service (NHS) in Britain faces a seemingly intractable conundrum. It is obligated to protect patients, visitors and staff from the effects of hot weather whilst delivering mandatory carbon savings at a prodigious scale.[1,2] If the 28 million m² NHS Retained Estate were to be sealed and mechanically cooled, the NHS Carbon Reduction Strategy (2010) would be derailed (NHS SDU 2010).[3] The NHS Carbon Reduction Target is proving to be intransigent even without this potential extra burden (Figure 6.2).

In 2010, the Department of Health calculated NHS England was responsible for 24.7 million tonnes of CO_2e ($MtCO_2e$), 30% of UK total public sector carbon emissions, 3% of all UK emissions, of which approximately 17% (4.07 $MtCO_2e$) was due to energy use in buildings.[4] To deliver even the required 28% reduction in carbon emissions by 2020, as compared to the 2013 baseline, let alone the 80% required by 2050, an annual incremental reduction of 4% in emissions, some 0.1628 $MtCO_2e$ annually, has been required from 2013.[5] Using a conservative allowance of a 9.5kg/CO_2/m² increment in emissions for adding efficient mechanical cooling across the estate, a minimum of an additional annual 0.266 $MtCO_2$/yr of emissions could be expected, in effect dismantling all attempts to save carbon by reducing building environmental energy and in practice it could be twice that figure, this allowance is derived from the 8.00 a.m. to 8.00 p.m. office world (Dunn *et al.*, 2006).

The conundrum will be further compounded by a changing climate. The impact of the predicted change across the UK, and its impact on the existing NHS Estate will be examined in Chapter 9. Figure 6.3 records the Graf *et al.* mapping of the distribution of the probabilities of temperatures occurring above 23°C, 25°C and 28°C across the UK for 1960–79 and 1980–99 taken from UK Climate Impact Programme data. Much of the southeast lies within a likely minimum of 30 days above 23°C in the later 20-year period.

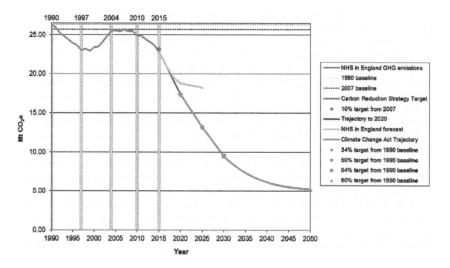

Figure 6.2
'NHS England CO_2e footprint 1990–2020 with Climate Change Act targets' as updated 2016, NHS Sustainable Development Unit. See www.sdu.nhs.uk (accessed 12 October 2016).

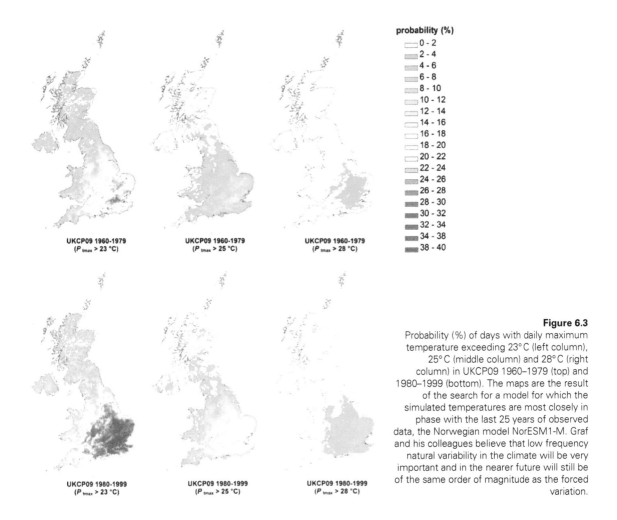

probability (%)
- 0 - 2
- 2 - 4
- 4 - 6
- 6 - 8
- 8 - 10
- 10 - 12
- 12 - 14
- 14 - 16
- 16 - 18
- 18 - 20
- 20 - 22
- 22 - 24
- 24 - 26
- 26 - 28
- 28 - 30
- 30 - 32
- 32 - 34
- 34 - 38
- 38 - 40

UKCP09 1960-1979
(P_{tmax} > 23 °C)

UKCP09 1960-1979
(P_{tmax} > 25 °C)

UKCP09 1960-1979
(P_{tmax} > 28 °C)

UKCP09 1980-1999
(P_{tmax} > 23 °C)

UKCP09 1980-1999
(P_{tmax} > 25 °C)

UKCP09 1980-1999
(P_{tmax} > 28 °C)

Figure 6.3
Probability (%) of days with daily maximum temperature exceeding 23°C (left column), 25°C (middle column) and 28°C (right column) in UKCP09 1960–1979 (top) and 1980–1999 (bottom). The maps are the result of the search for a model for which the simulated temperatures are most closely in phase with the last 25 years of observed data, the Norwegian model NorESM1-M. Graf and his colleagues believe that low frequency natural variability in the climate will be very important and in the nearer future will still be of the same order of magnitude as the forced variation.

Recovering 'the hospital'

Will this trend continue? Andrew Watkinson, Director of the UK Government's policy initiative 'Living With Environmental Change' (LWEC) has highlighted a general lack of 'transformative thinking' in national approaches to the changing climate.[6] A fundamental reinvention of the hospital form is urgently required to solve the conundrum in present conditions. It could be argued that the design of acute hospitals is atrophied in the UK, a paralysis induced by risk-averse procurement, not least the variants in which the contractor–developer builds and operates the new hospital. Contemporary designers do not have the understanding, the design time or the latitude under their appointment to invent innovative, validated passive designs. One has a sense that schemes for the UK are refashioned from projects designed to cocoon themselves from the Continental climate of the North American Midwest, deep-plan, compacted buildings with wholly artificial environments. Neither government nor industry are resourced to think 'transformatively' as they were in the 1950s and 1960s. The medical charities seem disinclined to fund this comprehensive review of hospitals as they once did (Nuffield 1955). In this chapter, however, a tentative start will be made.[7]

The key issue atrophying hospital design is anxiety about somehow increasing the risk of airborne cross-infection, the liabilities that would flow from that, and the belief that only mechanical ventilation can make hospital environments safe. Ironically, the Department of Health (HTM03-01) explicitly promotes the natural ventilation and passive cooling of hospitals to save energy and promote well-being but it is rarely adopted for new acute hospitals, a risk too far. The risks in poor hospital environments are very real. Excessively hot conditions are regularly reported in relatively recently built hospitals. The *Edinburgh Evening News* filed the following report at noon on 23 July 2014:

> Patients and staff are 'sweltering in intolerable conditions' at the Edinburgh Royal Infirmary Maternity Hospital as temperatures on wards reach 30°C. Nurses say they are close to collapsing on one of the worst affected wards at the Simpsons's maternity unit and new mums have been left in tears as they struggle to cope with the soaring heat.

The report blames, amongst other alleged factors, the restriction of window openings to four inches (for safety and security reasons) and the omission of air-conditioning from the design of the 2003 building to save cost.[8] It is extraordinary that at latitude 56° North it could ever be thought normal practice to seal and air-condition a public building to subdue summer overheating.

In the course of scoping the task of reinventing the hospital, albeit very roughly hewn, researchers discovered they were reinventing what was already known. The belief that effective ventilation would eliminate the communication of disease by removing its airborne agents drove much of hospital design through the 19th and early 20th centuries. Might there be clues locked in these forgotten designs? Is this not a sensible place to at least start searching for alternative strategies? The subsequent availability of mechanical ventilation and, in North America, air-conditioning, appears to have dissolved all interest in configuring healthy natural environments providing fresh safe air, natural light and even sunshine. There has of course been a systematic adverse reaction to these artificial environments orchestrated by Professor Roger Ulrich and colleagues (Ulrich, 2006). Earlier chapters discuss this collapse in knowledge and ambition in the general context of other non-domestic building types.

Learning from the 'dead hand of history': the original Johns Hopkins Hospital

However unlikely it may seem that useful learnings are available from the pre-antibiotic era of hospital design, it actually is illuminating to unravel historical precautions to reduce the incidence of airborne cross-infections through architectural form. An important moment in this history, perhaps contributing rather more than hitherto realized, was the invention of the original Johns Hopkins Hospital (JHH) in Baltimore between 1875 and 1879. Its shrewd benefactor's instruction to spend dividend income only to fund the works delayed completion of the first phase until May 1889 but the essential ideas were brokered out of highly productive consultations in early 1875 and published by the benefactor's trustees in that year.[9] It is by no means unknown as a project and was described at some length by Brieger (1965).[10] Subsequently, Thompson and Goldin (1975, pp. 175–193) published the JHH as it was built in some detail, Yanni (2007) discusses it briefly in the context of American asylums. More recently Fair (2013) discusses the circumstances of the final design. To reprise the history, briefly, Johns Hopkins was what would now be called a logistics entrepreneur, banker and railway company owner of such substance that he simply rode out the destructive financial 'Panic of 1873'. He developed the philanthropic imperative late in life in the manner of George Peabody, to whom he had been deliberately introduced with this end in mind. He had survived cholera. His last will and testament makes generous provision for founding a city hospital and a university.[11] He was an abolitionist and a Quaker, a very un-Baltimore figure, Baltimore being fundamentally Confederate but posing as Unionist in 1861 in an unsuccessful attempt to avoid being pillaged. His Letter of Instructions to the JHH Trustees of 10 March 1873 required an exemplary hospital, more advanced than any other in the known world, in which the indigent sick 'be received into the hospital without peril to the other inmates'. The Trustees' interpretation of this latter instruction was to use every means to eliminate 'hospitalism', the sudden onset of sepsis in the new building resulting from cross-infection between newcomers and incumbents. None of the Trustees being medically trained, the most authoritative advice on hospital design and operation available from five medical authorities was recruited and not, interestingly, from architects. Rather than concentrating on the effective containment and disposal of effluent from the new hospital, the Trustees believed 'hospitalism' would be prevented by vigorous ventilation of the wards.[12]

Their letter of invitation to respond with advice for this prestigious project, published as the 'Letter Addressed to the Authors of the Essays' (King, 1873),[13] states the Trustees' position:

> Certainly not second in importance to any of the matters you are invited to instruct us upon are those of ventilation and heating … as curative agents. The various problems of heating, combined with ventilation, form professional problems about which the most experienced and best informed minds seem to be far from being united.

Of the five, one proposal excels in the completeness of its vision, that submitted by Norton Folsom MD (1842–1903), Superintendent of the Massachusetts General Hospital pursuing a concept for ward design indigenous to the 'Mass General', and another in its accompanying text, its imperious grasp of hospital organization and management, that of Bvt Lt-Col and Asst Surgeon John Shaw Billings (1838–1913) (Garrison, 1915, p. 200).[14] The two experts promoted radically opposing plans for 'safe' wards, apparently anticipating by some years the coming debate in Britain between the proponents of narrow linear 'Nightingale' plans and 'centrally-planned' wards with rotational symmetry: square, octagonal or circular (Taylor, 1988; Marshall, 1878).[15,16] Billings was an unreconstructed supporter of Nightingale's views on hospital design and organization. The Trustees had already appointed their own architect for the project, John R. Niernsee (1814–85), sometime Greek revivalist, re-modeller of Johns Hopkins' own house, Clifton, a miniature Osborne House.[17]

The consultation was not intended by the Trustees to be competitive. There is no hint of remuneration, nor any kind of reward in their letter of invitation. A little naively, they planned to harvest the best ideas for Niernsee to blend and realize (Chalfont and Belfoure, 2006, p. 84). Knowing full well that this was a controversial field, they elaborated on their understanding of three fundamental positions then being propogandized in America and Europe (King *et al.*, 1875, p. xii):

> … whether ventilation should be accomplished by what is called the natural method, through doors, windows and unavoidable leakages, or through flues and ducts acted upon by the differing temperatures of the outer and inner air, or by enforced currents set in motion by fans, blowers or other mechanical contrivances, are points which have equally learned advocates and opponents.

What invisible menace was the immense forthcoming investment in ventilation designed to dispel? Chang and Jackson (2007) remind us that John Snow's 'germ theory', provoked by the 1854 London cholera outbreak, was not accepted by British Sanitarians, notably William Farr (1807–83), until the aftermath of the 1866 epidemic, so that in the mid 1870s the 'Contagion' and 'Environmental' theories of disease still prevailed in both Britain and America.[18] The two theories had elided through the contributions of Pettenkofer, held in high regard by Billings for his *Boden* theory.[19]

Pettenkofer associated communicable disease with 'miasmas', noxious gases, 'carbonic acid' and/or 'sulphurated hydrogen', issuing from the ground vertically or possibly, and more controversially, laterally. Foul smells were dangerous to human life for subscribers to this theory. It would be judicious to be 'lying in' somewhat above the miasmatic ground level, hence the preponderance of raised basements in the JHH proposals. Contagion might also occur through ingesting 'fomites', particles of concentrated pathogens, which it was thought could be diluted in fresh air to harmlessness, hence the perception that good ventilation was vital to survival in a threatening natural environment. 'Hospital fever' or 'Hospitalism' was frequently preceded by the horrifying symptom of putrefaction of the flesh. Florence Nightingale wholly subscribed to the theory to the point of arguing that excellent ventilation would remove any risk of direct contagion between soldiers the minimum distance apart:

> Natural ventilation, or that by open windows and open fireplaces, is the only means for procuring the life-spring of the sick – fresh air … as much fresh air to be admitted from without as suffices to keep the ward fresh. *No* artificial ventilation will do this.

Florence Nightingale explained this to the National Association for the Promotion of Social Science in October 1858 on her return from Crimea.

Responding to written questions from the Royal Commission into the Sanitary Condition of the Army, Nightingale (1858) explained further:

> … without the most perfect ventilation, there is always more danger of effluvia being driven by a draught till it accumulates in part of a very large ward, as was the case in the long corridors of Scutari.

Billings hedged his bid by explaining to the Trustees that, of course, 'miasmas' were problematic but 'rarely lethal in dosage' whereas 'disease germs' or 'contagia', including fungi, were lethal, destroyed only by high temperatures, chlorines, sulphurous acid, ozone and disinfectants, 'they are what we have to fear'. Dilution of the air, he insists, would not be enough to remove the danger. This is important to the design outcomes because if sufficient dilution was effective as a single measure, ventilation rates would have to be greatly increased with correspondingly unmanageable winter heat losses. As Billings says, employing a military analogy, if only ten marksmen are firing at you rather than 100, 'if one of them does chance to hit you, the practical difference will not be appreciable'.[20]

Tuberculosis mycobacteria, not identified until 1882, can infect the lungs on the principle of 'the unlucky particle' but other important airborne pathogens require higher concentrations to induce disease.[21] The answer, Billings explains forcefully, is very effective ventilation delivered through the configuration of the building, coupled with scientific cleanliness. The idea that the architecture itself could conjure up healthy airflows was not at all unprecedented by 1875.[22]

Although there is no mention of Joseph Lister or 'antisepsis' in the 1875 publication of the five essays, Garrison's 1915 biography of Billings claims he knew Lister well enough to have received a warm invitation, dated 5 December 1876, to join the Lister family for lunch during his official JHH Trustees' funded European expedition. He was cajoling Lister to put him up for a Fellowship of the Royal Society, ultimately unsuccessfully. Garrison (1915, p. 201) reported that:

> In the final plans, the wards were in single storey pavilions, and to ward off the miasms and malarial emanation which, under Pettenkofer's Boden theory were then supposed to emanate from the soil.

The Johns Hopkins Hospital, as built, on the cusp of the new understanding, was therefore braced for all eventualities (Barnard, 1972; Richmond, 1976).

Naturally driven, closely controlled, buoyancy-driven ventilation was thought to be the solution by the respondents Billings and Folsom, airflow driven solely by the pressure differences developing between a warming occupied interior and a cooler exterior, employing no mechanical fans, the effect accelerated by the natural reduction in barometric pressure by height. The other three respondents fell in with current practice.[23] Folsom and Billings conceived their proposed buildings as huge integrated ventilating instruments. Air moving devices were not merely bolted onto them, the building in each case was the ventilation machine, inducing and capturing naturally induced airflows by incorporating elaborate structural air supply and exhaust arrangements. Both used prodigious stacks to drive the airflows but differ fundamentally in the geometry of the supply and exhaust and thus the geometry of the wards.

Billings' case for edge-in/centre-out ventilated general wards

Billings' (1893) grasp of buoyancy-driven airflow was certainly solid. Chapter II in his *Ventilation and Heating* presents a diagram giving a reliable prediction of likely air movement in a space under steady state conditions (Figure 6.4).

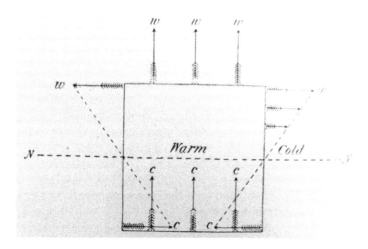

Figure 6.4
Billings' depiction of natural buoyancy and the concept of the neutral plane at which internal and external pressures align within the space, from Billings (1893).

He identifies a neutral layer below which air will flow into a building and above which it will tend to flow out as the decrease in pressure due to internal heat gains exceeds the diminution in barometric pressure with height. He derives his understanding, he writes, from the published work of General Morin, whom he much admires, and for good reason. From Chapter 5, readers are now quite familiar with Morin's thinking in the context of theatres.[24] Billings was probably also acquainted with the work of Dr Reid by 1875, most likely Reid's (1858) American work, *Ventilation of American Dwellings*. Billings cites Reid in his 1893 text, *Ventilation and Heating*.

Billings was proposing a climate-driven adaptation of what had become the standard English Nightingale hospital model. The throwing open of windows by an unsentimental matron to admit fresh air at the rates recommended would be very problematic due to the extreme cold of the Baltimore winter and subsequent excessive summer heat. Billings' point was that the new buildings would have to maintain adequate conditions in external temperatures ranging between 0°F and 100°F in the Maryland climate, classified as 'Cfa', 'humid subtropical', by Köppen (1884). Billings included tables of climate data for the four years from 1871 to 1874 obtained from the Chief Signals Officer of the Army as an appendix to his report to the Trustees. Peak summer temperatures in June and July are remarkably consistent across the four years and the preceding 53 years. The winter minima are rather more volatile, January lows between –4°F and 13°F across the four years and –10°F in 1835 to 19°F in 1869, the years 1817 to 1821 being exceptionally cold in winter. However, mean temperatures lie consistently between 53°F and 58°F throughout the century, suggesting cold winters were compensated by warmer summers, in part at least. This was a very much more extreme climate than that mediated by the contemporary Queen Anne style buildings in England emitting and enjoying the 'Sweetness and Light' associated with the movement, the phrase appropriated from Matthew Arnold to capture the style's pastel fragility (Collini, 2006).[25]

Billings' original entry was feeble as a design, neither very innovative nor well resolved. It relied heavily on Florence Nightingale's template (*The Builder*, 1858).[26] Our redrawing of his original Figure 2 of the two-storey ward cross-section (Figure 6.5), key to the design, transposes missing information from the more complete drawing for a single-storey ward onto the two-storey variant.

Billings is indifferent to the number of storeys but explains that the ceiling exhaust from the lower ward would be problematic in congesting the upper floor. It certainly is, and is conveniently omitted from his diagram. However, whilst referring warmly to Dr Reid's bottom to top 'displacement ventilation' scheme for the House of Commons in 1844, he commends the reverse, a 'top-in/bottom-out' flow of air driven by the stack effect created within a sizeable 'aspirating chimney'. It is a more expensive arrangement but puts the heat where it is required. It became the standard geometry for the delivery and exhaust of air in Victorian hospitals. Banham (1969, pp. 75–79) illustrates a similar arrangement at the Royal Victoria Hospital in Belfast, one of a family of large civic hospitals in Britain.

Recent experiments substituting one fluid for another, water for air, in transparent tanks at the BP Institute at the University of Cambridge suggest Dr Reid's 'bottom-in/top-out' system may be flawed in infection control terms.[27] Billings and his generation may have made safer environments, more efficient in flushing out airborne pathogens. Figure 6.6 shows particles of silicon carbide at 13.9 microns diameter, the size of the larger airborne pathogens, being introduced at low level, equivalent to a patient's tubercular exhalations, whilst fluid is continuously introduced at the base of the tank and extracted from the top.

The pattern is also typical of modern hospital mechanical ventilation systems. However, the system is unable to flush all of the particles from the notional patient room. The air is clear at the bottom of the room but there is a pronounced displacement layer, above which particles, proxy for the pathogens, persist in suspension. At the top of the room, close to the exhaust, is another clear layer.

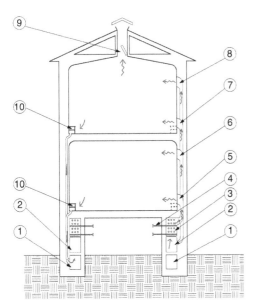

Figure 6.5
A redrawing of Billings' original proposal for the JHH, a two-storey Nightingale ward building with enhanced ventilation arrangements. The redrawing is informed by further and more persuasive detail in the original drawing of the single-storey version as published. The exhaust passages are connected to a substantial chimney beyond (exercise funded by the Sir Isaac Newton Trust, copyright author).
1. Foul air returns direct to connect to 'one central aspirating chimney' or to 'smaller ones for each building, which last will be the most satisfactory'.
2. Fresh air supply ducts. Billings implies that flows here will be fan-assisted, making this a genuine and early hybrid scheme.
3. Heating coils within.
4. 'Valves' to adjust airflow, supply to heating coil and supply from heating coil into vertical supply duct.
5. 'Fresh air register' low level. Note that this asymmetric section is intended to alternate bay by bay.
6. 'Fresh air register' high level.
7 and 8. Arrangement of lower floor supplies to upper floor.
9. 'Ridge ventilation' to be used in summer but Billings judiciously observes that on days when Ti or TE are similar, stack-driven flows may stall, in which case, the (10.) 'aspirating foul air closet' duct should be employed to gravity exhaust the spaces.

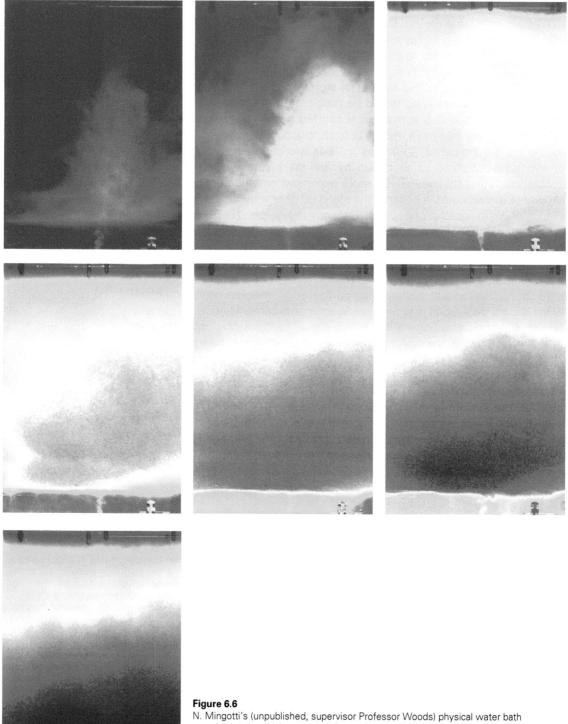

Figure 6.6
N. Mingotti's (unpublished, supervisor Professor Woods) physical water bath experiments in which a steady inflow at low level fails to flush out silicon carbide particles at 13.9 microns diameter through a high level exhaust at 1, 10 and 30 minutes. Dark blue represents clear cool fluid, lighter blues and greens a lower concentration of particles, more intense yellows and reds a greater concentration. Steady state is achieved at 40–50 minutes (BP Institute, University of Cambridge).

Gravity is playing a role and therefore the airborne particles are not being swept into the exhaust duct. After 30 minutes of flowing fluid, a discernible concentration is developing within the body of the tank representing the occupied realm of the room. Could it be that several decades of building and observing ventilation systems and the free exchange of practical experience had revealed this counterintuitive phenomenon to 19th century hospital builders, knowledge waiting to be recovered by their modern equivalents?

Recent measurements in redundant late 19th century Nightingale wards in the north of England confirm their efficiency in providing sufficient, reliable and well-distributed ventilation (Gilkeson *et al.*, 2013). Figures 6.7 and 6.8 show that in the final scheme for the JHH the air paths become indirect. Air is introduced into the occupied ward through ducts within the external wall, either directly in warm and mid-season conditions, or preheated via a 'diverter' through a steam coil supplied with steam at 150° F.

A combination of both was possible, the 'diverter' can be employed as a mixing valve controlled by a lever within the inlet grillage just above skirting level in the ward between pairs of beds, the principle means of control accessible to nurses, patients and orderlies (Billings, 1893, p. 202 and Figure 69 on p. 210). Air is exhausted, Billings explained, in winter and mid-season through domed mesh-covered openings in the floor and is drawn down into these exhausts because they are connected by iron pipes sweeping into a spine duct connecting to a 75 feet high chimney, 4 feet 2 inches in diameter, generating powerful stack-driven flows, assisted by a heating element at its base (ibid. Figure 69, p. 210). Out of the heating season a series of 2ft by 2ft shutters at the apex of the concave ceiling connect the ward volume to a tapering duct growing in girth as it progresses towards a high level connection to the same stack. It is not clear from Billings' descriptions whether the hatches were remotely operable, as they were heavy, and built of solid timber and galvanized sheet metal. One ward was equipped with a fan supplying the basement level duct. It helped considerably in summer. The final design devotes 55.8% of its cross section to voids required by the ventilation strategy, not including the stack. Only 44.2% is actually occupied, suggesting an investment broadly equivalent to half the construction cost in an attempt to eliminate the airborne transmission of disease.

Figure 6.9, a freeze frame view from the three-dimensional model, looking up into a slice of the ward, shows the air supply 'machinery' in the raised basement devoted to conditioning the air supply. Figure 6.10 shows how this arrangement appears within the ward.

Billings' account claims the supply of air, though variable, was intended in everyday mode to deliver one cubic foot per second. The author calculates this to be equivalent to 2.32 air changes per hour (ACH) supplied to the ward as background ventilation. Billings published a detailed drawing of what he describes as a 'bypass' arrangement in *Ventilation and Heating* (Billings, 1893) reproduced here as Figure 6.11.

The Billings papers in the New York Public Library include the record of an early monitoring exercise, 'Johns Hopkins Hospital. Thermometrical Indications February 1885', recording dry-bulb air temperatures taken on-site through the first two weeks of that February, in effect, records of the commissioning, reproduced here as Figure 6.12.[28]

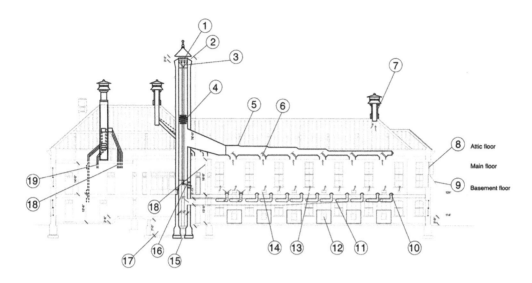

Figure 6.7

Long section through a Common Ward cut north–south at the Johns Hopkins Hospital (author).

1. Open Grillage assumed at stack termination to exclude birds; was there an insect mesh behind it? No evidence available.
2. Vertical free area at termination, 2ft 2 inches (651mm) height of openings, belfry louvres at all eight sides delivered 2226 inches^2 (1m^2) the cross sectional opening of the stack.
3. Centre folding damper, apparent umbrella type action deployable structure.
4. 'Accelerating Steam Coils' to force stack effect, as Billings describes them.
5. Longitudinal foul air duct in the attic, timber boarding lined in galvanized iron.
6. Hinged trap doors connect ward to exhaust duct. Were they manually operated individually or connected through mechanical linkages to a single control lever? Sadly we found no evidence but it seems likely, not least because of their inaccessibility.
7. Attic ventilation, 24 inches diameter dedicated stack.
8. Substantial sash windows along ward, collection of three at southern end to serve a sunspace. The free area varies from zero to full opening of 20ft^2 (1.85m^2), both sashes in line, the basement floor was served by a significant area of sash windows, providing a maximum of 12.1ft^2 (1.13m^2) opening area.
9. Double action casement and top hung louvred shutters shield glazing, clearly shown on contemporary photographs which also appear to indicate roller blinds internally. At the sunspace end of the ward three sets of shutters are indicated internally to add to the layers of sun screening.
10. Contemporary photographs and drawings show a domed grille beneath each bed, 1ft in diameter, intended to act as an exhaust outlet.
11. Foul Air Duct below ward floor.
12. Heating chamber 'heat coils' pumped hot water, preheats ventilation air for the ward above.
13. The air, warmed or un-warmed by virtue of a mixing damper within the duct below enters the ward through two sets of wall grilles, the primary inlet through grilles mounted below the large sash windows.
14. Principle delivery grilles above the 'heat coils' in the basement directly below were somewhat larger.
15. Below floor 'Foul Air Duct' connects to stack.
16. Direct connection to soil below drains away any rain or blown snow which might drop down the stack, the 'Central Ventilating Chimney', very much a contemporary problem in the design of naturally ventilated buildings. The Queens Building in Leicester employs deliquescent salts to absorb stray water in the stacks.
17. Manual Controller at waist height operates the umbrella type damper at the stack termination.
18. Pipe tunnel 7ft 10 inches × 10ft 0 inches connects all the buildings to the central boilers.
19. Vent pipe from private wards 18 inches × 22 inches (457mm × 558mm), a self-contained system.
20. Similarly the vent pipe from dining room 17 inches × 24 inches (432mm × 610mm).

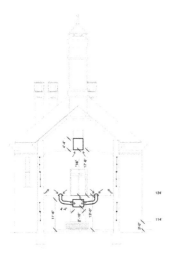

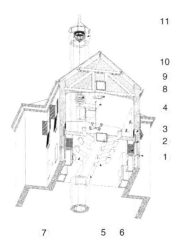

Figure 6.9

Freeze-frame isometric 'up-view' of the Johns Hopkins Hospital Common Ward (author).

1. Iron damper at air intake diverts incoming air up vertical duct within wall into the ward, one outlet per two beds with secondary outlets beneath windows. Manually operated from ward to enable mixing of warmed and fresh air to temper incoming air to the desired temperature.

2. Heating chamber within brick masonry structure with novel double layer iron plates with insulating felt trapped within to provide access to the pipework coil.

3. Iron damper coupled in some way with mechanical lever control to air intake damper below, connects heater chamber into air supply duct.

4. The air, warmed or un-warmed by virtue of a mixing damper within the duct below enters the ward through two sets of wall grilles, approximately 2.36ft^2 (0.22m^2) face area directly above the heating coils and a 1.39ft^2 (0.13m^2) mounted above the radiused skirting below the large sash windows. Staff would have had to crouch or kneel down to operate the large lever on the face of the grille to set the temperature of incoming air.

5. Tapered 'Foul Air Duct' below ward floor varies from approximately 2.05ft^2 (0.19m^2) to 12ft^2 (1.115m^2).

6. Domed grilles below each bed, 12 inches in diameter intended to exhaust air into the Foul Air Duct and thence up the stack.

7. Iron pipes, 12 inches in diameter, connect domed grilles to Foul Air Duct.

8. Longitudinal foul air duct in the attic, timber boarding lined in galvanized iron, scales of ⅛th inch drawings at 48 × 51 inches, 2448 inches^2 (1.579m^2).

9. Hinged trap doors connect ward to exhaust duct, Billings reports that they are 2ft^2 but they scale at 2ft 10 inches square (0.746m^2).

10. 'Accelerating Steam Coils' as Billings describes them.

11. Centre folding damper, apparent umbrella type action deployable structure, in eight linked moveable segments within the octagonal stack. Even the detailed drawings are unfathomable, the segments do not appear to interlock and seal the shaft when closed, could they be in a flexible material, leather or thin sheet material; it is manually operated from ward level via a long iron rod.

Figure 6.10
Freeze-frame downward-view of the Johns Hopkins
Hospital Common Ward showing the ward interior (author).
1. Air intake.
2. Diverter between fresh air and heated fresh air and/or a mixture of both.
3. Principle inlet grilles to ward with lever control for dampers below, between each window, secondary supply grilles below windows.
4. Domed exhausts below beds.
5. Secondary intake below windows.
6. Hinged trap doors connect ward to high-level exhaust duct.
7. 'Foul Air Duct' above ceiling intended to exhaust in summer.
8. 'Accelerating Steam Coil' in stack.
9. Folding damper at termination to stack.

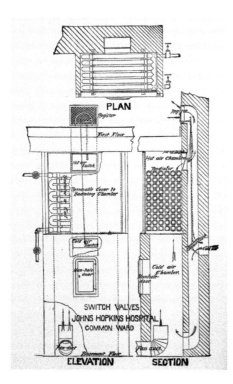

Figure 6.11
Air intake below the ward floor level showing
the bypass damper (switch valve) in position to
divert incoming air through the bank of heating
coils, from Billings (1884).

On 6 February, the weather was 'cold and clear'. At 8.00 a.m. the site was well below freezing at 26° F, the temperature falling to 25° F at 1.00 p.m. and to 23° F at 6.00 p.m. Internal temperatures in the three zones of the General Ward under scrutiny climbed variously from 53° F to, variously, bracing 60° F, 62° F and 64° F, consuming 5 US tons of coal that day. On the following day, learning from the experiment and despite even colder external conditions, 22° F at 8.00 a.m. barely rising to freezing point by lunchtime, temperatures between 70° F and 73° F were achieved. The paper records 'experiments in opening and

Figure 6.12

'Johns Hopkins Hospital. Thermometrical Indications. February 1886. Taken at about 8am, 1pm and 6pm.' Johns Hopkins Hospital and Medical School subject files. Astor, Lenox and Tilden papers, John Shaw Billings papers, Manuscripts and Archives Division, The New York Public Library.

closing doors to admit fresh air'. By 11 February in bitterly cold weather, 2°F at 8.00 a.m., internal conditions achieved a creditable range of 69°F to 72°F, rising to 76°F at one station by 1.00 p.m. as the temperature outside reached only 10°F but at a cost of 7 US tons of coal consumed. Nonetheless, this was a remarkable achievement, the internal volumes are substantial and the potential for air leakage at every construction joint potentially huge. The quality of construction, particularly the window joinery, must have been very high. A margin note records, with evident relief, 'expansion joints are apparently working all right', steam pressures were considerable. A second table in the same handwriting accompanies the thermometric record, untitled, but apparently recording achieved airflows in feet per second on selected days in a consecutive December and January (Figure 6.13). This is unlikely to have been December 1885 because ward temperatures are noted as slightly above 70°F. The implication is that the building must have been commissioned and in use but the table records an exploration of the effects of opening the cellar windows on ward airflows. Perhaps airflows had been disappointingly erratic during 1886.

Monitoring of contemporary buildings is time consuming, expensive and disruptive, and tends only to be undertaken to investigate disappointing environmental performance. The notes to the table record, 'The average velocity of incoming air when cellar was closed was 1.6 ft. per sec.', but when the cellar was open, a significantly higher 3.3ft per sec. was measured. There was no direct route between the spaces, a double lobby defended the stair down, this was probably additional supply air leaking up into the ward through the heater batteries and perhaps the floor.

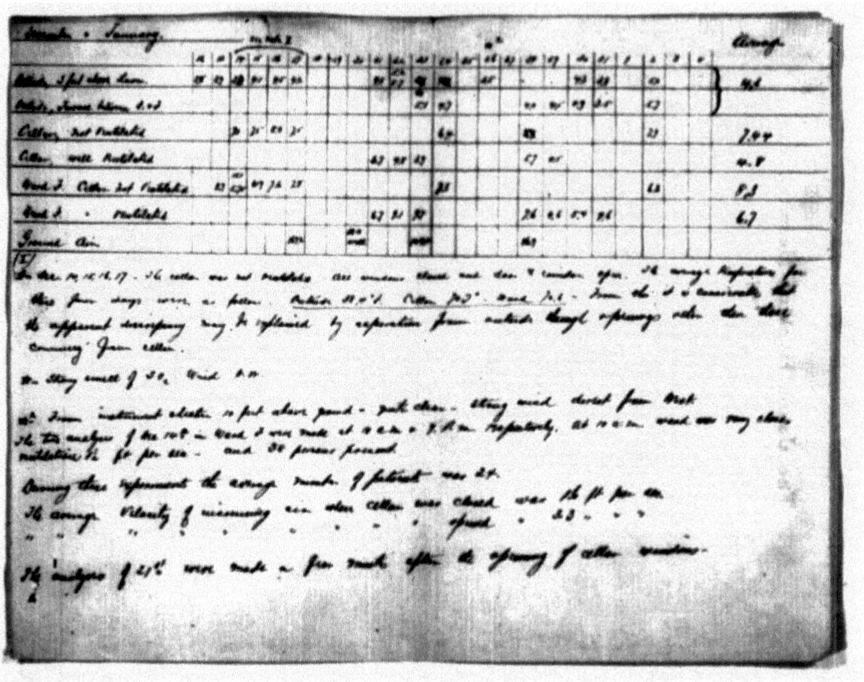

Figure 6.13
Billings' table of air velocities from Box 48.

The commissioning record is accompanied by a hand-written paper of February 1885, 'The Heating Apparatus of the Johns Hopkins Hospital', reflecting on the performance and economy of the system.[29] Billings notes that fuel use is higher than expected because it is an 'all fresh air system' with no recirculation, so that there was a complete renewal of air every 15–20 minutes, some 3 to 4 full ACH. There was a 700ft run of heating pipework losing heat along its length but the Hospital was maintained, he states, at 72–74°F, at a cost of less than 1000 US tons of coal per year. The papers include exhaustive hand-written schedules of heat emitters, each carefully sized to its situation, and similar schedules of gas lamps: 'combustion chandeliers' and 'combustion pendants'.

Sir Vincent Zachary Cope (1881–1974), the surgeon and honorary librarian of the Royal Society of Medicine, has reported on a correspondence between Billings and Florence Nightingale. Billings sent her the scheme in October 1876. His initial letter and his curt acknowledgement are retained in the Nightingale archive in the British Library. Billings' response suggests Nightingale, writing at length, urged caution, advising that just one ward be built first. Billings was not delighted at her 12 pages of comments or at the return of the drawings. Most surprisingly, her letter is not preserved in his archives either in the JHH or the NYPL, but his letters are found in her archive.[30] No record was found that his mentor's reservations on the design, arriving at a key moment in the debate, were ever communicated to the Trustees. Billings took a tactical decision at the 'Front'.

Folsom's alternative centre-in/edge-out proposition

In direct contrast, Dr Norton Folsom promoted a square ward plan in which beds were placed in a ring around a central stack, between very attenuated sash windows with multiple opening lights, prefiguring the equivalents at Norman Shaw's Melbury Road studio house by two years. Folsom explains his proposal drew on the design for the square ward building he had overseen at the Massachusetts General Hospital (Mass General) in 1872.[31] The hospital had developed a tradition of square pavilions deriving, Folsom explains, from Mr George M. Dexter's contributions to the 1844 enlargement. He cites this in his evidence to the JHH Trustees (Hospital Plans, 1875, p. 76): 'I do not think the attractive home-like character of such rooms, in comparison with long, narrow wards can be appreciated without seeing both in occupation.'

Folsom's scheme develops a sophisticated ventilation design around the Mass General plan. The fresh air feeding the whole system is drawn into two lateral ducts at undercroft level from both east and west orientations, a robust 'push-me-pull-you' configuration which can balance differential pressures deriving from changes in wind direction, guaranteeing some supply of air in all circumstances (Figure 6.14).

Fresh air is admitted upwards into the centre of the ward at low level, either directly, or via a diverter into and through a heating chamber before entry. The very substantial chimney climbs from the centre of the ward plan to well above the apex of the pyramidal roof. The air supply inlets into the ward straddle opposing sides at the chimney base whilst Folsom explained Franklin stoves or open fireplaces would occupy the other two faces. He predicted that warmed air will rise up to the higher cavity of the roof volume and circulate by tumbling down the pitch of the ceiling onto the patients' beds and thence out through exhausts positioned at bedhead level around the perimeter (Hospital Plans, 1875, p. 82). Some air, he explained, will be drawn back across the floor to the centre of the room before recirculating. The low level exhausts chime with Billings' strategy, adopting advanced practice of the time which current science may now validate.

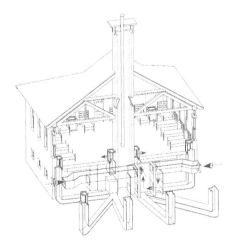

Figure 6.14
Freeze-frame isometric view of digital 3D model of Folsom's proposal showing air supply routes. Air is drawn in from east or west into two central mixing chambers, offset from the stack base, directed through heating coils or diverted directly into two entry points on the east and west sides of the stack (author).

The waste air is drawn down to, and then beneath the basement floor through a radial system of ducts congregating at the base of the central stack (Figure 6.15). Natural buoyancy will drive the warmer waste air up to the stack termination assisted by the heat of the stove or fireplace flues contained within (Figure 6.16).

Shuttered openings are provided into the stack at the highest internal point for additional exhaust in hot summer conditions. Folsom was sensibly concerned about the 'dog days' during which 'the air seems heavy and stagnant', the buoyancy effect would be very diminished and a coal fire was to be lit in the chamber below the base of the stack in the basement to maintain flow but without warming the occupied space. Folsom's guide to the physics of airflow in buildings was local, Mr A.C. Martin of Boston, an architect.

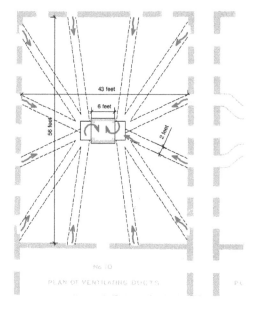

Figure 6.15
Folsom's diagram No. 10 'Plan of Ventilating Ducts', as reproduced in Hospital Plans (1875).

Figure 6.16
Freeze-frame isometric view of digital 3D model of Folsom's proposal showing air exhaust routes. Waste air is received through dampered grillages at bedhead height around the perimeter to be ducted down below basement floor level before congregating in two chambers adjacent to the aspirating stack base. The buoancy effect drives air up to the ambient lower pressure at the head of the stack, accelerated by the heat given off by the stove flues within the stack. In summer, Folsom provides direct high level openings into the stack just below the roof apex and his sections show further operable dampers connecting into the roof dormer voids connected to louvred gables, a third wind-assisted route.

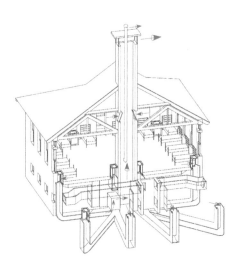

Folsom's master plan has the additional ingredient of a specific isolation ward building, similarly derived from a prototype at the Mass General, in which patients deemed to be in a state of malodorous putrefaction which would defy the capacity of the natural physics driving the environmental behaviour of the Common Ward, were confined to single occupancy rooms. Folsom's proposal remains a credible design strategy for a contemporary low energy single patient room hospital. The building is demolished and the author has relied on Billings' own record (1890) to produce new drawings and a 3D model (Figure 6.17 Cross-section, Figure 6.18 Part long section, Figure 6.19 Freeze-frame down view).

It was entirely comprised of single rooms to enhance infection control, very much as current UK NHS 'consumerist policy' requires, to enhance patient privacy and dignity but ultimately to reduce hospital (nosocomial) infection rates. This issue is discussed later in the chapter. Ironically the Isolation Ward becomes more of a potential model for a contemporary hospital than the Common Ward. The arguments currently being advanced for the retention of the open multi-bed ward (Short *et al.*, 2014), particularly for the care of older people with dementia, are considered in Chapter 9.

Billings writes that the Isolation Ward was 'designed for cases giving rise to offensive odors or in which a large amount of organic matter is thrown off, or in which for other reasons a large amount of air is desirable, the air supply is fixed at 2 ft^3/sec/head'. Billings calculated the volume of each single room to be 2145ft^3 with 162ft^2 of 'radiating surface'. For all wards the air is warmed in cold weather before it enters. All flues allow air to 'pass' with a velocity not exceeding 1.5ft/s under normal circumstances. Fresh air enters each room through a single register following the basic principle of the Common Ward diverter-mixing valve but the heating area provided by the coil below was greater,

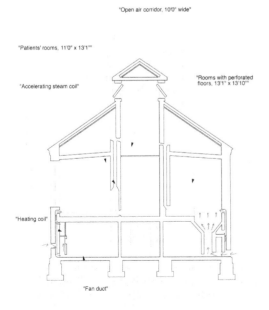

"Open air corridor, 10'0" wide"

"Patients' rooms, 11'0" x 13'1""

"Accelerating steam coil"

"Rooms with perforated floors, 13'1" x 13'10""

"Heating coil"

"Fan duct"

As built isolation ward scheme short section
Scale 1:100

Figure 6.17
Isolation Ward, short cross-section cut
north–south as built.

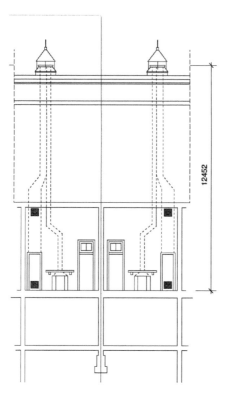

Figure 6.18
Isolation Ward, part long-section cut east–west as built.

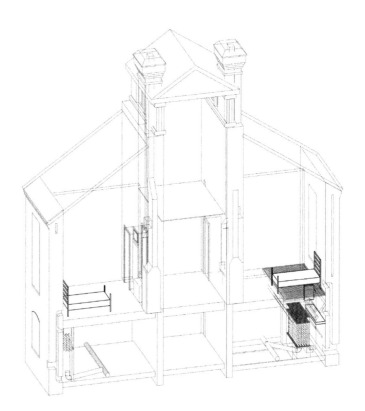

Figure 6.19
Isolation Ward, freeze frame isometric view looking down.

calculated to create 2ft^3/sec flow within each room. Each room is indicated to be 11ft × 13ft 1in, 144ft^2 (13.375m^2) with double doors into the rooms from the central corridor, intended to be an airlock enabling patients to 'sport the oak' in proper collegiate style. There was an open fireplace to provide instant radiant heat and to one side a cupboard accessible from the corridor side for the commode to be inspected and transported to the laboratory. The closet was lined with galvanized iron so that it 'can readily be cleansed with flame'. The closet door to the room had intake grilles at low and high level and the closet an exhaust connecting to a separate exit flue in which an accelerating steam coil was judiciously placed (Billings 1890, p. 95), perhaps a little low in the system, the foul air had to dislodge a considerable stack of cold air first thing in the morning. The iron flue from the closet passes alongside the flue from the fireplace, intentionally or unintentionally but no doubt helpfully enhancing flow by warming the stack. The flues extend to the top of the octagonal stacks, one per room projecting just above the ridge level (Billings, 1890, p. 96).

In addition, three slightly larger rooms in the Isolation Ward are provided with perforated floors below which a funnel form connects at its base to a heating coil to provide double the airflow rate, 4ft^3/sec, with, Billings reports, the capacity to double this again if desired. The perforated floors comprised ¼ inch holes for first 7ft from the outside wall which Billings reports provided 50 holes/ft^2, a total of 5000 holes per room. The extension of the module to accommodate these is effortlessly subsumed above so that the rhythm of the stacks is uninterrupted.

The central corridor was open at both ends, not uncommon in Europe at the time, for example at the Winchester College Sanatorium 1884–93 by William White, a Silver Medal winner at the World's Columbian Health exhibition, all rooms gave onto the exterior. The Oundle School sanatorium is very similar. The 10ft wide central corridor is shown with a clerestory along its whole length, 'with glass louvres', presumably the centre-pivot windows depicted in contemporary photographs. Isolating the top half of the building and venting it vigorously would have helped calm temperatures in the attics and reduce the effect of prodigious summer solar gains. Billings emphasized that 'the Aim is for air to pass constantly upwards so that it is not re-breathed'.

The Trustees' conundrum

Niernsee was wholly persuaded by Folsom's 'centrally planned' alternative to the linear Nightingale ward plan. He was granted two appendices to the publication of the five responses. In the rubric to the first he addresses the Trustees as 'Architect to the Board of Trustees', reminding them he is their architect, 'Member of the Austrian Institute of Architects and Civil Engineers, Member of the US Commission of Science and Art and of the Jury of the Exposition of Vienna in 1873', in other words a figure of some substance, with international experience and recognition, not a provincial figure (Hospital Plans, 1875, p. 235). The Vienna Hofoper had only recently been completed in 1873.

To this effect the Trustees allowed the inclusion of his own radical scheme of octagonal pavilions as the second appendix, a reinterpretation of Folsom's proposal intended to resolve the simple geometrical problem of juxtaposing beds in the corners of a square plan (Figure 6.20).

It is not a particularly convincing arrangement; beds are stranded mid-floor with not quite enough space behind to walk comfortably around the window walls. Nonetheless, one octagonal pavilion was actually incorporated into the built scheme, but only because, Billings tells us in the celebratory publication, ungraciously, there was insufficient site area to accommodate a final and superior linear ward building. This was manifestly not the case as can be seen in the plan of the first phase (Figure 6.21). The former Union Army officer Billings had, perhaps, taken against Niernsee, who had served, almost circumstantially, as a major in the Confederate Army.

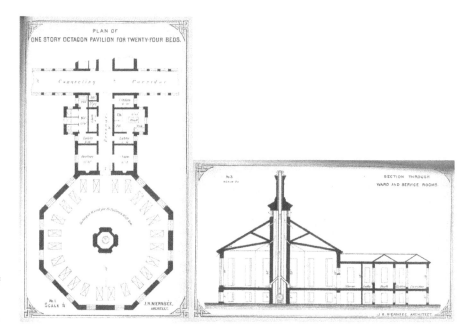

Figure 6.20
Niernsee's alternative proposal for a hospital of octagonal wards described in Appendix 2 to Hospital Plans (1875).

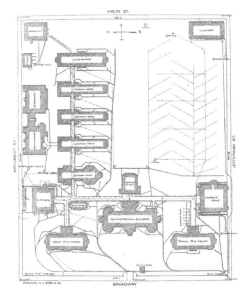

Figure 6.21
Site plan for construction showing the extent of the first phase of JHH in 1885. The Octagonal Ward is closest to the central Administration Building (Billings, 1890).

Niernsee argued forcefully for his proposal, clearly feeling on the back foot. He claims to have visited 'the best and latest hospitals in Europe', he rejects the usefulness of the medical experiences of the Civil War, a dig at Billings, and commends Folsom's scheme. He had clearly had sight of all the entrants' drawings. Billing's high pitched voice (as reported in the guide to the JHH Archives), misogyny and domineering 'military' manner must already have been heard in the Trustees' room, to Niernsee's alarm, lobbying hard to take control of the project. Niernsee writes further, referring to the 'Herbert's Commission Report' and other 'high medical authorities' (*Hospital Plans: Five Essays Relating to the Construction, Organization and Management of Hospitals, Contributed by their Authors for the Use of the Johns Hopkins Hospital of Baltimore*, New York: William Wood and Co., 27 Great Jones Street, 1875, Appendix II p. 337):

> ... the advantages of both the square and octagon ward over the oblong shape in that respect are so apparent and decided that I was led to the investigation of the further development and possibilities of the octagon ward, stimulated by the inspection of Dr. Folsom's temporary one square pavilion erected in connection with the Massachusetts General Hospital in Boston.

One suspects the late spring and early summer of 1875 saw the Trustees struggling between the Folsom–Niernsee proposition and the rather less imaginative but well established position on ward design adopted so stridently by Billings.

The argument prefigures the debate in Britain between the proponents of linear Nightingale and radial Circular Wards. George Godwin, the editor of *The Builder,* having promoted Nightingale's ward plan, subsequently argued fiercely for circular plans in his journal in late 1878, the idea deriving from a lecture by John Marshall, The Professor of Surgery at University College London and Royal Academy Professor of Anatomy, to the Social Science Association in Cheltenham in 1878. This was three years after Folsom's innovative plan was formed, and two years after it was published (Taylor, 1988).[32] By 1885 the Nightingale–Marshall opposition had become intensely partisan. Saxon Snell, formerly an enthusiastic circularist, denounced circular wards in *The Lancet* in September 1885. He declared that the first proposition for the popular 'circular system' in Britain was made in late 1878, corroborating the originality of Folsom's square 1875 scheme and Niernsee's octagonal design of later that year, at least to British eyes.[33]

Billings persuaded the Trustees against Folsom's plan in a similar spirit and thereby discredited and marginalized Niernsee and Folsom. Niernsee wrote an open letter to the *Baltimore Sun* newspaper on 7 February 1877 expressing his deep unhappiness at being omitted from the reported authorship of the JHH (Chalfont and Belfoure, 2006, p. 110). Nonetheless, Garrison records from Billings' records that Niernsee drew the hospital from Billings' sketches such that they were incorporated into Billings' report to the Trustees as approved on 15 July 1876 before Billings departed for Europe on an apparently fruitless Trustee funded reconnaissance trip and that the Boston Arts and Crafts architects Cabot and Chandler were appointed merely to dress the volumes in Queen Anne style elevations.[34] Billings credits them in his celebratory publication on completion but does not mention Niernsee who was tempted back to the construction of the South Carolina State House, thereby forfeiting his official role advising the JHH

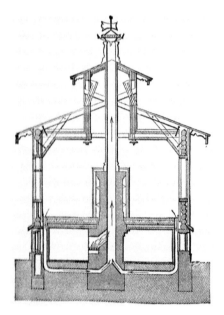

Figure 6.22
The Roschdestvensky City Hospital in St
Petersburg, 1871–72, favoured by Billings but
more reminiscent of Folsom's Common Ward
design (from Billings, 1884).

Trustees. Despite Billings' disillusionment with contemporary hospital design
abroad, environmental design ideas in both Folsom's and Billings' schemes are
strikingly reminiscent of the then recently completed Roschdestvensky City
Hospital in St Petersburg (Figure 6.22) built between 1871 and 1872, three years
before the Trustees' consultation process. Billings eventually visited it in 1881
and was impressed. The section and basement plans show a very similar
arrangement to Folsom's design, an apparently 'edge-in' and clearly an 'edge-
out' scheme, looping exhaust ducts back under the basement floor level in a
radial plan connecting to the base of an exhaust shaft. Were Folsom and Mr A.C.
Martin aware of the design in 1875? In effect, it compresses the Mass General
arrangement into a truncated, linear Nightingale plan, hence, perhaps, Billings'
enthusiasm. Extracts are located in the floor between pairs of bedheads much as
Billings' as-built scheme. How effective would the two Common Ward
propositions have been in venting away pathogens?

An indication of the comparative performance of the two ward types

Computer-based fluid dynamics analysis has been evolved recently to hazard a
prediction of the path of airborne pathogens, calibrated against a growing body
of observed data. The technique is applied here to the two schemes. A fine
mesh of cubes is defined within both three dimensional models, generating
some 3.2–3.8 million cells. The model solves standard fluid dynamics equations
at each cube interface to predict where a particle within each would move to
next, as the software calculates the natural buoyancy forces likely to develop.
This potential motion is depicted by vectors. The solutions reproduced here are
steady-state snapshots and depict a stable winter condition in which the ward
temperature is maintained at 64°F (18°C) whilst the external temperature is
39.2°F (4°C), assumptions based on the commissioning records and Billings'

own note of temperatures achieved in a built ward, 67.3–74.5°F, in the December before the opening ceremony, during which external temperatures varied between 33.3°F and 50.1°F, a mild winter period.[35] The building was performing relatively well three years after commissioning despite the intensity of human occupancy and manipulation of more than 300 manual controls.[36]

For both wards the effectiveness of the stack in drawing up air is calculated using the standard UK Chartered Institute of Building Services Engineers' stack effect equation. The driving force generated is considerable for both, delivering 12.6 ACH, twice the current UK Department of Health guidance.[37] Figure 6.23 depicts the likely path of particles in the air arriving through each inlet in the two ward types. In A, the Billings ward, the flow circulation is largely confined to the slice of the ward volume served by the particular inlet and outlet, there is little lateral mixing, whilst ward type B, Folsom's design, enjoys a swirling, mixing regime at similar velocities. Flow up the Folsom stack is particularly healthy.

Figures 6.24 and 6.25 attempt to model the flow of particles emanating from a particular prone patient's mouth. At the chimney end of Billings' ward, Figure 6.24 suggests the suction developed by the stack of the chimney is so strong that the exhaled air barely enters the main ward volume but disappears down the exhaust beneath each bed. However, the effectiveness of the exhaust diminishes markedly as one progresses southwards away from the stack. Folsom's ward maintains the swirling mixing distribution of exhaled air before

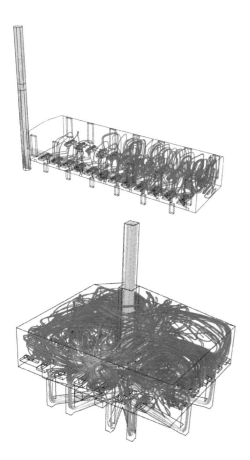

Figure 6.23
Views of the airflow structure using pathlines released from patients' mouths only for (a) ward A (Billings) and (b) ward B (Folsom). Note that the pathlines are coloured by velocity magnitude with dark blue colours denoting slow airflow currents, green highlights medium speed flow and yellow/orange signifying relatively fast airflow.

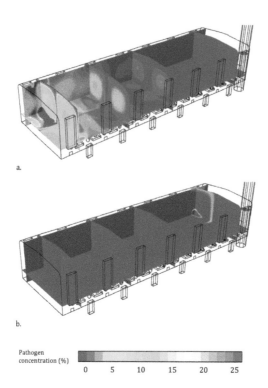

a.

b.

Pathogen
concentration (%)

0   5   10   15   20   25

Figure 6.24
Contour plots of the pathogen
concentration for (a) pathogen release
from the south end of ward A
(Billings) and (b) release from the
north end adjacent to the chimney for
a global ventilation rate of 6.3ACH.

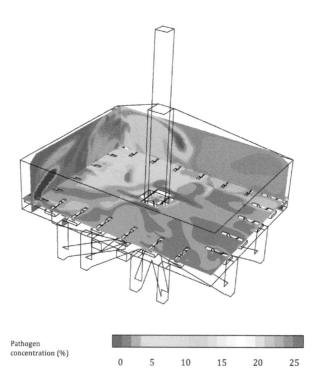

Pathogen
concentration (%)

0   5   10   15   20   25

Figure 6.25
Contour plots of the pathogen concentration ward B (Folsom) for a global ventilation rate of 6.3ACH.

it disappears down the perimeter exhausts despite healthy flows along those exhaust ducts (Figure 6.25). Is this a symptom of inadequately sized exhaust, which is unlikely, or the fundamental geometry of the ward? Certainly Folsom's own explanation, given in his bid, of a slow vortex developing in the two dimensions of the vertical plane is inadequate, the three-dimensional geometry of the ward volume dominates.

It is possible to make an approximation of the likely pathogen concentrations in each ward type and these are given for Billings' 'as-built' design in Figure 6.24. A patient lying opposite another infected patient coughing liberally would be markedly less vulnerable to cross-infection at the north end, close to the stack, some 0.1% of pathogens landing on his/her bed, whilst his or her confederate at the south end of the ward would receive some 7.8% of the source concentration. However, the spread of airborne infection longitudinally across adjacent beds is markedly low at a maximum of 0.2%.

In contrast, the airflows in the Folsom ward would appear to be much more three dimensional, achieving greater mixing. Figure 6.25 shows there is a more evenly distributed but higher exposure across the beds. Intriguingly, the peak concentration is measured in the seventh bed along from the source, here taken as the occupant of the corner bed. It reaches 12.9% at the unlucky bed but the average pathogen concentration across all 23 beds is 4.25%, creditable but more than double that of the Billings scheme. The modelling suggests a more uniformly distributed air supply would lower this risk. However, conspicuously infectious patients would have been confined to the isolating ward building. Figure 6.26 compares averaged ventilation rates across each of the two wards for increasing air change rates.

For a creditable and compliant 6.3 ACH there is very little difference between the schemes. The difference widens at 25 ACH but this is a truly prodigious rate of flow, a rate prescribed for contemporary operating theatres. This is a partial analysis which suppresses certain key variables. The model considers air turbulence resulting from airflow being disrupted at the inlets and exhausts, but not the effects of turbulence created within the ward as nurses, doctors

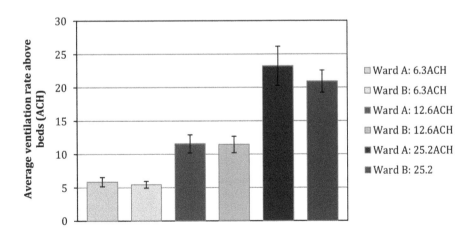

Figure 6.26
Plot of the average ventilation rates observed across all beds for both wards.

and patients move about. Recent work at the University of Cambridge BP Institute reveals that this can be significant. There is no allowance as yet for the plumes of heat rising from the human beings or other heat sources in the spaces. However, the possibility of generating naturally driven airflows at this impressive rate suggests the designs may well be resilient to hot summer periods under a sensible control regime.[38]

A recovery of lost understanding

To Folsom's disappointment and Niernsee's despair, Billings was appointed as one of the first 'project managers' for a major project in the modern sense, delivering both the building and its future organization. How could the non-architect Billings have led successive architects and contractors through the complex and precisely dimensioned design? None of Billings' sketches for JHH survive at JHH or in the New York Billings archives but the New York Public Library holds the very competent sketch plan he made of their Central Building (Figure 6.27).[39]

It is broadly to scale, dimensioned, symmetrical, with key architectural components and uses indicated. The architectural detail is clearly of little consequence.

The competition was, intellectually, a draw. The stakes were low. By 1929 a perspective drawing of the hospital site proposed much higher density arrangements, three-storey wards would soon replace Billings' buildings, indifferent to miasmatic emanations. North American post-war hospital design, liberated by air-conditioning from any concern for the prevailing climate, pursued a very different deep plan type, multi-storied and sealed, discovered subsequently to have profound energy and carbon implications. The type migrated latterly to Britain with developer–constructors well versed in the mechanism of Private Finance Initiative schemes.

However, public buildings of the relatively recent past capable of sustaining successful internal environments with minimal energy demand could well be of interest again to policy-makers and designers.[40] Both Billings and Folsom used the very stuff of construction, the architecture, to make very effective low energy buildings, coupling thermal mass with judicious ventilation. Might their environmental design principles be relevant today? Billings' commissioning

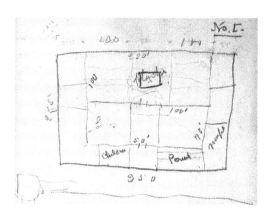

Figure 6.27
Sketch plan by Billings for the future New York Public Library. New York Public Library NYPL catalog ID (B-number): b11524053, IMAGE ID: 465480.

exercise recorded 4 US short tons of coal being consumed daily to sustain a ward in very cold weather, outside temperatures hovering around freezing point. For an occupied ward volume of 3015 cubic metres (106 464ft^3) the daily energy use would be 0.86 giga Joules (GJ) per 100m^3, the archaic metric collected by the UK Department of Health.[41] Billings' climate data for the four years 1871–74 show a five month heating season. The 1961–90 data presents an annual average of only 11 days continuously below freezing. Even if Billings' 1880s heating system was running at full pace for 120 days it would consume 103.2 GJ/100m^3 per annum, equivalent to an NHS acute city centre hospital in a temperate climate. For the whole volume, including the roof and basement voids, the figure would be a very respectable 50.67 GJ/100m^3, well within the UK Department of Health best practice guideline for new hospitals. Modern heating technology would dramatically reduce this figure.

Could this work of the last quarter of the 19th century in North America and its European antecedents possibly inform the design of the next generation of ultra low energy, low carbon hospital buildings in a changing climate? An immediate clue lies in Billings' inclusion of a borrowed figure on page 169 of *Ventilation and Heating* (1893) depicting the Mr W. Briggs' scheme for the school at Bridgeport Connecticut (see Figure 6.28), apparently built and published in the *Sanitary Engineer* on 1 December 1881 (Billings, 1893).

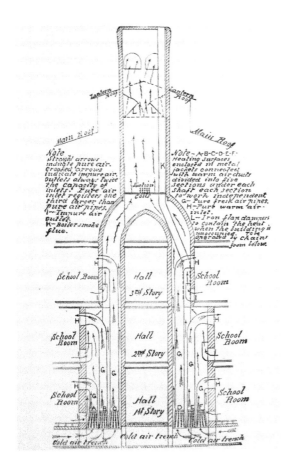

Figure 6.28
Proposed 'Vertical Section of School Building', Bridgeport School, from Billings (1884, p. 169 Figure 50).

Briggs had a contrary idea to the prevailing notion of supplying air at low level at the perimeter, to supply it in the centre of the plan two-thirds the height to the ceiling, some 8ft up at the inner corner. A platform sat over the low level exhaust and Briggs' theory was that the high level supply would cool and descend across the whole room. Briggs' figures suggest an air change rate of just under 6 ACH, the UK NHS strongly suggests 6 for hospitals, regardless of volume. Yet more interesting is its extension as a system vertically through a multi-storey school building served by multiple supply and exhaust stacks. Such diagrams are being drawn today (Lomas and Ji, 2009). Briggs supplies air through a below-ground plenum, it rises up through 'onerating' chains, apparently important when the building was unoccupied, into sectionalized flues sitting within a brick construction in which the free volume is used as exhaust. The exhaust is uncompartmented, heating coils at the neck of the stack above induce 'suction' but outlets are twice the size of inlets suggesting ample scope for reversing flows. Dampers in the termination in the horizontal plane sit within a belfry top to the stack. Briggs published good figures in the *Sanitary Engineer* on 11 January 1883 and Billings reproduces an experimental afternoon's figures, 22 December, with 50 child 'volunteers'. To our sensibilities, a variation between 65°F and 108°F may seem excessive but 59–67°F at desk level would be consistent, if spartan. Billings berates school architects for ignoring fresh air supply, no other class of building in the United States being 'in such an unsatisfactory condition' (ibid. p. 157). He reports there has been 'a good deal of growling at architects lately on the core of their neglect in sanitary matters'. He speculates on their liability and failure to grasp that carbonic acid, 'the specially dangerous impurity that is to be gotten rid of', will fall to the floor as it cools but exhausts are designed at high level. This argument is as relevant today, downdraught cooled buildings lend themselves to low-level exhaust in summer.

As noted, wholly artificial health environments are now subject to sustained attack. Ulrich's (2006) comprehensive literature review, which exposed penalties in recovery times and outcomes, was enthusiastically adopted by the UK Department of Health. Ulrich was seconded to the UK.

Where did this cocooning of hospitals start? The 1955 Nuffield Provincial Hospitals Trust Report, 'Studies in the functions and design of hospitals', includes the casual aside that operating theatres were tending to be mechanically ventilated so that ceiling heights could be reduced to enable their incorporation into new multi-storey hospitals, what were to become the 'matchbox on muffin' type, ward towers on podia of outpatient cubicles and theatres (Provincial Hospitals Trust, 1955). This suggests an element of serendipity in the evolution of the type. Did air-conditioning entrepreneurs drive this development or simply react opportunistically to a popular image of a 'modern' health environment? Fear of cross-infection without all-enveloping air-conditioning took hold. Deep plans for hospitals, close-packing clinical departments, are not going to deliver this target, quite the reverse. The Queens Hospital in Romford, UK, a PFI Project of the Year, is 50 rooms deep between one window wall and that on the opposite side of the building at ground and first floor levels.

There is a particularly urgent need for Watkins' 'transformational thinking' in hospital design. How might new and existing hospitals be reimagined as robust, passively conditioned, low energy buildings?

The potential for importing low energy ventilation and cooling strategies from other non-domestic building types into new-build hospital design was investigated by an interdisciplinary research team drawn from the universities of Cambridge, Loughborough and Leeds reinforced by medics and NHS staff, funded by the National Institute for Health Research (NIHR) (Short and Al-Maiyah, 2009).[42] The vehicle for the investigation was the invention of an outline strategy for an acute hospital to meet the Department of Health (DH) best energy target for new-build. Without assistance from renewable energy devices, the scheme did not quite achieve the DH best target of 35GJ/100m³, indicating the severity of the challenge to practitioners.

Barriers within the service and industry, perceived and real, exist to stall more passive strategies despite the failure to stem energy usage. The principle barriers include increased intensity of use, increased use of high-technology equipment, growing implementation of the single room policy, increased use of mechanical cooling in response to heatwaves and the diversion of funds earmarked for energy efficiency measures (Department of Health, 2006, p. 8). In addition, current environmental performance indicators appeared to be self-fulfilling in leading to the choice of sealed mechanical environments. More responsive adaptive comfort models as enshrined in BS EN 15251 and the ASHRAE Adaptive Comfort model have been recently included in the author's redrafted HTM 07-02. But clearly the most pervasive barrier remains anxiety about propagating airborne infection. Researchers reviewed the state of the literature as of 2009: finding there was no clear agreement on the relative significance of the airborne route available to certain pathogens and not others. The project simply concluded, on the advice of project co-investigators Dr Catherine J. Noakes and her colleagues at the University of Leeds Pathogen Control Engineering Research Centre, that all air supply and exhaust routes in the notional schemes serve one space only and do not combine, an entirely viable stiffening of the Briggs strategy, whilst low wattage fans operated by airflow sensors maintain flows.

The project identified five potential environmental design approaches to the principal hospital space types from Simple Natural Ventilation to Advanced Natural Ventilation employing controls, hybrid systems, full mechanical systems to full air-conditioning, filtering, humidifying, cooling and warming. It distributed them across the accommodation required to build a quadrant of the theoretical hospital and then a complete plan.

A five-category coding system was developed to denote type spaces by proposed environmental control strategy:

A. **Simple Natural Ventilation (SNV) all the time (opening windows).** Ventilation of outdoor air directly into the space through occupant-controlled windows. Flow of air out of the space may be through the same window, other windows or via stacks.
B. **Advanced Natural Ventilation (ANV) with passive cooling.** Outdoor air supplied via stacks fed from below ground, concrete-lined plena providing passive cooling and/or warming by season. Air leaves the space via ventilation stacks. Airflow rates are controlled by Building Management System controlled dampers at the inlet and outlet locations to each space.

C. **Hybrid Ventilation, combined natural and mechanical ventilation including Passive Downdraught Cooling (PDC) and Mixing Ventilation strategies.** Supply of outdoor air directly into the space via damper-controlled inlets. Flow of air out of the space is via exhaust stacks. During peak load (warm) conditions, fans are used to increase ventilation cooling. Passive Downdraught Cooling (PDC) as at the SSEES building, encouraging air to fall through chilled water pipes at high level, may also be used to provide additional cooling where and when necessary. In PDC mode, air leaves the space via damper-controlled openings at low level or stacks, depending on external air temperature.

D. **Full Mechanical Ventilation.** Airflow into and out of the space is driven by variable speed fans providing full control over ventilation rates, but with no mechanical cooling. The system enables heat recovery via an Air Handling Unit (AHU).

E. **Full Air Conditioning and Filtration.** Air is supplied to and exhausted from spaces via High Efficiency Particle Arrestor (HEPA) filters, driven by a central AHU controlling temperature and humidity according to the requirements of each space.

The environmental expectations of type spaces were investigated in these terms as identified within the Department of Health Activity Database, from waiting areas, patient rooms, examination, treatment and imaging spaces to an operating theatre. Having evolved space type strategies, these were aggregated into a quadrant of a hospital plan at three storeys. Figures 6.29 and 6.30 summarize a basic planning unit in which occupied spaces are gathered around the south and west facing sides of courtyards attenuated on their east–west axis below each of which is formed a shallow pre-conditioning labyrinth.

Consulting/examination/treatment rooms are arranged around an external courtyard, 7.2m × 21.6m, attenuated east–west, to offer predominantly south-facing elevations. The 'L'-shaped accommodation to the north and east of the courtyards fits into a 'slipped' tartan grid of circulation routes, directly adjacent to the courtyard on the south and west sides. These circulation routes thus enliven what will inevitably be long corridors and aid navigability. The plan yields 'dark' locations for services, support rooms, and other largely unoccupied spaces. To the north of the rooms adjoining the courtyard lies a further range of rooms, facing the next courtyard to the north across a lateral circulation route. Circulation routes are naturally ventilated directly from the courtyards. Rooms adjacent receive supply air ducted within a deep façade and exhaust back into further ducts within the façade. The façade depth shields the south-facing glazing from summertime solar gain. The inboard rooms receive supply air from the courtyard to the north across a lateral circulation route. Supply air enters the rooms through an acoustically attenuated transfer duct. The exhausts are coupled together via a lateral high-level extract duct, connecting into exhaust stacks at regular intervals. As the tartan pattern builds, envisaged on three storeys, the 7.2m and 10.8m planning module develops 14.4m deep floorplates on the north-section axis, 21.6m deep on the east–west axis. Whilst North American hospital planning tends to develop a minimum of 35m deep packets of floorplate, much contemporary UK hospital planning is achieved in 25m plan depth. The cross-section Figure 6.30 depicts supply air passing through a concrete labyrinth lined

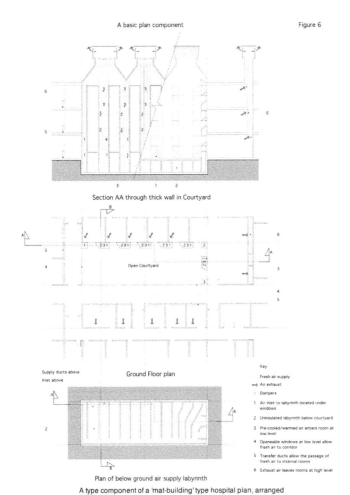

A basic plan component

Figure 6

6

5

Section AA through thick wall in Courtyard

Open Courtyard

Supply ducts above
inlet above

Ground Floor plan

Plan of below ground air supply labyrinth

A type component of a 'mat-building' type hospital plan, arranged
around a courtyard attenuated East-west.

Key

Fresh air supply
Air exhaust
Dampers

1 Air inlet to labyrinth located under
   windows
2 Uninsulated labyrinth below courtyard
3 Pre-cooled/warmed air enters room at
   low level
4 Openeable windows at low level allow
   fresh air to corridor
5 Transfer ducts allow the passage of
   fresh air to internal rooms
6 Exhaust air leaves rooms at high level

Figure 6.29
A component of a 'mat-building' type hospital plan
arranged around a courtyard attenuated east–west:
section through deep elevation, Level 1 plan and
plan of below grade labyrinth.

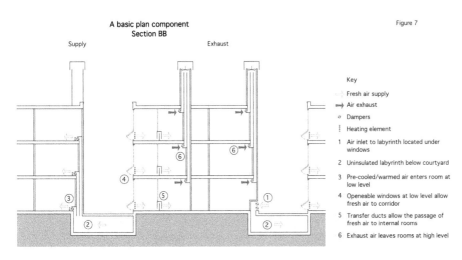

A basic plan component
Section BB

Figure 7

Supply   Exhaust

Key

Fresh air supply
Air exhaust
Dampers
Heating element

1 Air inlet to labyrinth located under
   windows
2 Uninsulated labyrinth below courtyard
3 Pre-cooled/warmed air enters room at
   low level
4 Openeable windows at low level allow
   fresh air to corridor
5 Transfer ducts allow the passage of
   fresh air to internal rooms
6 Exhaust air leaves rooms at high level

Figure 6.30
Section cut north–south through one module of the mat plan hospital.

in anti-fungicidal surface treatment below the courtyard, feeding supply ducts within a double façade, delivering to three floors. Exhaust is provided by stacks spliced onto the supply ducts below, a simple 'edge in/edge out' strategy delivering pre-cooled (or pre-warmed) supply air. Across the courtyard, air is admitted in the circulation zone and across adjacent rooms, exhausting through a central duct, connecting, as described, to stacks. All the stacks indicated are provided with fan assistance operated by flow sensors to prevent flow reversal.

Quadrant plans are assembled space by space, mapping the relative distribution of the five available strategies in five tones, the lighter the tone, the less mechanical the environment (Figures 6.31, 6.32 and 6.33).

Figure 6.31
Air supply and exhaust within ground floor, northwest quadrant:
1. Administrative Offices.
2. Consulting/Examination Room.
3. Treatment Room.
4. Operating Theatre.
5. Ancillary.
6. Plant.
7. Void.
8. Courtyard.
9. Waiting.
10. Laboratory.
11. Short-term Recovery.
12. Preparation.
13. Scrub.
14. Anaesthetic.
15. Sluice.
16. Critical Care (One Bed)/Cubicles.
17. Relatives Room.
18. Isolation Rooms (Critical Care).

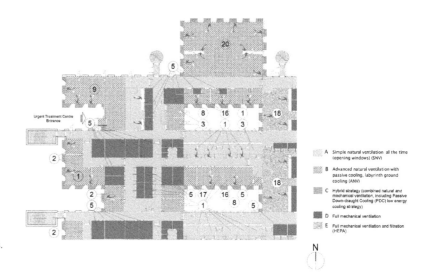

Figure 6.32
Air supply and exhaust within first floor, northwest quadrant (see Figure 6.31 for key to room types).

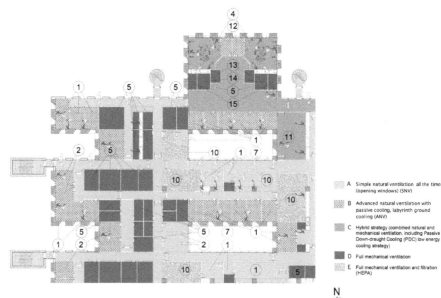

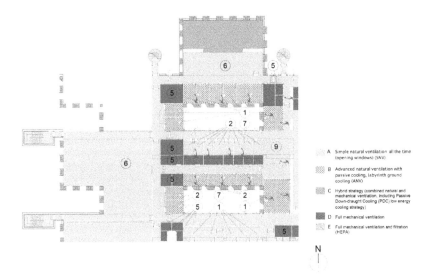

A Simple natural ventilation: all the time
(opening windows) (SNV)

B Advanced natural ventilation with
passive cooling, labyrinth ground
cooling (ANV)

C Hybrid strategy (combined natural and
mechanical ventilation, including Passive
Down-draught Cooling (PDC) low energy
cooling strategy)

D Full mechanical ventilation

E Full mechanical ventilation and filtration
(HEPA)

N

Figure 6.33
Environmental strategies, air supply and exhaust modes mapped onto a Level 3 plan for a northeast
Quadrant, including operating suites (see Figure 6.31 for key to room types).

Simulations were conducted on the northwest quadrant. It has a higher proportion of spaces requiring controlled mechanical ventilation and cooling than is the case for the rest of the design; the results gained may be pessimistic in the context of the whole hospital. The level two quadrant revolves around the operating theatres, recovery and support spaces. Operating theatres are particularly intriguing. The post-1945 theatre arrangement is both accidental, according to the Nuffield report, profligate in energy and unpleasant to work in for long periods. The Department of Health invited a particular study of the potential to design out current very high operating theatre energy consumption. Figure 6.34 shows a notional hybrid environmental strategy developed for a theatre for smaller scale, elective, routine surgery.

The proposed room is double height, north-lit but guarded to prevent solar penetration whilst providing daytime general background illumination, similar to a contemporary museum gallery. It is reminiscent of late 19th and early 20th century German and British theatres.[43] There is certainly no intention to substitute the 50 000 lux required on the operating cavity and LED lighting is available as colour rendering issues subside. The theatre is supplied with pre-cooled air at the perimeter from three sides. A portable HEPA filter, equipment in general use, delivers air across the operating cavity.[44] Instruments are placed in an open-sided, glazed cabinet receiving a constant stream of HEPA filtered air in preparation for use. The scale of the problem is dismantled into containable sites. The NHS Activity Database describes a relatively low space (3m ceiling height) maintained at 22°C in winter and 20°C in summer. Discussions with surgeons, hospital administrators (via the NIHR project Sounding Panel) and the National Patient Safety Agency suggest, anecdotally, a tendency to overheat, although deeper, more invasive operations do require higher temperatures to ensure patient safety, because hypothermia is a risk. High air-change rates are called for, in part, to dislodge micro-organisms and skin

NHS Environmental Requirements

Proposed Environmental Design Strategies
Space Type 6: Operating theate

Room Type	Operating theatre	
	Maternity	
Room Layout	Area (m2) = 40.00	
	Height (m)= 3.00	
Room Environmental Data:		Information Sources:
Design Temperature (winter)	22 Deg C	DoH-Activity Database-March,2008
Design Temperature (summer)	20 Deg C	
Design Temperature (F/C)	62-80 Deg F	AIA (2001)
Minimum Fresh Air Exchanges per Hour	15 ACHR	ASHRAE Handbook (1999)
Minimum Total Air Exchange per Hour	15 ACHR	ASHRAE Handbook (1999) AIA (2001)
Relative Humidity (% RH)	40-60%	DoH-Activity Database-March,2008
	45-55%	HTM 03-01
Pressure Relationship to Adjacent Areas	Positive	DoH-Activity Database- March, 2008 ASHRAE Handbook (1999)
Lighting Intensity- General	500 lux (Floor)	DoH-Activity Database-March,2008
Lighting Intensity- Local	50,000 lux (Operating cavity)	DoH-Activity Database-March,2008

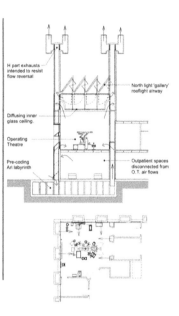

Figure 6.34
Ventilation modes
mapped onto Level 1 of a
complete notional plan of
the theoretical hospital.

squamae from settling in the open wound. The standard fully mechanical approach is familiar, a low ceilinged, windowless space within which air is discharged through a HEPA filter within a hood down onto the operating area at some 25–45 or more ACH and exhausted from at high level and part recirculated through the filter, with fresh air make-up. Papers critical of this custom and practice model have been published (Pereira and Tribess, 2005). The research team questioned the need to change all the room air at the specified rate to secure the necessary condition within the relatively small operation site. NHS microbiologists explained to the NIHR research team the advantage of a constantly turbulent environment in protecting instruments laid out in preparation on trolleys placed within the general volume of the room. Something needs to replace that.

The component plan elements were then developed into a full 35 000m², 180 bed, acute hospital and similarly coded (Figures 6.35, 6.36 and 6.37).

Figure 6.38 depicts the network of below-ground labyrinths required to modify air supply temperatures.

It is immediately evident that not all of each plan area is toned black as requiring a sealed artificial environment, although this is the industry standard approach. The plans indicate the tartan configuration of the floorplan, which is used to accommodate the principal departments and ancillary functions of a small to medium-sized acute hospital. As discussed, 'compact' planning is favoured by clinicians and managers over shallow, linear planning because it offers a greater frequency of closer medical adjacencies. However, this theoretical deep plan will be liberally punctured to offer fresh air, using garden courtyards as day-lit navigation landmarks. There is no need for the courtyards

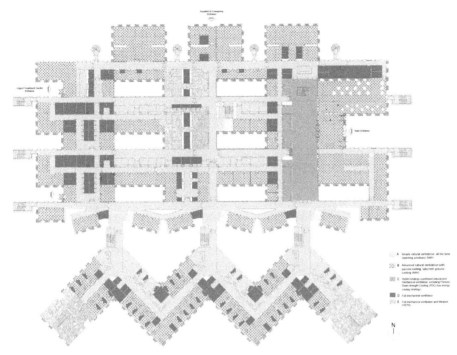

Figure 6.35
Level 2 of the notional hospital.

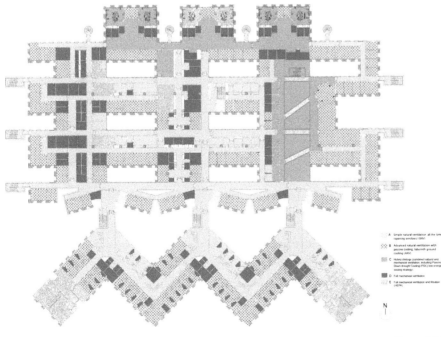

Figure 6.36
Level 3 of the notional hospital.

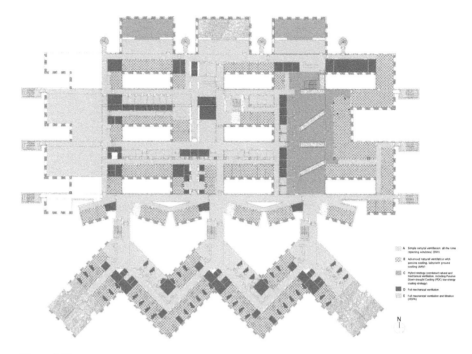

Figure 6.37
Plan of below-ground labyrinths beneath the notional hospital ground slab.

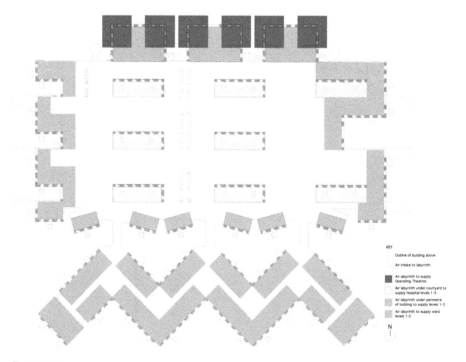

Figure 6.38
Relative distribution of the environmental strategies across the notional hospital by gross internal floor area.

to be similar or for the plan to be orthogonal throughout, although extended west-facing elevations are avoided to mitigate solar gains. Again the plan is coded to indicate the distribution of the five environmental strategies to cope with the present-day climate of the south of England, which may become the future climate of the English Midlands, as the simulations indicate. The pie chart Figure 6.39 indicates that some 70% of the plan may be naturally conditioned in some way, a further 11% may be hybrid environments and only 19% sealed and mechanically conditioned. Internal heat loads are estimated but here the team reports a serious dearth of data.

A perspective section gives some impression of the types of space generated, ceilings are appreciably higher than the norm with exposed concrete soffits to calm temperature change (Figure 6.40).

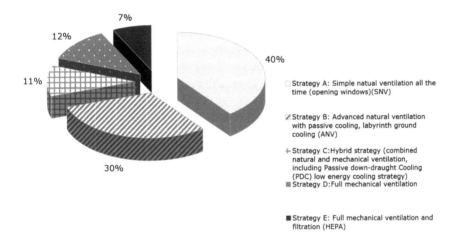

Strategy A: Simple natual ventilation all the time (opening windows)(SNV)

Strategy B: Advanced natural ventilation with passive cooling, labyrinth ground cooling (ANV)

Strategy C:Hybrid strategy (combined natural and mechanical ventilation, including Passive down-draught Cooling (PDC) low energy cooling strategy)

Strategy D:Full mechanical ventilation

Strategy E: Full mechanical ventilation and filtration (HEPA)

Figure 6.39
Relative distribution of the environmental strategies across the notional hospital by gross internal floor area.

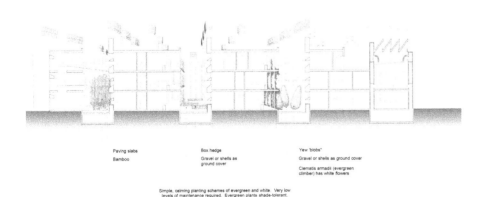

Paving slabs
Bamboo

Box hedge
Gravel or shells as ground cover

Yew 'blobs'
Gravel or shells as ground cover
Clematis armadii (evergreen climber) has white flowers

Simple, calming planting schemes of evergreen and white. Very low levels of maintenance required. Evergreen plants shade-tolerant.

Figure 6.40
Perspective section through the notional hospital looking west.

Computer simulations were carried out to investigate the likely comparative energy consumption of the proposed design, using Integrated Environmental Solutions software (IES VE, version 5.8.2). Simulations were carried out based on current (CIBSE 2005) and predicted climatic conditions for the years 2020, 2050 and 2080.

A model was built comprising different proportions of each of the space types determined by the relative proportion of volumes within each of the five environmental strategy types. These included high heat gain spaces requiring continuous mechanical ventilation and cooling, the spaces requiring mechanical ventilation due to their isolated location within the core of the building and spaces proposed to be SNV or ANV. The spaces modelled included both continuously occupied spaces, such as patient rooms and wards and the high dependency unit, and spaces only occupied during the daytime such as consulting rooms. The rooms modelled were distributed over all floors with a variety of orientations.

The model results suggest this outline scheme is a $38GJ/100m^3$ building against 2005 climate data, 38.5GJ by 2020 but reducing thereafter to 37.6GJ in 2050 and 36.7GJ in 2080. Renewable sources could then make a meaningful contribution once the fundamental demand reduction is achieved. These results suggest a near halving of typical achieved figures. As the study from which they are taken was calibrated against energy-use predictions based on current hospital designs and construction, it is clear that significant energy savings can be achieved. However, the exercise reveals that delivering the lowest best target of $35GJ/100m^3$ will be very demanding. Medical equipment heat loads will be required to be significantly reduced at source. There is little evidence that this is a procurement priority.

Could renewable energy technologies solve this conundrum, restricted to those technologies which do not burn fossil fuels on-site? The base case prediction on 2005 data of 376.7 kWh/m^2 yields a total energy consumption of 467 485 kWh. Even a 10% contribution of 46 749 kWh requires a total of 234 wind turbines with a 3.2m blade diameter, rated at 1.5 kW and assuming an average wind speed of 2m/s. This is infeasible in development control terms alone. Environmentally responsive design is much more effective.

Cost consultant Davis Langdon AECOM report that capital construction costs of £3448/m^2 compare with the then average new-build hospital construction cost of approximately £3300/m^2. Figure 6.41 graphs comparative energy costs as net present values over 40 years, revealing the value of climate resilient design.

Although the avoidance of future expenditure does not figure in public sector business case calculations it is important to note that the 'business as usual' base hospital must be refitted at 20 years with additional cooling capacity and thereafter becomes increasingly more expensive to operate than the notional scheme.[45] This prospective redundancy deriving from poor resilience is endemic to 'business as usual' non-domestic building types.

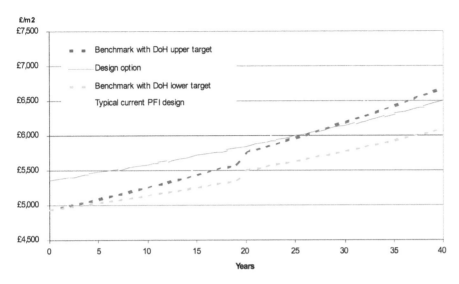

Figure 6.41
Energy costs compared to capital costs over 40 years for the notional design, hospitals achieving the Department of Health upper and lowest best energy targets and a typical PFI hospital from Davis Langdon AECOM's building cost database (commissioned from DL/Aecom).

Notes

1. Results from the September 2010 UK National Health Service census. 1.4 million staff, 5% of the UK workforce and one million patients every 36 hours. Online at www. ic.nhs.uk/web!les/publications/010_Workforce/nhsstaff0010/Census_Bulletin_March_ 2011_Final.pdf; 22 March 2011 (accessed 19 January 2011).
2. The Climate Change Act 2008, Office of Public Sector Information (OPSI), UK Statute Law Database.
3. NHS Sustainable Development Unit, NHS England CO_2e footprint 1990–2020 with Climate Change Act targets. See www.sdu.nhs.uk (accessed 20 April 2015).
4. DEFRA, *Climate Change, The UK Programme, Department for Food and Rural Affairs*, March 2006, p. 202.
5. Recorded data and performance benchmarks are given in 'Making energy work in healthcare (HTM 07-02)', Short, Guthrie, Soulti, MacMillan, 25 March 2015. See www.gov.uk/government/publications/making-energy-work-in-healthcare-htm-07-02 (accessed 12 October 2016). Another insight into hospital energy performance comes from 'An Analysis of Display Energy certificates for Public Buildings, 2008–2012. A Report to the Chartered Institute of Building Services Engineers Oct 2013' prepared by Sung-Min Hong and Philip Steadman, Energy Institute, University College London. A sampling of Design Energy Certificates yields a median figure of $423\,kWh/m^2/yr$ for clinical and research hospitals or $243\,kWh/m^2/yr$ where the building environmental control is all electric.
6. Watkinson, A. Keynote talk, Adaptation and Resilience to Climate Change Conference, Oxford, 6 April 2011.
7. Through work funded by the National Institute of Health Research (NIHR), 'Design strategy for low energy ventilation and cooling of health buildings' NIHR Project B (06/03) and the Department of Health and the Engineering and Physical Sciences Research Council (Design & Delivery of Robust Hospital Environments in a Changing Climate (De^2RHECC) Council Ref. EP/G061327/1).

8. *Edinburgh Evening News*, 24 July 2014, 'Soaring heat at maternity ward "intolerable"'. Seewww.edinburghnews.scotsman.com/news/health/soaring-heat-at-maternity-ward-intolerable-1-3485678 (accessed 20 April 2015).

9. *Hospital Plans: Five Essays Relating to the Construction, Organization and Management of Hospitals, Contributed by their Authors for the Use of the Johns Hopkins Hospital of Baltimore* (New York: William Wood and Co., 27 Great Jones Street, 1875). The volume consulted is that presented to the British Medical Association Library on publication, imported into Britain by Sampson Low, Marston, Low and Searle, Publishers and Importers.

10. Brieger explains the then current understanding of antisepsis in North America. Lister's 'system' for protecting patients during and after surgery was being discussed in America as the JHH was being conceived. Brieger points to articles in the *British Medical Journal* by T. Spencer Wells, 'Some causes of excessive mortality after surgical operations' (1864, Vol. 2, 384–388) and in *The Lancet* by John Eric Erichsen 'American Surgery' (1974, Vol. 2, 717–720).

11. The Last Will and Testament is reproduced as Appendices I and II in Thom, H.H. (1929) *Johns Hopkins: A Silhouette*. Baltimore: Johns Hopkins University Press. Thom was Johns Hopkins' great-neice. A more accessible facsimile edition of Thom's biography is available, edited by the Archivist of the Sheridan Libraries, James Stimpert: *Johns Hopkins: A Silhouette* (Baltimore, Johns Hopkins University, 2009).

12. Maryland was an unhealthy state with one of the highest state mortality rates. Baltimore enjoyed very little mains drainage and was backwards in relation to other major East Coast cities. It is surprising that the disposal of excrement was not the primary concern of the Trustees, particularly as it was thought by Pettenkofer and his followers to be the principle constituent of miasmas, as we shall see shortly. A 1936 paper recording the history of the city's sewerage system by a University of Maryland fraternity aspirant, Alexander Lopata, survives: see Lopata, A.A. (1936) *History and Development of the sewerage system of Baltimore up to 1916* (records of Phi Mu, Special Collections, University of Maryland Libraries. See http://ia700603.us.archive. org/6/items/HistoryAndDevelopmentOfTheSewerageSystemOfBaltimoreUpTo1916/ LopataAlexander-univarch-014627.pdf (accessed 14 November 2012). Lopata reports that progress was stalled by political infighting, the cost and pressure from the Chesapeake Bay oyster farmers who typically recovered 20 million bushels of oysters annually and who believed a centralized sewerage system would destroy their businesses. Effective political lobbying stalled a succession of various commissions' reports so that, by 1887, the city had invested – chaotically – more than $4 million on 33 miles of drains to little effect, 'the Basin had become a gigantic cesspool with a very objectionable odour' (Lopata 1936).

13. King, 'Letter addressed to the Authors of the Essays', reproduced in *Hospital Plans...*, xii, refers to ventilation, heating, light and sunshine as 'curative agents', almost exactly as Professor Roger Ulrich and colleagues at Texas A&M University have been insisting in their many publications (see for example, Ulrich *et al*., 2008, accessed via https:// smartech.gatech.edu/bitstream/handle/1853/25676/zimring_HERD_2008_researchlit review.pdf?sequence=1 on 6 February 2013).

14. The author consulted the scanned facsimile published by Bibliolife, digitized for the Microsoft Corporation by the Internet Archive 2007 from University of California Libraries. A concise but illuminating portrait of Billings as a figure of great esteem in American medical circles is given by J. Fraser Muirhead in *The New England Journal of Medicine*, 'Doctors Afield, John Shaw Billings' (Vol. 268 No.14, 778–779). Dr John Cameron records Billings as having been tasked with dismantling the Unionists' field hospitals, which enjoyed a good contagion record, and then with inspecting all 27 less health-bestowing Marine Hospital Service hospitals in 1869, experiences which provided his expertise in hospital construction, see Cameron, J.L. (2001) 'Early Contributions to the Johns Hopkins Hospital by the "Other" Surgeon: John Shaw Billings.' *Annals of Surgery*, **234**(3), 267–278. By 1870 Billings was promoting the French style pavilion hospital in Circular No. 4 from the Surgeon General's Office, 'A Report on Barracks and Hospitals' and subsequently Circular No. 8 'A Report on the Hygiene of the United States Army' appeared. Billings was a commissioned officer

throughout the JHH exercise, cited for exemplary surgical performance at Gettysburg, retiring in 1895. It appears that Billings was, in effect, seconded to the JHH exercise at the Army's expense. Dr Cameron paints a rosy picture of the outcome of the competition and Billings' relationship with Niernsee. However, as we shall see, the proffered schemes were by no means similar. Billings could clearly be officious and irritating, even the archivists at JHH refer to his bizarre high-pitched voice, his peremptory treatment of his own family and colleagues and his blanking of any contributions made by women.

15. Godwin's editorial appeared in *The Builder* on 2 November 1878, Volume 36, 1140–1441.
16. Marshall (1878) discussed in Taylor (1988).
17. John Rudolph Niernsee (1814–85), born in Vienna, emigrated to the US in 1837, apprenticed to Benjamin Latrobe, enlisted as a Confederate army major by default, his then clients were Confederates so there was, perhaps, little empathy with Billings. He practised with James Neilson (1816–1900), a similarly cosmopolitan architect, in a spirited Greek Revival. Niernsee had lived in England and Brussels. See Chalfant and Belfoure (2006).
18. A full discussion of the then competing theories of the spatial spread of disease appears in Chang and Jackson (2007).
19. Morabia reports on how Pettenkofer (1818–1901) envisaged a cycle in which the cholera germ was deposited on the soil in excrement, transformed in the soil into a cholera miasma and released through the putrefaction of vegetal matter into the atmosphere. If inhaled, cholera would follow (Morabia, 2007). Low-lying damp soil was thought to be particularly dangerous but the water table had to be below the surface to enable the miasmic reaction to occur. In this theory Pettenkofer married the 'Environmental' and 'Contagion' theories of disease transmission but, in denying the role of water in transmitting cholera directly, he inadvertently failed to prevent many of the 8606 cholera epidemic deaths in Hamburg in 1892, whilst the contagionist Robert Koch saved the population of neighbouring Altona through filtration and quarantine. Pettenkofer retired unceremoniously in 1894. The JHH project spanned this period of conflicting theory and evidence. Billings was particularly interested in Pettenkofer's deduction that the level of carbonic acid is constant vertically to the outer limits of the atmosphere. He set that concentration at about 400 parts per million, measurable using Pettenkofer's portable apparatus, which Billings illustrates (1883). Four hundred parts per million atmospheric carbon dioxide concentration prefigures the Intergovernmental Panel on Climate Change (IPCC) 'disaster scenario'.
20. Billings' theory of contagion was probabilistic. He had privileged access to the national statistics as they were emerging. Having been Medical Statistician of the Army of the Potomac and appointed to the Office of the Surgeon General, he directed the collection of Vital Statistics within the Census for 20 years, the exercises in 1880 and 1890, and lauded as 'perhaps the ablest vital statistician the country has ever produced' (Willcox, 1933). Billings' own paper on the subject (1883) is reproduced in Brieger (1972). The Vital Statistics record historic mortality rates back to 1800. Not all states were registered but Maryland had registered by the time of the first thoughts for the Johns Hopkins Hospital.
21. See explanation of tuberculosis transmission in internal spaces in Noakes *et al.* (2006).
22. The idea that a public building could defend itself from attack by dangerous and pervasive invisible vapours simply through its architectural configuration (Corbin 1996) was popular with Enlightenment architects: 'The building must be designed so as to separate putrid exhalations from currents of fresh air, in the same way that currents of fresh water had to be divided from used water. The idea was that the shape of the building itself would ensure satisfactory ventilation, thus rendering traditional methods redundant.' But what were the traditional methods. Corbin explains his interpretation derives in part from Francois Beguin, in Foucault *et al.* (1979, p. 40), '[Planners aimed] to use nothing but architectural resources to capture the air, cause it to circulate, and expel it'. This concept, embodied in the JHH designs, could not be more relevant today.

23. Miasmatist Dr Casper Morris of Philadelphia would rely on 'ventilating fireplaces' after Captain Douglas Galton, perhaps supplemented by cavity flues; Dr Stephen Smith, New York surgeon and designer of the Roosevelt Hospital, on simple cross-ventilation for which he provides a full page but dully repetitive diagram; and Professor Joseph Jones of the University of Louisiana in his declared anxiety to avoid controversy, recommends an unlimited supply of air as the 'cardinal consideration' that should govern the construction of hospitals, entering the wards through doors and a generous supply of windows augmented by piping air to each bed, a minimum of 72 square inches free area per patient.

24. Morin, as noted in Chapter 5, was a high-ranking military officer whilst serving as the first Professor of Mechanics at Metz, an impeccable provenance perhaps for Surgeon Major Billings.

25. Matthew Arnold's 1869 'Culture and Anarchy' lecture at the Sheldonian, Oxford University, discussed by Collini (2006).

26. Article anonymous, 'Hospital Construction – Wards', *The Builder*, 25 September 1858. Thought now to have been written by Scottish surgeon John Roberton, see King (1966).

27. See Chapter 3, 'Architecture and Ventilation', sub-section 'Ascending and Descending Atmosphere' in Reid (1844, p. 85). Reid proposes in para. 156 what has come to be known as 'displacement ventilation' in the era of air conditioning, '… in this country, air, vitiated by respiration, tends invariably upwards … vitiated air collects above in any apartment … and that an ascending movement should be given to air which enters…'. He maintains this view throughout *Ventilation in American Dwellings*, with a graphic depiction in Figure 2, p. 19. In Chapter 3 he considers the problem of a radiantly heated room in which the heat escapes too rapidly through the outlet at ceiling level as shown in Figure 32. He argues in Figure 33 that by moving the outlet to floor level the air quality will deteriorate and using pink and blue washes he suggests that the displacement layer between fresh and vitiated air will lie diagonally across the height of the room. Figure 34 proposes a perforated zinc floor above a below floor plenum discharging warmed air evenly across the plan providing a perfectly ventilated space, very much as appeared in the Isolation Ward at JHH. In Figure 37 he depicts a safe ventilation regime for the domestic bedroom of a yellow fever patient preventing the spread of the disease, a high-level outlet discharging into a duct falling to floor level and then doubling back vertically to form a stack, the J-tube configuration used by both Billings and particularly Folsom to strengthen the stack effect.

28. The handwritten 'Thermometrical Indications' table is in Billings' own personal archive at the New York Public Library (NYPL), see J.S. Billings, single manuscript page, February 1886, John Shaw Billings Papers, Box 48, Manuscripts and Archives Division, The Brooke Russell Astor Reading Room, NYPL, New York. The authors are particularly grateful to Weatherly Stephan, Manuscripts Specialist at NYPL for suggesting Box 48 may be revealing. The handwriting is perhaps too well formed for Billings' own hand, the capital letters a little too florid, unless Billings was trying to make a polished document, but the subsequent table of airflow velocities is very much in his hand.

29. See 'Johns Hopkins Hospital Ventilation per Ward and Service Building', handwritten manuscript apparently in Billings' handwriting in John Shaw Billings Papers Box 48 at NYPL. He calculates boiler sizes for a steam temperature of 230°F at 26lbs of pressure, assuming the temperature in the wards should be 70°F and that the average summer external temperature would be 70°F and in winter, 40°F.

30. Sir Vincent Zachary Cope spotted this correspondence in the British Museum and reminded readers of the *Medical History*, News and Queries pages; see Zachary Cope, *John Shaw Billings, Florence Nightingale and the Johns Hopkins Hospital* (*Medical History*, News Notes and Queries, medhist00185-0086.pdf, pp.367–368). For Sir Vincent Zachary Cope see Hamilton (2004), www.oxforddnb.com/view/article/30968 (accessed 28 July 2014). Cope wrote, 'It is not generally known that Billings, in his endeavors to perfect the plans for the hospital, travelled on the continent of Europe and visited England, where he consulted Miss Florence Nightingale. Among the Nightingale manuscripts at the British Museum the following two letters from Dr. Billings to Miss Nightingale are included. They sufficiently explain why they were

written and indicate the response the first letter obtained from Miss Nightingale. The first letter is dated 23 October 1876 and runs as follows:

> I take the liberty of sending to you through Mrs. Wardroper, who has very kindly consented to forward the package, a set of sketch plans for a hospital to be constructed at Baltimore, U.S. under the terms of the Johns Hopkins Trust. With this hospital are to be connected a training school for nurses, a convalescent hospital, an orphan asylum and some other things. I am now on my way to the Continent and shall return towards the end of November, spending a day or two in London on my way to the U.S. Knowing as I do the great interest you take in such subjects I shall consider it as a great favour if before my return you will, if your health permits, examine these plans and the two pamphlets which accompany them and let me know what you think of them. They are only sketch plans and I desire criticism before going further. I am with great respect very sincerely yours John S. Billings'

The second letter is dated 4 December 1876 and shows to what good purpose Miss Nightingale examined the plans:

> I have the honour to acknowledge the receipt of your letter of Dec. 2. enclosing 12 sheets of notes on the Johns Hopkins Hospital plans, and I desire to express my sincere personal thanks for this favour. Your remarks shall be laid before the Trustees as soon as I return to America and I feel sure that they will be greatly interested in and influenced by your criticisms.
>
> The labour and expense of conducting an hospital built in this general plan I fully appreciate as also the complications which arise in trying to cut off all service rooms from the ward. I infer from your note that you may not have received a copy of a book containing five different plans for this hospital published by the Trustees about 9 months ago. I am quite sure that a copy was sent to you. The first of the plans in that book I prepared and in it I think many of your objections were avoided. If you have not seen it I shall make sure that a copy is sent to you.
>
> That at first but one or two blocks should be built is precisely what I think. But I will not attempt now to comment upon your notes nor indeed do I think it probable that I should do otherwise than agree with them. The copies of plans sent you were intended to be kept by you and other plans will be sent to you hereafter. If you have not the volume containing the five plans and will let me know it I will have it sent. For the next 9 days my address will be care of Brown Shipley and Co, Founders Court, Lothbury, London. After that Surgeon Generals Office, Washington D.C. U.S.A. I leave London by the 5th inst for Leeds, Edinburgh etc and sail on the i6th. Again thanking you for your criticism.
>
> Miss Florence Nightingale
> very respectfully and truly yours John S. Billings, M.D. U.S.A.

31. Thompson and Goldin (1975) reproduce an interior view of a square ward of 1845 in the Bullfinch Building as Figure 113, p. 104. Folsom was following what had become a standard formula for ward planning at the Massachusetts General Hospital.
32. There was a European manifestation of the circular, i.e. non-linear, plan which is closer in date to Folsom's submission. Baeckelman had completed Antwerp's City Hospital entirely comprised of circular ward pavilions in 1878. It was not published in *The Builder* until 7 July 1883, but as a major investment in city infrastructure the scheme may have been 'in the ether' during its conception. Billings does not record a visit, he travelled directly from London on 30 October 1881 to Amsterdam, arriving in Leipzig on 2 November although it is just conceivable that he turned southwards before connecting to 'quaint and strange' Cologne en route. Given his agenda, he may have wished to suppress the then apparent success at Antwerp, even in letters home to his wife. Antwerp, however, became controversial, condemned by the Council of Public Hygiene in Brussels in 1875.
33. See Saxon Snell (1885), reporting his talk at the Congress of the Sanitary Institute of Great Britain held in Leicester on 24 September 1885. Saxon Snell declared Antwerp

was a 'failure', reminding readers that Nightingale had decreed too few patients in a ward would lead to a collapse of discipline. Saxon Snell showed only 22 beds could be accommodated in a circular plan of the same area as a 30 bed Nightingale ward at the same density. His Figures 1 and 2 purport to prove this and demonstrate that circular wards are problematic in achieving effective cross-ventilation, the 22 opposing windows, as opposed to 34 in a parallelogram-shaped ward, would be 60–70ft apart, whilst in a 30 bed version, windows would be an alarming 87ft 9in apart. The nurses in the centre was at the epicentre of bad air. Quite simply circular wards were profligate in cost, demanding more wards, more nurses and more cleaning.

34. Longfellow is thought to have drawn the scheme, at least in part, between his MIT education and his departure for Paris in spring 1879; see Henderson Floyd (1974, p. 38). The irony may be that the relatively vast and unfeasibly attenuated windows Shaw provided for what Saint terms the Painters' Houses of the mid 1870s, Nos. 8 and 11 Melbury Road, for racy clients with slightly questionable practices, informed the JHH ward building elevations as much as any other examples of the Queen Anne with perhaps the immense brick stacks of the Queen Anne 'palazzo' Lowther Lodge of 1873–75. The inspired act at JHH was the transformation of smoke stacks into conduits to drive natural ventilation.

35. The author is very grateful to Dr Alistair Fair for sharing this archival discovery of post-occupancy measurement made at the JHH and discussed in his paper (Fair 2014). It is interesting to compare this post-occupancy snapshot of conditions which one imagines was more diagnostic than experimental, with the original commissioning record Thermometric Indications held by the New York Public Library. Although the temperature range is wider, the wards were clearly reasonably resilient.

36. Both buildings required much manual adjustment and 'trimming'. This appears to have been the responsibility of nurses and perhaps orderlies at the mechanical lever interfaces provided, one set for every two beds in the common wards and one set per single room, a huge number of control points. We count 44 on the main floor of a common ward alone, perhaps 400–500 control points in the first phase alone. At five potential positions for each valve, there would have been 27^5 possible combinations of control settings, not including the ceiling shutters, which is over 14 million. One wonders how long it took staff to be confident in 'setting' and then regulating the ward environment, how long before, to use Shove's developed concept of 'social practice', staff became 'willing carriers' of unusual and unprecedented interactions with an advanced environmental strategy, a social practice we may well need to recover to empower occupants of a future generation of more naturally conditioned significantly lower energy buildings. See Shove *et al.* (2012).

37. See UK Department of Health (2007), *Health Technical Memorandum 03-01: Specialised Ventilation for Healthcare Premises* (London, HMSO, Department of Health).

38. The research team is seeking additional funding to rework the modelling under summer conditions. This is potentially a rich and revealing exercise.

39. For the 'Billings sketch' see New York Public Library NYPL catalog ID (B-number): b11524053, IMAGE ID: 465480, Shelf locator: MssARC RG10 5928. b. 34 f. 21. Construction Correspondence – William Ware, re: Billings's original sketch of the Central Building Feb–May 1909.

40. The *ASHRAE 170 Natural Ventilation Task Group* was (2014) reviewing the evidence base for the safety of natural ventilation in clinical environments with a view to issuing new guidance for designers in the United States.

41. Base data on energy output of coal varies depending on source, but here we adopt 7.2GJ/ton as recorded by the US Energy Information Administration (eia). See www. eia.gov/energyexplained/index.cfm?page=coal_use (accessed 28 July 2014).

42. National Institute for Health Research (NIHR) funded research project 'Design strategy for low energy ventilation and cooling of health buildings' (NIHR B (06) 03). The project outcomes are described in Short and Al-Maiyah (2009).

43. Des Neues Stadtischen Krankenhauses, Nurnberg 1898, for example. Very interesting research awaits into the transformation of theatres from anatomical amphitheatres, as

at JHH in the 1880s into these handsome, airy, well lit rooms and their likely environmental performance.

44. Information on portable HEPA filters can be obtained from: www.toulmeditech.com/en/home/home/index.html (accessed 12 October 2016).

45. This exercise should be seen as a scoping exercise for a series of more detailed investigations and certainly not as a blueprint for a new hospital. The project won the RIBA President's Commendation for Outstanding University-located Research 2009, for the project 'Design strategy for the low energy ventilation and cooling of health buildings'. The judges' citation reads, 'Its originality lies in the adaptation and development of existing knowledge to meet the varied and demanding requirements of large healthcare buildings. The work is highly significant, addressing strategies to meet the very demanding energy and carbon reduction targets set for the health care sector.' The work is cited in the 2012 national Climate Change Risk Assessment (CCRA) and DEFRA's National Adaptation Programme (NAP).

7

Mediterranean climates

Figure 7.1
'Caprice architecturale' originally attributed to Leonardo Coccorante (1680–1750). A mythical coastal landscape set in the southern Mediterranean, attributed to a Neapolitan artist and probably intended for English connoisseurs enjoying the Grand Tour. The painting is heavily varnished in the manner of Claude Lorraine to convey the characteristic balmy evening atmosphere sustained by the hot radiant temperature of the stone surfaces after a day of intense sun (author's collection).

Wherever the Air is extremely rarify'd by Heat, the neighbouring cooler Air rusheth in to restore the Balance. The nights in hot Countries are often very cold, and upon that Account extremely dangerous to the Health of such as expose themselves to it.

(John Arbuthnot, 1733,
Concerning the effects of air on human bodies, p. 85)

The sun, still far from its blazing zenith on that morning of the 13th May, was showing itself the true ruler of Sicily: the crude brash sun, the drugging sun, which annulled every will, kept all things in servile immobility, cradled in violence and arbitrary dreams.

(G. Di Lampedusa, 1958, *Il Gattopardo* ['The Leopard'], p. 48)

... the range of projects Di Fausto completed in both settings ... (Libya and the Dodecanese) ... attests to Italian modernism's engagement with contextual idioms in the making of colonial architecture and urbanism. Unlike many of the public structures built during the French colonial campaign in North Africa, Di Fausto's built and unbuilt projects refine the vagaries of traditionalism into a medium that is difficult to categorize. His architecture stands between histories. It is neither modern nor traditional ... the multiple temporalities of the Mediterranean basin saturated Di Fausto's architecture.

(Sean Anderson, 2010, *The Light and the Line: Florestano Di Fausto and the Politics of Mediterraneità,* abstract)

Hotter dryer climates, the southern Mediterranean (Köppen's C type regions: 'warm temperate' tending to B 'arid') would seem to present an intractable challenge to passive design. Chapter 2 presented its protagonists' claim that air-conditioning has revivified enervated populations and economies in these regions and is essential to continue their induction into civilized Western systems of commerce and all the benefits and comforts that flow into a society with that. This, however, is not the case. It is possible to draw the temperature within a building below the peak shade environmental temperature if the building is designed to achieve this. In the preceding chapters this phenomenon has been recorded in hot spells in the English Midlands and in the urban heat island of Central London. Similar design strategies can be successfully transferred to the hotter climate of the central Mediterranean. An example is presented: a zero carbon industrial building for an export brewery in Malta, Simonds Farsons Cisk, more popularly known as Farsons. The provenance of the brewery's name had a bearing on the design of its new building. The proposition here is to repeatedly pre-cool a heavy masonry structure at night to increase its resilience to the following day's heat, to 'beat the climate' by some 12 hours or perhaps multiples of 12 hours.

Thirty years of climate data for Malta's Luqa airport reveal the latent potential for a successful night cooling strategy. Mean monthly values, as provided by the US National Oceanic and Atmospheric Administration (NOAA) for this location show a high temperature of 28°C in May, 32°C in June, 36°C in July, 35°C in August and 32°C in September, which correspond with mean monthly values low temperature for those months of 10°C, 15°C, 18°C, 18°C and 16°C (Table 7.1). The designers' hourly analysis of the preceding 11 years of Luqa data showed that significant diurnal variations occurred frequently enough to hold a night-time ventilation cooling strategy. Figure 7.2 reveals, perhaps surprisingly, that although IPCC climate zone 13 is generally predicted to enjoy a considerable change in diurnal temperature range, the central marine modified Mediterranean climate is likely to retain its current high daily variation with little change. The strategy described here will be effective for some time to come. Köppen specifically designated Malta and Sicily Csa type, 'warm temperate with hot dry summers', and the recent updating of the original map with IPCC

Table 7.1 The data used in devising the strategy for Farsons Brewery. US National Oceanic and Atmospheric Administration (NOAA) data for Luqa airport 35° 51′ N NOAA Station ML16597, 1961–90, originally gathered for the World Meteorological Organization (WMO), then processed by the National Climatic Data Center (NCDC), and finally provided by the National Oceanic and Atmospheric Administration (NOAA) via NOAA Global Climate Normals 1961–90.

		Jan	Feb	Mar	Apr	May	June	July	Aug	Sep	Oct	Nov	Dec	
0101	Temperature mean value	c:	12.2	12.4	13.4	15.5	19.1	23.0	25.9	26.3	24.1	20.7	17.0	13.8	18.62
0201	High temperature mean value	c:	15.2	15.5	16.7	19.1	23.3	27.5	30.7	30.7	28.0	24.2	20.1	16.7	22.31
0218	High temperature mean monthly value	c:	18.8	19.3	21.4	24.2	28.9	32.9	36.5	35.1	32.2	28.5	24.9	20.8	26.96
0301	Low temperature mean value	c:	9.2	9.3	10.1	11.9	14.9	18.4	21.0	21.8	20.1	17.1	13.9	11.0	14.89
0319	Low temperature mean monthly value	c:	4.4	4.5	5.6	7.8	10.8	15.0	18.0	18.8	16.0	12.2	8.9	6.3	10.69
0615	Precipitation mean monthly value	mm:	89.0	61.3	40.9	22.5	6.6	3.2	0.4	7.0	40.4	89.7	80.0	112.3	46.11
1101	Relative humidity mean value	%:	79.0	79.0	79.0	77.0	74.0	71.0	69.0	73.0	77.0	78.0	77.0	79.0	76.00

Figure 7.2
Predicted change in diurnal temperature range, IPCC 4th Report, Chapter 10 Figure 11. The Mediterranean may experience some of the most intense changes predicted globally.

predictive data by the University of Veterinary Medicine in Vienna and the Carinthian Institute for Climate Protection, retains Malta and Sicily in this category in its climate zone projections for the period 2076 to 2100 (Rubel and Kottek, 2010).[1]

It is, surely, important that architects should experience anxiety when invited to design a building outside their accustomed milieu. Their 'antennae' should be sensitive to the prospect of a fundamental misunderstanding of the social, cultural, visual, political and religious contexts as well as the physical environment, not least the climate. This does not seem to trouble those corporate designers for whom a brief overseas escorted visit suffices to guide the pencil back at headquarters, for whom, unwittingly, the historic vision of international Modernism provides some self-affirmation, nor those clients who are repelled by their own regional contexts in pursuit of the so-called 'advanced' environments within which international commerce can prosper.

Not so with this client. Malta lies 100 miles north of the Libyan coast, an intense environment, socially and culturally as well as thermally. It was once a vital part of the Kingdom of the Two Sicilies and a survivor of the Great Siege of 1565 (Figures 7.3 and 7.4). It is a complex post-post-colonial society which has pre-occupied various social anthropologists. One such was Boissevain who published *Saints and Fireworks, Religion and Politics in Rural Malta* in 1965, describing the fierce rivalries between brass band clubs, the foci of 'partiti' devoted to the patron saint of the village or an emerging secondary saint, a phenomenon originating in the 1850s. One instance is recorded in the village base, Naxxar, of the co-client for the Farsons Building, the Scicluna family, thinly disguised in the text (Boissevain, 1965).

Figure 7.3
Giacomo Gastaldi, 'Fundamental map of Sicily 1545', issued in Venice by an unknown publisher. One sheet, 37.5 × 54cm. Reproduced by courtesy of Bayersiche StaatsBibliothek, Münchener DigitalisierungsZentrum Digitale Bibliothek, http://daten.digitale-sammlungen.de/bsb00092105/image_1 (accessed 28 December 2015).

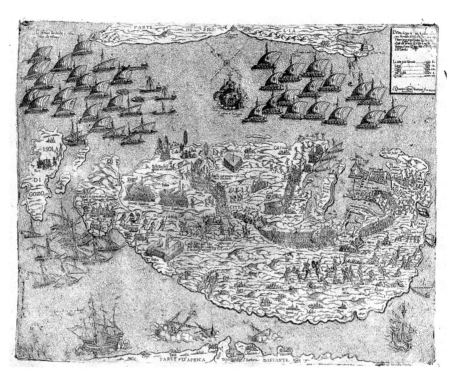

Figure 7.4
Domenico Zenoi, 'Siege of Malta 1565'. One sheet, 34 × 43cm. Taken from a nearly contemporary map by Antonio Lafreri, south coast of Sicily at top. With Figure 7.3, reproduced in Meurer (2004). Reproduced by courtesy of Bayersiche StaatsBibliothek, Münchener DigitalisierungsZentrum Digitale Bibliothek, http://daten.digitale-sammlungen.de/bsb00092051/image_1 (accessed 28 December 2015).

Architectural principles and lessons

The occupation by the Order of St John connected Malta directly to the Vatican, so that unlike the southeastern quadrant of Sicily where art historian Anthony Blunt (1975) described a wild and apparently unsupervised variant of Counter Reformation church architecture, Malta appears to have received its architectural direction from Rome. The churches adopt the strict Counter Reformation template devised by architect G.B. Vignola (1507–73) in the form and embellishment of Il Gesu (1568) (Blunt, 1975). Wettinger (1989) observed:

> ... influences from abroad were penetrating at least at the level slightly higher than that of the actual peasantry – at that, for example, of the extremely numinous clergy. These had come increasingly under the influence of the Counter-Reformation even before the arrival of Monsignor Duzzina here in 1575. Clerical discipline had been tightened up.

Extraordinarily, in the late 1980s, we found the stone cutting diagrams in Vignola's treatise being used to draw templates by Maltese masons. Vignola's type configuration is ingenious in creating a cool, comfortable environment in the summer heat. Realized in massive masonry construction, the design is highly judicious in its admission of direct sunlight so that the internal fabric is little charged by direct solar radiation. The only sunlight permitted to enter bores through tightly radiused lanterns above the vaults of side chapels and clerestorey windows high above the nave arcades set deep into cross vaults and thick walls. Light equals heat. Architectural historian Pevsner (1960, p. 381) explains:

> ... in the Gesu ... certain important features are introduced into the composition exclusively to make light effects possible. The nave is lit from windows above the chapels – an even subdued light ... the last bay before the dome is shorter, less open and darker than the others. This contraction in space and lightness prepares for the majestic crossing with its mighty cupola ... floods of light streaming down from the windows of the drum.

The injection of top light is barely heating up the congregation's layer of the internal atmosphere. The mass and volume of the church can absorb and neutralize the re-radiated heat of a burning disc cycling slowly on the stone floor, even in mid-summer. Contemporary architects can learn from this. The mass of the vaults is buffered by a ventilated over-roof in timber and pantiles. This layering around the occupied volume provides the opportunity for some immediate understanding.[2]

Similar principles seem to have driven the development of contemporary secular architecture, perhaps to protect paintings and furniture as much as human occupants. Syracuse's 18th century palaces – Figure 7.10 shows the Palazzo Beneventano del Bosco of 1779 – are massive stone vessels, punctures are few and heavily shielded from direct sun. Heavy damask curtains are drawn behind the shuttered windows. In the 18th century palace now housing Malta's National Museum of Fine Arts in Valletta, the glazing is carefully contrived to distribute reflected sunlight in manageable quantities through a heavy masonry envelope.

Figure 7.5
The carefully framed image of the dome of Il Gesu, Rome in Pevsner
(1960). Image source unidentified and so generally credited to 'archive of
author or publisher'.

Figure 7.6
External temperatures measured on-site
at Sta. Caterina, January to May 2009,
show internal May temperatures
consistently peaking at about 24°C but
dropping to about 16°C at night with
occasional lows of 5°C at intriguingly
regular intervals, perhaps the opening of
the main doors for early morning
processional events or dawn
downloading. Reproduced with grateful
thanks to Maria Roberta Galea, Professor
Jo Ann Cassar, Built Heritage, Faculty for
the Built Environment, University of
Malta.

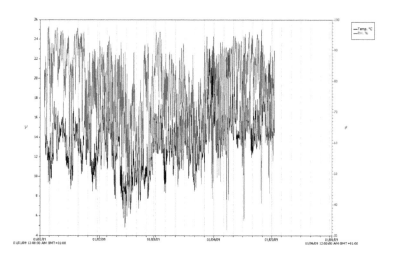

Figure 7.7
Location of the B series of loggers in Sta.
Caterina d'Italia, Valletta, Malta (Professor
Cassar and Maria Roberta Galea,
University of Malta, based on a drawing
included in De Lucca and Carapecchia
[1999]).

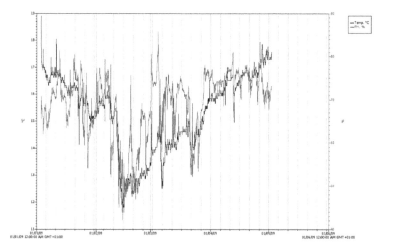

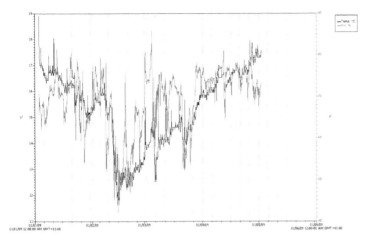

Figure 7.8
Sta. Caterina, Valletta, data logger B4
records internal temperatures in May at the
occupied level, cycling around 16.5°C as
external temperatures regularly peak at
24°C. Sadly the university calendar
curtailed further measurement into June,
July and August (Professor Cassar,
University of Malta).

Figure 7.9
Courtyard of the Cathedral Museum, Mdina,
Malta, the former Diocesan seminary.

Figure 7.10
One of Tim Benton's iconic 1967 plates in Anthony Blunt's *Sicilian Baroque*,
showing Palazzo Beneventano (Plate 89) showing the heavy defensive form of
the urban palace in the Southern Mediterranean (Tim Benton).

An incremental layering of the building envelope developed. Occupants could escape the massive enclosure by slipping through fissures in the deep masonry shell into a lightweight, timber cross-vented projection. These appear to have been devised as supernumerary devices providing circulation to short-circuit sequences of rooms *en serie* but evolved into places of observation. Subsequently, earlier 20th century apartment developers in Valletta extended the projecting bay principle into a vital component of everyday *de minimus* living in this climate (Figure 7.11).

The seamless continuity of masonry building and these relatively simple environmental strategies in Malta was disrupted by Giuseppe Psaila's Balluta Building of 1928 in St Julian's Bay, the first apartment building intended for the emerging upper middle class in Malta in a southern transformation of the Liberty style, the Italian Art Nouveau (Figure 7.12).[3]

The architect punctured the heavy envelope by drawing major external spaces deep into the block, making externalized internal rooms so that occupants can dwell indoors at external shade temperature, in some magnificence. It was developed speculatively by the Marquis John Scicluna, who was instrumental in the development of the Cisk Brewery. Similar contemporary buildings are reported as having been developed along those sections of the Mediterranean coast under Italian influence: Libya, the Dodecanese and Egypt in particular. Evidence of Italian architectural ingenuity in an extreme climate is well documented in Alexandria. Awad (1990) explains the concessions, 'capitulations', which enabled Europeans to settle and build in

Figure 7.11
A vestigial lightweight occupied layer projecting beyond a heavy envelope (author).

Figure 7.12
The Balluta Building, St Julian's Bay (author).

Alexandria, a sophisticated physical 'Occidentalizing' of Alexandria importing and reinventing more familiar neo-Gothic and neo-Baroque elements, clearly climate driven to some extent. Post-Risorgimento Italian entrepreneurs developed a monopoly in the provision of urban infrastructure in Alexandria, introducing reinforced concrete and terracotta blocks. Awad further explains that the King had been educated in Italy and maintained warm relations with the 'House of Savoia'. An Italian architect, Lasciac, was Chief Architect to the Royal Palaces, Italian contractors dominated institutional and public construction whilst Malta maintained close commercial connections with Egypt, under the auspices of a constant flow of Royal Navy vessels.

Khalil (2009) discusses the elements in the Italian contribution to Alexandria's architecture in more detail and the metamorphosis of its clear ethnic constituents. He demonstrates how Italian architects from the 1880s coped with their new context by initially transforming neo-Gothic Venetian architectural configurations, citing Count Zizinia's Palace on Place des Consuls by Antonio Lukovitch, and the astonishing Haramlek-Montazah palace by Ernesto Verrucci, built 1923–28, derived in part from the exuberant Castle Mackenzie in Genoa.[4] These buildings attempt a folding in of external space within their envelopes, apparently newly invented for the more southerly climate. Perhaps this is an example of the process of 'mechanical selection' which Le Corbusier claimed to have identified, a 'will to form' guided by custom and practice, and simple common sense.

Khalil (2009, p. 37) comments:

Alexandria became the commitment and practices [sic] of Italian Venetian professionals, who found their clientele within their own community, and among others, specially the Jews and also increasingly among Egyptians. This was a perfect opportunity and setting for Giacomo Alessandro Loria 1879–1937 to reproduce his Little Venice in the Alexandrian context. Loria's S. Salem, M. Douak, and the El Nokaly's apartment blocks in Ramleh Station 1926–1928 carry strong Venetian references, such as the Gothic detailing borrowed from its Palazzo Ducale ... the Italianated facade won for its architect the Municipality Honorary Prize for Best Facades.

All this occurred under the British mandate. Further Italianization yielded attempts to transplant the top-glazed 'galleria' of Milan, but to much tighter dimensions presumably to restrict solar access, Antonio Lasciac's 'Gallery Menasce' of 1883–87. It is not recorded what level of success was achieved in this gambit, one might reasonably be sceptical (Khalil, 2009). Nonetheless, despite the inventiveness and scale of the Alexandrian Italian architectural community, the Balluta Building in Malta emerges as being unusually innovative in its transformation of a radical prototype into a new proposition in an unfamiliar context.

The Italian architectural diaspora extended to Libya, a mere 100 miles to the south. Italian modernism in Libya is documented by McLaren but Anderson draws particular attention to architect Florestano di Fausto's programme (c.1923–40) of eliding Italian, Modernist and local architectures, 'arabisances' which emerges as a particularly compelling approach for a displaced European (McLaren, 2006; Anderson, 2010). Anderson writes of the speculations on

Mediterraneità, an argument in which 'the Mediterranean' is an Italian invention, with a natural empathy between north and south as culture flowed across the sea but one that was progressively eclipsed by colonial and subsequently Fascist racial biases. Di Fausto appears to have drifted 'off message'. His one explanatory text (Di Fausto, 1937) explains: 'Not a stone was placed by me without filling myself with the spirit of the place, making it mine.'

Anderson (2010) claims:

> Di Fausto merged the surfaces of Italian modernism with the visual remains of regional histories. Such adaptations became a source for and stage through which he enacted modern corporeal structures. In this manner, the modern architecture of Di Fausto was no longer an extension of the colonial power but, like the Mediterranean itself, a broad space governed within the dictates of Italy's appropriated metaphysical language of self-preservation.

Di Fausto appears to have achieved what the design team for the Farsons building aspired to against the background of another historical colonial presence (Anderson 2010):

> ...Di Fausto's built and unbuilt projects refine the vagaries of traditionalism into a medium that is difficult to categorize. His architecture stands between histories. *It is neither modern nor traditional.'* [original author's italics]

This is something indeed for the displaced designer to aspire to.

The Malta Brewery

The author's involvement with the Simonds Farsons Cisk brewery commenced during the Summer School for Young Architects held at the Old University in Valletta in August 1985, by chance during a fiercely contested election campaign. The School's outcomes were broadcast by the defending Minister of Education and published in the *Architectural Review.* Dom Mintoff's Malta Labour Party ruled still, but under siege conditions. Malta only enjoyed friendly foreign relations with China and North Korea. This political isolation had starved architecture students at the University of Malta of international contact. The summer event, supported by UNESCO and organized with the support of the unlikely, aristocratic Foreign Minister Alex Sciberras Trigona, was designed to break this involuntary exile. The event raised public perceptions of the very rich but distinctly vulnerable architecture that was still intact on the island. Equally, it brought ideas of environmental sustainability as applied to buildings in this Mediterranean climate.

Farsons had already commissioned a design for its new process block, in effect the laboratory, the yeast propagating, filtration and testing centre of the brewing process. A lightweight 'wriggly tin' shed was proposed, sealed and air-conditioned. The project managers and suppliers from the brewing industry clearly saw the intended building as a simple shroud for the very highly crafted stainless steel vessels, pipework and valves comprising the plant. In contrast, the existing brewery buildings, designed before the First World War but built shortly thereafter (Figure 7.13), pursued the architectural aspirations of the line

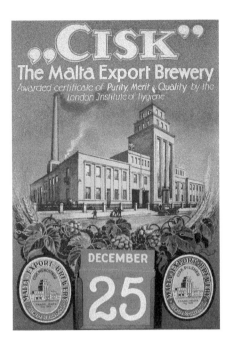

Figure 7.13
The original building designed by British
consultants in London, completed 1912, a
typical proto-Great West Road factory in an alien
climate (Farsons Brewery archive).

of industrial headquarters along the Great West Road leading out of London towards Slough, the Firestone factory in particular. Indeed, these brewery buildings had been designed in London. Surprisingly, there was little acknowledgement of the rather different climate.

This is evidenced by the maintenance workshop in which the continuous array of angled northlights was quickly blacked out. The designers had failed to account for the fact that the June sun reaches 80° altitude which had clearly led to overheating. The immediate remedy, to cover these rooflights, meant the building was artificially lit in perpetuity, 24 hours a day (Figure 7.14).

The beer was fermented in eight substantial open baths, approximately 10m × 3m, in reinforced concrete, stacked on two floors, the whole building environment being refrigerated to below 7°C. This emerged as a common mid-century brewing process strategy in the hot lower latitudes before sealed refrigerated cylindroconical fermentation vessels were introduced (Figure 7.15). Debate rages as to whether open fermentation produces a finer product. However, the energy demand entailed by this technique is prodigious. In Malta it is remembered, anecdotally, as having threatened the continuity of electrical supply to the neighbouring town of Birkikara.

Three families owned the brewery, the Maltese–Sicilian Farrugia-Micellis, the British Simonds's of Courage (absorbed subsequently into the Imperial Group), and the Sciclunas, represented by the eccentric Marquis Scicluna, practising Surrealist painter, creator of the Madliena Merz Bau as head of the Scicluna Testaferrata-Moroni-Viani family, who were bankers in origin. The proposition for an exemplary sustainable new process block emerged out of the realignment in ownership, hence the reconstituted Board's irrevocable commitment to the idea. It is important to understand the complex drivers for what appeared to external commentators to be an act of extraordinary corporate

Figure 7.14
The 1947 'northlight' workshop designed by British architects in London. All of its northlights were blacked out, probably within a week of opening, mid-summer sun angles are almost vertical, the space has been permanently artificially lit ever since (author).

Figure 7.15
Open fermentation baths, designed in the late 1930s, an industry standard around the globe. The entire space is chilled to 7–8°C encased in 100mm of uninsulated concrete (author).

selflessness in the global interest, hardly the product of 'nudging'. The UK Department of Energy and Climate Change may wish to take note as it expands its interest in behaviours and the drivers of behaviour.

The client's intent was to create a protoype for the food and drink processing industry beyond Northwest Europe and the USA, in order to export products back to those economies in compliance with all western hygiene regulations. Equally, there was a desire to have a building that was strongly characterized to the specific place and culture. The architecture that emerged is charged with the identities of the two families, a little reminiscent of those recorded in *Il Gattopardo*, the Marquis Scicluna being a contemporary equivalent of figures within the Duke of Salina's circle.

The new building is the antithesis of the lightweight, steel portal-framed, 'wriggly tin' wrapped shed, the gift of the post-war Anglo-American process and logistics industries which is characteristic of the contemporary brewing industry globally. The new building is heavy and substantial, made from local globigerina limestone, in an idiom with meaning for those who inhabit and visit it. It is configured to drive substantial airflows through itself when it is judicious to do so, to recharge its thermal mass to dissipate the effects of the inward passage of heat through direct radiation, conduction and convection (Figure 7.16). Figure 7.17 shows how simple opening vents are set and reset on a simple daily cycle with virtually no energy expended. Figure 7.18 is photographer Peter Cook's dawn view of the north elevation enclosing the perimeter free-running buffer zone around the Process Hall within.

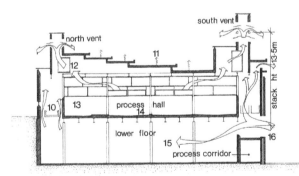

Figure 7.16
Perspective section of the proposed building (drawn by the author and rendered by Professor David Dernie).

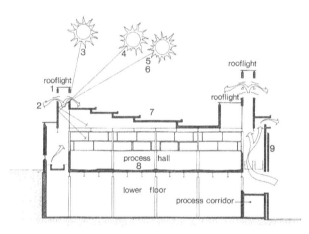

Figure 7.17
Summer day (top) and summer night modes:
1. Cast glass troughs on solar stack above perimeter jacket indirectly light Process Hall.
2. Windows to solar stacks permanently open except in storms to freely vent jacket, probably a mistake.
3. Solar altitude 78° at 12.00 a.m. on 21 June.
4. 43° at 12.00 a.m. on 21 October.
5. 30° at 12.00 a.m. on 21 December.
6. Low altitude winter sun reflected off solar stack rear face.
7. Insulated heavy stone roof construction delays heat entry by eight hours.
8. Process Hall should remain below 27°C even as peak external rises to 39°C.
9. Freely respiring stone diaphragm wall convects away solar gains reducing radiant transfer to technicians on walkways suspended in stack driven flow.
10. Stack flows in perimeter zone permitted either side of flying corridor bridges (Pugin's innovation at Scarisbrick Hall).
11. Clear night radiation cools roof.
12. Windows connecting upper cavity of Process Hall to solar stacks open to enable buoyancy forces to drive airflow with no mechanical assistance whatsoever.
13. Internal faces of massive enclosure exposed to internal air volume.
14. Dawn temperatures in Process Hall drop to 24°C and below.
15. Basement volume prodigiously stack vented throughout.
16. Air admitted at low level to basement.

The internal floor to ceiling heights are 6m, the main windows are 6m high, cathedral scale, all in unreinforced stone masonry, stone joinery in effect, the lower openings being the view windows at eye level. Cast glass troughs, a single clip at each end, present a serrated edge to the solar stack tops. Dramatic solar gains at the terminations accelerate the buoyancy-driven flows up and through the building so that the buffer zone freely ventilates throughout. This is probably a mistake; it would be prudent to stem this flow through periods of

Figure 7.18
Peter Cook's 4.00 a.m. view of the north elevation, mouldings just catching the rising sun. The small lower windows provide views out for the technicians, ceiling heights are 6m, a relatively high glazed area, limestone envelope, solar stack above, disguised plant access doors, storm water gargoyles, cast glass troughs to the solar stack tops to boil the exhaust terminations, simple centre-pivot windows (Cook, plate 878/001).

peak temperatures when the airflow may actually be reheating the masonry shell. The political situation demanded that imported materials were kept to the absolute minimum, the spectre of 70% import duties and interminable delays prompted a search across the island for reusable materials. The simple centre-pivot windows were made up from first generation aluminium sections found in storage. Chic, pre-war, French reeded-glass blocks were discovered in a barn and built into the solar stacks. Slender stone piers and upstands are stabilized and buttressed by loading them with stone blocks; they are not cosmetic (Figure 7.19). Figure 7.20 shows the minimal arrangement of punctures and cast glass presented to the sky.

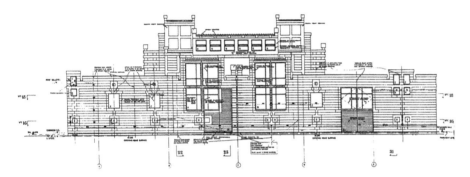

Figure 7.19
Setting out drawing for the north elevation (author).

Figure 7.20
The north buffer space with solar stack perched above on steel grid (Peter Cook).

The load-bearing stone envelope

The record image of construction in progress (Figure 7.21) reveals what was blindingly obvious in retrospect: it was essential to define every stone block for the masons. Traditionally, the form of individual stones was suppressed by the liberal use of flush pointing, subservient and sometimes contrary to the articulation. For example, a less inconsequential elevation of the presbytery of the cathedral in Mdina recorded in Figure 7.22 shows this treatment.

The 19th century American architect, H.H. Richardson's Glessner House in Chicago (1885–87), visited and scrutinized by the designers, reveals a very deliberate and manifest solving of the compositional puzzles for the building façade arising around interruptions, the pattern-making part of the architecture, familiar in polychromatic work but not in monochromatic masonry (Figures 7.23 and 7.24).

Figure 7.21
The building under construction. The globigerina limestone is quarried 2 miles away and delivered to one of three stone yards set up around the building. Rough-cut blocks were tipped out and then dressed very accurately to precise dimensions. Every block was drawn, an unexpected task. It was explained to the architects that God provided the raw material en masse. The masonry coursing of the blocks is clearly expressed and enters into the architectural expression, a modern idea, not at all apparent on the Baroque, in which joints are flush to maintain a contiguous surface. Richardson clearly delights in solving the coursing panels on the Glessner House. Vignola's constituency may have taken warmly to mass concrete (author).

Figure 7.22
The modest side elevation of the presbytery at the cathedral in Mdina. The stone blocks are suppressed whilst a simple grid of stone bands organizes various types of opening in this working elevation (author).

Figure 7.23
H.H. Richardson's stone detailing of cellar openings at the J.J. Glessner House, Chicago, 1885–87 (author).

Figure 7.24
H.H. Richardson's public façade of the J.J. Glessner House, Chicago, 1885–87. Every stone block is very deliberately sized and placed. This is an un-modern architecture of thermal mass drawing as much on the future as the past (author).

The Mdina building also shows a modest device for establishing some order on the façade, nets of raised stone banding. Perhaps this level of articulation might be appropriate for a working industrial building a long way down the honorific hierarchy of building types? Borromini deploys this device on the Presbytery adjacent to San Carlo alla Quattro Fontane, thoroughly eclipsed by the church but nonetheless compelling. It delivers a high level of freedom in puncturing the elevation, a framework in the vertical plane, a popular contemporary gambit. Stone cannons on the brewery building fire water away from the elevations in seasonal flash storms (Figure 7.25).

Above the occupied level, house-sized constructions manage and exhaust air, sitting on steels above the main stone enclosure. This device for propping masonry in the air was suggested by the bizarre 19th century, stone trussed hothouse at the Palazzo Parisio (Figure 7.26), saved and restored as a byproduct of the brewery project. The soffits of the north stack climb towards the exhaust chamber (Figures 7.27 and 7.28)

The roof of the building is active, as this approach demands. The general arrangement is depicted in Figure 7.29. Direct view of the sky is permitted, in the traditional manner, through relatively very small openings, the design refined using physical models in the artificial sky (Figure 7.30). The upper surface of the roof is limestone, the continuous lines of cast glass vaguely reminiscent of the water bodies and irrigation channels of a central Mediterranean formal garden. The south facing solar stacks borrow from the architecture of wellheads, not least from the examples at Palazzo Parisio of 1723 (Figure 7.31).

Figure 7.25
Stone storm water gargoyles hand carved on-site. Heavy downpours come in September and February, too heavy for the conduits to the water cisterns to cope (author).

Figure 7.26
The discovery of an 1870s hothouse at the Palazzo Parisio roofed in stone trusses supported on iron I-beams, an extraordinary construction evolved on an island with an acute shortage of timber. The stone members have dovetail joints, joinery in masonry defying the orthodox explanation of the evolution of stone building out of timber prototypes. It delivered the solution to the problem of perching very large and heavy solar stacks high above the principle volume. In poor condition, over 110 years the iron was corroded in the saline environment, it was restored as a byproduct of the brewery project (author).

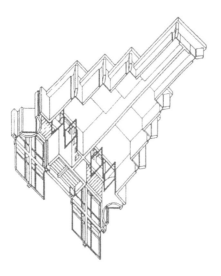

Figure 7.27
Choisy-type up-view of the north solar stack sitting above and within the north elevation on steels, an unusual juxtaposition, showing the soffit above the Hall cavity stepping up to encourage warmed air to drift towards the exit vents into the stack (author).

Figure 7.28
View up into the north vent, profiled glass troughs admit
toplight into the less critical jacket space, freely venting
(author).

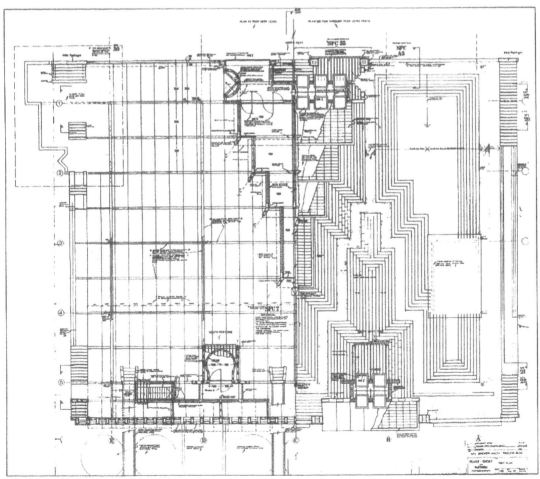

Figure 7.29
The general arrangement of the roof (author).

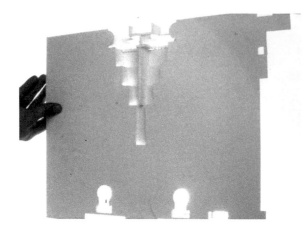

Figure 7.30
A model of the roof as tested and modified in an artificial sky, showing the strategy of injecting very small quantities of direct sun into the deep plan Process Hall. This is potentially hazardous, because with the light comes solar gain (author).

Figure 7.31
A well head in the gardens of the Palazzo Parisio 1723, drawing water up from the vaulted cisterns below and storing it at two levels for gravity distribution through stone channels across the garden. A south solar chimney imagined as its airborne equivalent (author).

The south solar stacks grow out of the diaphragm wall (Figure 7.32). They comprise an intermediate chamber to capture air as it rises through the temperature sensitive process hall (Figure 7.33). The two stack constructions address the countryside beyond in the direction of Quormi (Figure 7.34). The stacks were constructed from simple pencil drawings dimensioning each stone, prepared in the heat of the day before the afternoon's stone laying commenced (Figure 7.35)

The south elevation is a 14m high unreinforced masonry diaphragm wall (Figure 7.36). Gravity is key in its elemental unreinforced construction (Figure 7.37). The parallel cells enable convective currents to rise throughout the day to reduce heat transfer to the interior of the buffer space, also convectively cooled. The stone labyrinth formed at the top openings intercepts Saharan sand blown across the Mediterranean in the mid-season commencing in September. It is bi-axially symmetrical, useful in setting up the fields of stainless steel tanks and

Figure 7.32
Looking directly up into a south solar stack from within the Process Hall, the vents eventually opened briefly during daylight hours for the photographer by unwilling technicians familiar with the strategy cajoled into overriding the controls (author).

Figure 7.33
Choisy-type view of a south solar chimney showing the chamber above the Hall, the opened vent, the stone labyrinth at the head of the diaphragm structured south wall to defend the interior against wind blown Saharan sand (author).

Figure 7.34
View across the roof towards the south (Peter Cook).

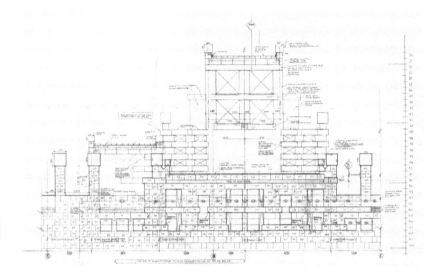

Figure 7.35
On-site pencil masonry setting out drawing for a south-facing solar stack. Setting out occupied the intense heat in the middle of the day when stone-laying ceased until 4.00 p.m. (author).

Figure 7.36
The south elevation, subject to prodigious solar attack, venting continuously through its open diaphragm cells (Peter Cook, plate 878/CT006).

Figure 7.37
The archaic cellular diaphragm construction of the
14m high, load-bearing, self-supporting, respiring,
limestone south wall (author).

in addressing the distant horizon, attempting a little 'calculated ambiguity of expression' as architect and writer Robert Venturi (1977, p. 20) reminds us of Edwin Lutyen's design of Nashdom at Taplow, UK very much a programmatically driven duality, less so in Moretti's highly sophisticated apartment building in the Via Parioli in Rome.

Could it work? Does it work?

Figure 7.38 is the output graphed by Professor Brian Ford from Dr Nick Baker's early simulation model FRED, derived from an electrical network model, which contained an algorithm to simulate the effects of natural stack ventilation. Judiciously, Baker modelled the onset of a 40 year heatwave from historic data. Typically, the external night-time temperature lows remain depressed even as highs rise to 40°C plus; the model predicts that the night ventilation will remove the previous day's delayed incoming heat pulse and maintain peak internal temperatures well below peak external. This was a familiar strategy for the local board members whose parents and grandparents operated their homes in a similar way but unintelligible to the English project managers. The histogram shows predicted rates of respiration climbing to 12 air changes/hr in the early hours of the morning at peak temperature difference.

Actual measurements show how the building works in hot conditions. Figure 7.39 shows a section of the continuous data collected by the Head Brewer, recording an unpleasant September (1991) in which night-time lows do not fall as far as hoped for. Nonetheless, the building maintains remarkably stable conditions, surface temperatures cycling around 23°C whilst external temperatures exceed 38°C at zero energy. The passive cooling is calculated to avoid 213 841 kWh of mechanical cooling load, which would have resulted in emissions of some 69.7 tonnes of CO_2 annually.

The opening ceremony on 12 July 1990 was an important moment in the economic history of the island. The President, the new Christian Democrat Prime Minister and the Archbishop attended. It was televised. Unfortunately, for the preceding week the doors were all flung wide open as 400 seats, podia and stage lighting were installed. The interior heated up to ambient external

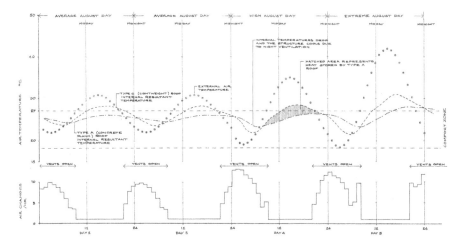

Figure 7.38
Nick Baker's FRED model prediction of likely conditions within the process hall during a once in 40 year heatwave. The model predicts a natural respiration rate climbing to 12 air changes per hour driven entirely by temperature/pressure (Dr Nick Baker and Professor Brian Ford).

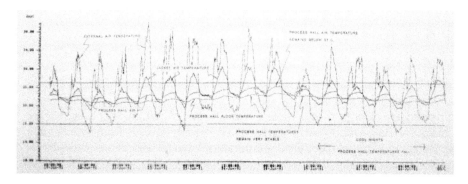

Figure 7.39
Real observed data, temperatures annotated at 4°C intervals, late June. Some 14°C+ depression in peak temperature by early morning.

temperatures. The guests had been promised an environmental miracle. Despite tipping 5 tons of ice into the building the night before, temperatures on the day remained as they were the day before, the mass of the building had buffered any useful cooling. Before the evening ceremony, Short and Ford opened all of the vents fully. As the audience was seated and the powerful lights were switched on, a gentle and then an increasingly vigorous breeze developed through the building as the internal heat loads built. Figure 7.40 shows the guests discharging into the courtyard after the ceremony, wholly persuaded of the 'miracle'. This experience persuaded the designers that it should be possible to naturally vent and passively cool a densely occupied auditorium. Chapter 5 records the subsequent investigation into the prospects for driving densely occupied theatre spaces naturally.

Figure 7.40
The opening ceremony. An economy outside the
United States and Northwest Europe receives an
industrial plant to re-export western products back to
the west, earning foreign exchange, whilst
conforming to the necessary EU/US hygiene
requirements but with none of the attendant energy
implications, saving some 70 tonnes of carbon dioxide
production per year (author).

Reception history

The reviews were universally kind. The building was regarded as an exotic oddity and received high profile coverage and recognition. *Newsweek* commented:

> Projects like Farsons Brewery will redefine architecture in the decades to come. Wealthy nations will want to decrease operating costs, while developing nations will not want to waste precious resources.

The building was reviewed in *New Scientist*, the *Observer*, filmed by the EU, subject to thorough publication by *Architecture Today*, written up by the *Architects' Journal* as 'An alternative to air-conditioning', and visited by a UN scientific mission. Ineligible for almost all prizes, being in Malta, the Process Building was awarded the High Architecture Low Energy Award by judges Colin (Sandy) St John Wilson, Bill Bordass and Robin Spence with the citation, 'exemplary in its integration of structure and services' but hinting at some reservations about its detail and its stylistic aberration. More importantly the Maltese public were enthusiastic, a foreign intervention had seemingly caught the *genius loci*, its designers ever conscious of potential intrusion (Boissevain, 1965; quoting from the 1962 Manifesto of the Malta Labour Party):

> Since time immemorial our islands have been dominated by foreign powers. Our people have not therefore been the custodians of these fundamental human rights … on the contrary our social, political and economic institutions have been at the mercy of foreign masters who have fashioned them to suit their own requirements.

Notes

1. Open Access Article, Gebruder Borntraeger 2010 (published online, accessed June 2013).
2. It has been surprisingly difficult to find measured data collected in Baroque church interiors but fortunately research undertaken by graduate students at the University of Malta under the supervision of Dr Jo-Ann Cassar have collected data on outdoor and

indoor thermal conditions in Sta. Caterina d'Italia, a centrally planned church by the Maltese architect Girolamo Cassar, built between 1572 and 1580 (remodelled by M. Blondel c.1682, façade by Romano Carapecchia 1713) frequented by the Italian community on the island (Figures 7.6, 7.7 and 7.8). It has contained a major painting by Mattei Preti since its completion and so the environmental stakes have remained high.

3. National Inventory of the Cultural Property of the Maltese Islands. See www. culturalheritage.gov.mt/filebank/inventory/01212.pdf (accessed 8 July 2013).

4. Khalil was Assistant Lecturer, Department of Architectural Engineering, Faculty of Engineering, Mansoura University, Egypt.

8

Continental climates

Figure 8.1
Deutscher Meister, Düsseldorf School, first half of the 19th century. Signed C. Herzog. As members of the Düsseldorf School migrated to America, they introduced the exceptional clarity of atmosphere and extreme attention to detail of the Düsseldorf artists into the Hudson River School to record natural North American landscapes as yet untouched by human presence (author's collection).

A confused murmur of talk and the shuffling of many feet arose on all sides, while from time to time, when the outside and inside doors of the entrance chanced to be open simultaneously, a sudden draught of air gushed in, damp, glacial, and edged with the penetrating keenness of a Chicago evening at the end of February.

(Frank Norris, 1903, *The Pit*, Chapter 1)

This old house with the leaky weatherboards was a very different thing from their cabins at home, with great thick walls plastered inside and outside with mud; and the cold which came upon them was a living thing, a demon-presence in the room. They would waken in the midnight hours, when everything was black; perhaps they would hear it yelling outside, or perhaps there would be deathlike stillness – and that would be worse yet. They could feel the cold as it crept in through the cracks, reaching out for them with its icy, death-dealing fingers; and they would crouch and cower, and try to hide from it, all in vain. It would come, and it would come; a grisly thing, a specter born in the black caverns of terror; a power primeval, cosmic, shadowing the tortures of the lost souls flung out to chaos and destruction. It was cruel iron-hard ...

Later came midsummer, with the stifling heat, when the dingy killing beds of Durham's became a very purgatory; one time, in a single day, three men fell dead from sunstroke. All day long the rivers of hot blood poured forth, until, with the sun beating down, and the air motionless, the stench was enough to knock a man over; all the old smells of a generation would be drawn out by this heat – for there was never any washing of the walls and rafters and pillars, and they were caked with the filth of a lifetime.

(Upton Sinclair, 1906, *The Jungle*, Chapters 7 and 10)

The later 19th century 'Hudson River School' of painters captured North American landscapes deep within the continental landmass in the moments before any imported human occupation. Key figures in the school had migrated from Germany, from the Düsseldorf School, bringing the highly characteristic panoramic super-realist Düsseldorf 'gaze', constructing an overwhelming clarity of atmosphere well beyond the human eye's perceptual capabilities. Every growing thing depicted across the canvas is given a botanical identity in sharp focus. The eye has to travel back and forth across the paintings to capture the vivid undifferentiated detail. What is now the eastern part of Germany, well to the east of Düsseldorf, lies on the periphery of the Eurasian Continental landmass extending to Beijing, with a climate every bit as ferocious as that of the American Midwest. The climates are surprisingly similar. How could the citizens of both continents possibly survive before the era of the artificial environment? They clearly did but not without some difficulty.

By 1906 the human-induced environmental conditions Upton describes in the meat-packing district of Chicago were every bit as appalling as those so meticulously catalogued by Corbin (1996) in the industrial areas of 18th century Paris. Responses to Sinclair's novel contributed to the pressures for the first US Meat Inspection Act.[1] The Chicago wind-rose suggests that the stench of the West side would have been discernible in Mid-town on a seasonal basis. The wind blows in from the west-southwest in November and December, veering to the west in January, February and March, its reach unimpeded across the prairies.[2] Perhaps a collective memory of this corrosive airborne environment, well within living memory, reinforced the later 20th century Chicagoan middle class environmental 'social practice', the unquestioning expectation of a wholly artificial environment. At their most advanced, these environments were created within hermetically sealed interiors wrapped in a seamless film of glass. Does the climate actually demand this? How did the better resourced survive before mechanical cooling was available?

Köppen-Geiger's 1901–25 map covering the period of Upton's account, records Chicago as being at the volatile transition between the Cfb region, 'Warm Temperate, fully humid, warm summers' to the south, and the Dwc climate, 'Snow, winter dry, cool summers' to the north (Rubel and Kottek, 2010). The original 1884 map actually inserts a narrow intermediary zone at this latitude.

Meyer (2000) explains that summer heat had become the principle climatic problem in North America by the Civil War. Meyer quotes John W. Draper (1868), a New York physician and historian of the War who wrote ruefully: 'In the infancy of humanity, cold was man's antagonist: his more perfect civilisation struggles less successfully with heat.'

Draper interpreted the Civil War as a conflict between people who had become adapted to very different climatic regions. He promoted migration across the latitudes through his writing, bucking at the theory that migrants tended to relocate along the same latitude. During the 1853 Northeastern heatwave, the expected weekly death rate in New York City had doubled to 600. Many of the fatalities were amongst male adult manual labourers. More recently, the July 1995 heatwave affected Chicago particularly badly, some 200 additional deaths being recorded. Surprisingly, the number of male adults affected through excessive physical exertion was still significant in the 1995 casualty list.

What did and does it mean to be comfortable in a Continental climate?

The central protagonists in Frank Norris's 1903 novel about commodity trading on the Chicago Exchange, *The Pit*, the aggressive and hugely wealthy grain speculator Curtis Jadwin and his wife Laura enjoy their new and lavish mansion through the summer extremes of climate with no recorded attempt at mechanically assisted cooling (Norris, 1903):

> It was a July evening … it had been so warm that she had not changed her dress after dinner – she recalled that it was of Honiton lace over old-rose silk, and that Curtis had said it was the prettiest he had ever seen. It was an hour before midnight, and the lake was so still as to appear veritably solid.

Laura Jadwin is surviving a warm and very still night in layers of silk and lace. A century later, the National Renewable Energy Laboratory (NREL) database presents normally expected weather conditions for Chicago, the Class A Typical Meteorological Year (TMY2) weather data, for which the measurement site is Midway Airport in the urban fringe close to Lake Michigan (Marion and Urban, 1995). Notwithstanding a changing climate, the TMY2 provides a little insight into Laura's environment that evening. The data suggests the temperature could have still been 25°C but possibly falling towards 15°C the following evening. Perhaps, customarily, Laura put on a more substantial dress after dinner but desisted that night. There is no mention of the use of an electric fan but they were available. The Jadwins could certainly have afforded such conveniences. Arsenault (1984) reports that in the Southern states in 1902:

> Ownership of an electric fan quickly became a badge of middle-class respectability in the urban South. The 'whirligig' was a luxury item beyond the reach of most working-class families, although by 1902 small table fans could be purchased from Sears Roebuck for ten dollars (*The 1902 Edition of the Sears Roebuck Catalogue,* New York, 1969, p. 663). In the rural South, where electrification was almost unknown and where farmers and villagers were left to the mercy of natural forces, the electric fan became a symbol of urban opulence – or in some cases decadence.

The American National Standards Institute (ANSI) and the American Society of Heating, Refrigeration and Air-conditioning Engineers (ASHRAE) Standard 55-2004 places a recommended upper limit on the 'allowable' operative temperature of about 28°C for American citizens and their diaspora, which reduces towards 27°C as the moisture content increases to the limiting value of 12g/kg of dry air (ASHRAE, 2004). De Dear and Brager (1998) have constructed alternative guidance for naturally conditioned buildings within the standard based on adaptive comfort observations, raising the bar a little as the recommended temperature range follows the movement in mean monthly temperature. In theory, it would permit a 'hybrid' building, a building able to cycle between natural and mechanical modes of environmental control, to remain in its natural mode for longer, as shown in Figure 8.2.

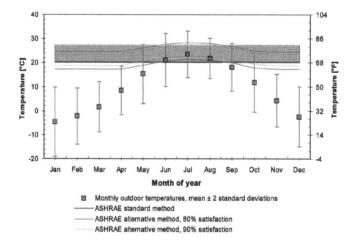

Figure 8.2
Eleven years of Midway Airport temperature data with an indication of corresponding heating and cooling demand in the Chicago area, following ASHRAE (American Society of Heating, Refrigeration and Air-conditioning Engineers) guidance. Very surprisingly, mechanical heating and cooling only appear to be essential, theoretically, for about half of the calendar year. From Short and Associates team Competition Report to Judson College 2001.

The Midway data records approximately 300 hours per year for which the ambient temperature exceeds 28.2°C, the 80% acceptability rate temperature limit and the permitted peak temperature of the alternative method. One might then safely, perhaps too safely, determine that the design criterion to be adopted for a new building is that the achieved temperatures should meet the indoor temperature targets of the Standard 55-2004 method, even in spaces with operable windows. Given the well documented harsh extremes of Chicago winters and perhaps 50 excessively hot days or more in summer, is there any possibility of a naturally ventilated, free-running low energy design surviving here? Surprisingly, there is every chance for a significant part of the year and, more surprisingly even, the predisposing conditions may become more and more benign as the century unfolds (Norris, 1903, Chapter 5):

That year the spring burst over Chicago in a prolonged scintillation of pallid green. For weeks continually the sun shone. The Lake, after persistently cherishing the greys and bitter greens of the winter months, and the rugged white-caps of the northeast gales, mellowed at length, turned to a softened azure blue, and lapsed by degrees to an unruffled calmness, incrusted with innumerable coruscations.

It certainly would have been possible to free-run a building through the Chicago spring of the early 20th century.

Playing the hybrid hand

More recently the author's design for the new Academic Center for Judson University in Elgin to the west of the city departs from business-as-usual designs conforming to current Illinois Building Codes. It brokers an altogether different and significantly lower energy environmental design approach through a close examination of the historic and present climate data. This is an exercise which almost certainly yields surprises, even for those born and bred in the region. The TMY2 weather data for Midway Airport is close to Elgin (Marion and Urban, 1995). Elgin is roughly 64km (40 miles) from each site and at a similar vertical elevation, approximately 200m. The data proposes typically 28 hours below −20°C and 167 hours above 30°C. There were no discernable differences in the diurnal temperature swings or the solar radiation data. The Chicago Midway TMY2 data was used in all the simulations for the Judson project (Lomas *et al.*, 2006).

The data confirms that the extremely cold winters and high summer temperatures preclude 'simple' natural ventilation all year round for contemporary occupants following our assumed rather benign criteria. It is likely to be an economic necessity to employ heat recovery and, potentially, to maintain comfort to this standard guidance, humidity control in winter and summer although millionaires once prospered without it.

The Chicago weather is susceptible to unannounced swings at all times of the year. In January, for example, the TMY2 data shows daytime temperatures between −22°C and −14°C early in the month, then 3–10°C mid-month and back to very cold conditions again at the month's end, so that even in January there are days in which a hybrid building could adopt the natural mode if it could

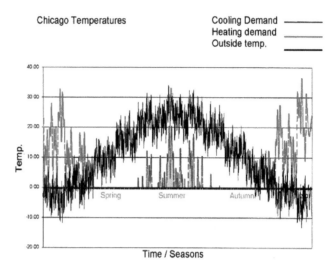

Figure 8.3
Mean monthly Chicago temperatures with ASHRAE (American Society of Heating, Refrigeration and Air-conditioning Engineers) Standard and Alternative (climate adapted) method derived comfort bands superimposed (ASHRAE 2004). First published in Short and Lomas (2007).

change mode quickly. During the summer there is similar evidence of oscillating weather patterns. In July, the hottest month, two- to five-day spells with warm humid conditions develop with southwesterly winds from the Gulf of Mexico, peaks of 30–35°C and moisture contents of 8g/kg to 11g/kg are recorded, interspersed with more moderate periods induced by cooler, dryer airstreams originating from the north.

The diurnal temperature range is also variable, with swings of 15K or more being reasonably common in April, May and September, but not as reliable as found in the central Mediterranean as discussed in Chapter 7. On some days the swing is below 5K, so that night ventilation cooling cannot be the only tactic available to defend the internal environment from excessive heat.

Figure 8.3, compiled from 11 years of the Midway weather data from hourly temperatures and moisture contents, suggests that the Chicago year is divided into three type periods: winter, from the beginning of November to the end of March, in which peak daily ambient temperatures almost never exceed 15°C and the mean monthly daytime temperatures are below 4°C; a full summer period from mid-May to mid-September in which the peak daily temperatures invariably exceed 20°C; and the mid-season periods, the spring and autumn, in which temperatures oscillate around the band in which the successful use of 'Advanced Natural Ventilation' could be imagined.

The design mechanics of hybridity: Judson Academic Center

A simple exercise taking a notional building designed to the current Illinois codes demonstrates that by shading all of its exposed glazing from direct solar gains, enabling effective daylighting and ventilating the interior through the cooler nights, the energy demand could be halved (Figure 8.4).

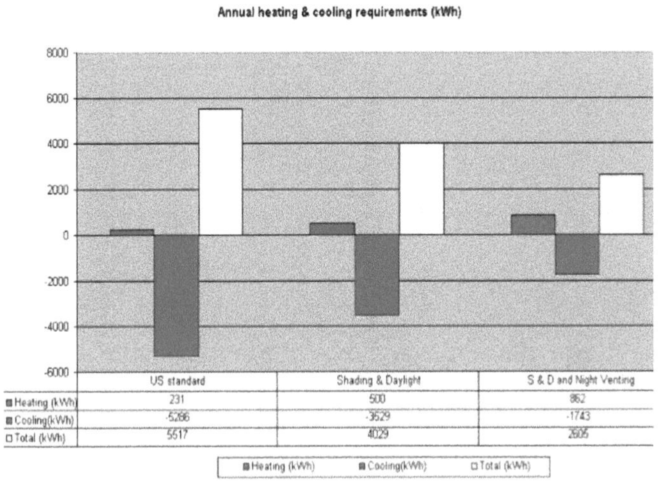

Annual heating & cooling requirements (kWh)

	US standard	Shading & Daylight	S & D and Night Venting
▣ Heating (kWh)	231	500	862
▣ Cooling(kWh)	-5286	-3529	-1743
▢ Total (kWh)	5517	4029	2605

▣ Heating (kWh)   ▣ Cooling(kWh)   ▢ Total (kWh)

Figure 8.4
A preliminary modelling exercise reveals that by shading the glazing from direct solar gain whilst admitting adequate natural light for normal tasks and then ventilating the building at night it may be possible to halve the energy consumed by a US Standard Building designed to early 21st century Illinois codes (courtesy of Professor K. Lomas).

A 'hybrid' environmental strategy, cycling between natural and mechanical modes, could cope with conditions extending beyond the recognized comfort envelope. It could yield significant energy and carbon savings and, seasonally, actually reintroduce its occupants to the natural world outside. The prospect of a building configured to cycle through these modes raises many interesting questions: cost, in the potential duplication of system; controls, how quickly could/should the transitions occur; what does it mean to be comfortable in a hybrid environment; what would the occupants' expectations of the environment likely be, and would they vary by mode; would they be more tolerant of the building in free-running mode; would they know what mode, should they know what mode; how might one stem the potential profligacy of occupants selecting mechanical cooling and heating when natural ventilation is in full flow?

The Level 2 plan of the Judson building as built shows that it is functionally diverse, housing the institution's library, faculty offices, lecture rooms, classrooms and the art, design and architecture studios. Its functional complexity, along with a scale of room sizes ranging from single occupant rooms to large densely occupied teaching spaces, taxed the base environmental strategy heavily. This is an informed client and, as one of the many Christian colleges in North America, it maintains a strong ethical position. The Trustees were prepared to pay a premium to deliver the sustainable intent (Short and Lomas, 2007).

Their 2001 competition brief asked:

... in what ways may both digital and environmental technologies, issues of sustainability, and effective stewardship of resources become design positions in architecture in general and in these facilities in particular.[3]

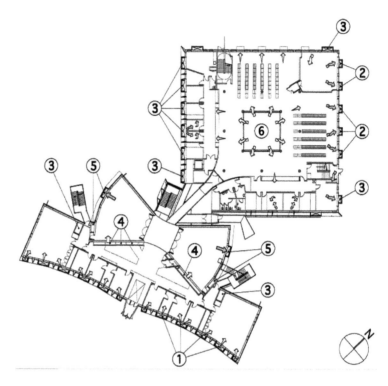

Figure 8.5
Judson University Academic Center Level 2 plan (Short and Associates).
1. Supply air ducts embedded in façade.
2. Exhaust air ducts embedded in façade.
3. Return air duct from roof plenum.
4. Riser ducts supply classrooms.
5. Exhaust air ducts connected to roof level plenum.
6. Central supply to library.

The environmental design strategy must allow close juxtaposition of deep (for the institution's library) and narrow sections (for faculty offices and teaching rooms). Large buildings until the early part of the 20th century were variants of 'mat' buildings, apparently deep overall but, in effect, woven ribbons of narrower elements interspersed with voids. The lessons from the 1866 Law Courts competition for the Strand site in London reveals this entirely understandable gambit, grids of terraces of very large houses connected by bridges and, in some cases, by glazed arcades (Wilson, 1985). As observed in Chapter 2, Le Corbusier's Moscow Tsentrosoyuz building is a mesh of narrow buildings spiralling around an auditorium building. But the Swedish master Gunnar Asplund (1885–1940) devised a fascinating variant for his 1939 competition design, his last, for the Stockholm City Archives, in which a narrower inhabited bar of administration building, oriented almost due south, clips the edge of a very large northeast–southwest oriented square book repository, embedded by the appropriation of a square bay of the deep square book repository (Holmdahl et al., 1950, p. 220). The National Association of Swedish Architects' rather mean-spirited text fails to recognize the ingenuity and elegance in the invention of a 'New Formal Device'.

James Stirling may have known of this arrangement when he invented the plan for the proposed but unbuilt Olivetti building in Milton Keynes, UK. Between these works many industrial sites were developed as thin administrative buildings glued face-on to a large shed. This was the formula of the Great West Road Firestone factory in London, with all the social implications of the divide along the junction, a type Gropius and Meyer replicated at the Fagus factory in Germany and pursued by the Ford Motor Company's highly inventive architect Albert Kahn, from the early Ford plant at Highland Park, Illinois 1909–10 to the Curtis-Wright Corporation Airport plant of 1941 (Hildebrand, 1974).

In the hybrid Judson building, the proposition is that the mechanical system will operate in displacement ventilation mode. As the driving pressures in a buoyancy-driven natural system are very much lower than those of mechanical systems, the cross-sectional area of the air supply and exhaust routes to deliver a given volume of air must be significantly greater, with lower resistance than a typical mechanical air distribution system. The ducts, plena and shafts will be big. Big enough to become architecturally significant. By using the air paths of the natural ventilation (NV) system to also distribute mechanically-conditioned air, fan energy consumption is reduced and this doubling up of the infrastructure for both modes is the innovation explored here. The plant is centralized to avoid the parasitic energy consumption associated with the pumps, valves, small fans and the other paraphernalia of distributed local cooling systems such as fan coil units and chilled beams. However, there is a penalty: a loss of flexibility. This is an interesting conundrum and important to solve as hybrid buildings should become the norm in our 'recovery'.

Early schemes placed a north-lit architecture school above the library and pursued a hybrid, top-down, 'centre-in/centre-out' scheme, proposing what came to be known as 'fresh-air fountains' at points across the floors, places for congregation discharging cool air which the building's occupants could gravitate toward (Figures 8.6, 8.7 and 8.8).

They could sit around the source of coolth on an Ottoman. This is an archaic but sophisticated form of furniture allowing for subtle social interactions, much to be commended as type furniture for a passive building. The elevational studies retract the glazing back onto the inner line of the double façade, the residual volume between them being used to enhance the basic Lanchester Library diagram by providing a perimeter air supply and exhaust infrastructure contained within a thick double façade. This is certainly not a wholly glazed 'doppelfassade', the popular, expensive and counterproductive leitmotiv of contemporary office buildings that was considered in Chapter 2. The New York architects Mitchell Giurgola employed a deep façade for their Columbia University laboratory building to organize live services, an idea derived, perhaps, from Franco Albini's 1961 La Rinascente store in Rome. The original northlight roof scheme was met with amused scepticism by seasoned Chicagoans who explained to the European designers how the build up of snow would result in leaks over the books and drawings before collapsing the structure. Nonetheless, the first unbuilt scheme presents an interesting alternative strategy for negotiating the Continental climate and later in the chapter it will be applied to the situation of Beijing.

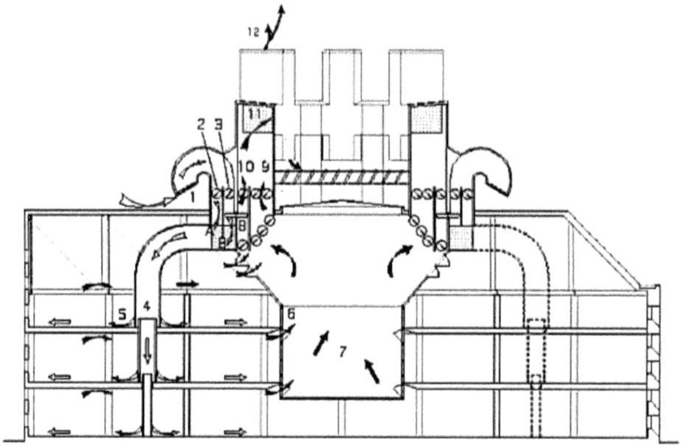

Figure 8.6
An early, unrealized, but promising top-down, centre-in/centre-out scheme for the Judson University Library. The section shows rooftop intakes supplying 'fresh-air fountains' in the centre of the floors either directly or through dessicant wheels. All the air is exhausted through a central lightwell. The Future House building in Beijing, discussed later in the chapter, takes this form but inverts it (author).

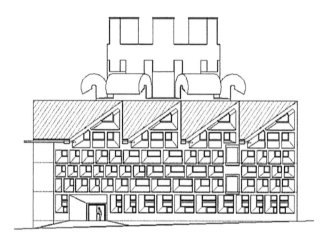

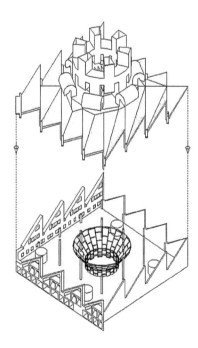

Figure 8.7
An early study for the centre-in/centre-out scheme for the Judson building, pushing all glazing deep into the elevation to avoid all direct solar gains in the 'cooling season' (author).

Figure 8.8
Axonometric view of the early centre-in/centre-out scheme for the Judson building (author).

Das Ding an sich: *Judson as built*

The built building is as heavy as is practicable, as at Leicester, Coventry, Malta and Bloomsbury, but more so. Two pre-cast concrete entrepreneurs, who had recently purchased a 400ft casting bed 100 miles to the north, under-bid the blockwork masons to make the external shell, an ever-present possibility in this fluid construction management arrangement in which the design is free to absorb opportunities encountered *en route*. The building was reconceived rapidly in precast concrete panels, every opening determined and finalized at very short notice. All concrete is left exposed internally, including the concrete plank floors. Open steel trusses support the planks across the 32m spans to avoid damming the warm stale air accumulating below the soffits. 3.4m high ceilings enable deeper penetration of daylight from the lightwell and a second layer of high level daylighting windows.

The built scheme preserves the 32.9m × 32.9m deep plan element of the early version, but explores a perimeter supply and exhaust infrastructure, so that the built building enhances the basic 'centre-in/edge-out' (C-E) approach as shown in Figure 8.5 (Lomas and Cook, 2005). Figure 8.9, a diagrammatic

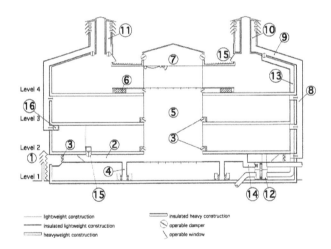

Figure 8.9
Library building, abstracted cross-section to show airflow routes, points of control and relative positioning of thermal mass and insulation.
1. Low level air intake sized to compensate for loss of free area through insect mesh.
2. Air inlet plenum insulated from building interior.
3. Air inlet with open hospital radiator behind to provide reheat.
4. Air supply route to Level 1.
5. Lightwell and air supply plenum.
6. Acoustic attenuation in Level 4 air supply path.
7. Ventilated buffer space between air supply plenum and exterior.
8. Extract air stacks from Levels 1, 2 and 3 incorporated within façade construction.
9. Insulated plenum within depth of roof construction.
10. Exhaust termination connected directly to roof plenum, incorporating rooflight. Outlet dampers behind belfry louvres.
11. Exhaust termination dedicated to Level 4 incorporating rooflight.
12. Air handling plant.
13. Return air duct connects exhaust air plenum to air handling plant.
14. Mechanical supply to air inlet plenum.
15. Air intake to cellular office with acoustic attenuator box, damper and reheat.
16. High level air exhaust via attenuator box to minimize cross-talk between offices.

cross-section, shows fresh air delivered to a centrally located, glazed but uninhabited atrium via a low-level plenum, from which fresh air is directed to each floor. It is sealed top and bottom as is the void in the SSEES building, the vessel which actually 'manufactures' the environment, to pursue the analogy with Boyle and Huygens' vacuum pumps (Figure 8.10).

The ventilation infrastructure

Figure 8.11 shows the plenum plan which is now very much more sophisticated than that of the Lanchester Library. Stacks, arranged around the perimeter, run the whole height of the building and exhaust the air to ambient, without any mechanical assistance. This arrangement will be familiar from the Lanchester Library.

In Natural Ventilation (NV) mode, air enters through the wide louvred inlets into the plenum (Figures 8.9 and 8.11), resisting wind-driven rain and snow. Dampers control the airflow and sensors close against reversing flows to prevent conditioned air escaping in mechanical mode. Raked heating coils pre-warm the air before it passes deeper into the building. The plenum is well insulated to the line of the heating coils and more lightly insulated thereafter. Insulation is necessary because, in winter, the incoming air, even after pre-warming, is likely to be cooler than the temperatures of the spaces above and below. On summer days, the warm incoming air should not warm the night cooled concrete floor of the space above or the ceiling of the floor below. Supply air is fed down to the spaces on level one and discharged into the lightwell through a plenum between levels one and two. The lightwell is sealed top and bottom by horizontal glazed planes (Figure 8.10).

There are two levels of airflow control on both the inlet and the exhaust routes. The dampers at the plenum inlet provide coarse control, whilst the dampers at the inlet to each floor provide fine control in response to air

Figure 8.10
View across the Judson air supply lightwell, glazed lenses above and below, low level opening lights to each floor.

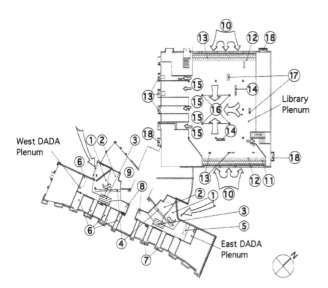

West DADA
Plenum

Library
Plenum

East DADA
Plenum

Figure 8.11
Lower plenum level plan.
1. Air intake to DADA wing for natural and mechanical modes.
2. Air rises to mezzanine platform at main plenum level.
3. Mechanical air-handling plant.
4. Riser to bowtie classrooms for supply in all modes.
5. Return air path from roof level air collection plenum.
6. Supply route to riser ducts in DADA wing façade.
7. Vertical supply routes within façade.
8. Supply to atrium.
9. Supply to Level 1 studios.
10. Air intakes to library.
11. Insect mesh folded to increase available free area.
12. Heating elements.
13. Risers to supply offices on Level 2.
14. Mechanical supply to plenum from air handling unit.
15. Supply to double façade distribution ducts servicing Level 2 offices and Level 3 teaching rooms.
16. Glazed base of central supply atrium.
17. Supply to Level 1.
18. Return air ducts.

temperature, CO_2 and humidity sensors. Air entering any space is reheated to the final desired temperature by hospital radiators located after the inlet dampers. The flow of air up each of the 20 exhaust stacks is controlled by dampers at the point of inlet, which operate in tandem with the air inlet dampers under the control of the building management system (BMS). In the perimeter offices, sensors will relay the state of the windows to the BMS, which can then close the air inlet and outlet dampers and thus avoid simultaneous passive (open window) and active (from air handling unit) air supply: enjoyable but irredeemable. Brokering between optimized control and occupant freedom is a black art. Occupants rarely behave as expected, which, ultimately, must bode optimistically for the future of humankind. Control strategies require a lot of time and not a little clairvoyance as we shall see and there must be a formal opportunity in the contract to review them after a year or more in operation.

The perimeter library offices connect into the double façade so that they operate on the 'edge-in/edge-out' (E-E) principle (Lomas and Cook, 2005). Figure 8.12 shows that the Department of Art Design and Architecture (DADA) wing is almost wholly configured as E–E.

The direct air supply, via the up-feeds from the plenum, avoids the need for 'transfer ducts' to enable air from the library to pass into the offices. In the Judson building, acoustic attenuators are used in the stacks to prevent 'cross-talk' between the perimeter offices and the library. The detailed wall section (Figure 8.13) shows some of the key features and illustrates how the perimeter stacks create a 1.2m deep recess, which naturally provides lateral solar shading to perimeter windows.

This recess is fully exploited on the southeast and southwest facing elevations where horizontal shading is created above the vision and clerestory windows. The gloss white finish to the window sills and reveals, redirects and diffuses direct sunlight into the building, thereby lifting the daylight levels (Figure 8.14). The design was subject to close analysis using Radiance software, undertaken by the Institute of Energy and Sustainable Development (IESD) at De Montfort University. The defensive geometry of the southeasterly window

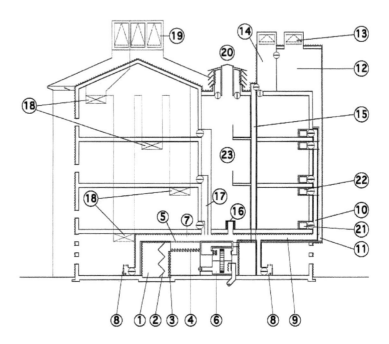

Figure 8.12

Diagrammatic cross-section through the narrower section Division of Art, Design and Architecture (DADA) building to show airflow routes, points of control, the relative location of thermal mass and insulation.

1. Low level air intake for all modes of operation, natural and mechanical.
2. Insect mesh, folded to maximize available free area.
3. Dampers mark effective line of enclosure to building.
4. Dampers control natural air path into plenum.
5. Heating elements.
6. Air handling unit.
7. Distribution plenum below Level 1 soffit for all modes of operation.
8. Air supply to Level 1.
9. Feeder ducts to south facing façade for supply air to upper levels.
10. Supply ducts within façade feed Levels 2, 3 and 4.
11. Return air ducts (shown in same plane for clarity, but actually alongside supply occupying full depth of façade).
12. Exhaust air collection in roof plenum.
13. Exhaust termination with dampers behind louvre blades.
14. Dedicated exhaust plenum for Level 4 with own exhaust termination.
15. Return air duct connects roof plenum to air handling unit.
16. Supply to atrium.
17. Dedicated supplies to Level 2, 3 and 4 classrooms (shown in same plane for clarity, but actually in line forming deep double wall to classrooms).
18. Air extract from classrooms shown dotted, within plane of end wall beyond, developed into a deep façade.
19. Exhaust termination to classroom extracts above double façade.
20. Exhaust termination to atrium incorporating rooflight.
21. Air intake to cellular office with acoustic attenuator box, damper and reheat.
22. High level air exhaust via acoustic attenuator box to minimize cross-talk between offices.
23. Atrium.

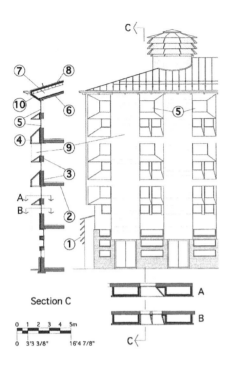

Section C

0 1 2 3 4 5m

0   3'3 3/8"   16'4 7/8"

A

B

C

Figure 8.13
Detailed wall section, part plans and elevation of the library double façade.
1. Belfry louvre air intake.
2. Precast hollow core floor planks.
3. Precast wall panels, 3.07m overall.
4. Deep façade, light steel frame holds shafts (9), and deep window reveals (5).
5. Glazing set back into precast panel, defended against solar gain by deep white finish reveals.
6. Precast soffit to Level 4.
7. Void of roof exhaust plenum.
8. Lightweight insulated roof deck.
9. Shaft houses air extract stacks.
10. Prefabricated connection to plenum at eaves to ensure air-tightness at vulnerable change in direction.

Figure 8.14
Detail of southeasterly elevation at 10.00 a.m. on 21 June.

cavities is shown in Figure 8.15a, and in operation at 10.00 a.m. on 21 March in Figure 8.15b. Very fine tuning of shading geometries is possible, potentially a rich and meaningful design driver.

The infrastructure of the deep elevation is unfurled across the DADA wing. Figure 8.16 shows the punctured concrete plane against which will be mounted the system of interlocking supply and exhaust ducts around the grid of window openings. The full façade is clad in copper, another unexpected contribution from an under-cutting supplier. Figure 8.17 shows this system in construction and Figure 8.18 the completed building across its new Fenland setting.

The system enables each office to have its own dedicated supply and exhaust. A natural symmetry develops around this arrangement. As the volume of air being carried up the supply shafts diminishes floor-by-floor, so the volume of air which is being carried by the exhaust shafts increases. Thus, the two separate shafts can be neatly interleaved, giving a composite vertical supply/exhaust route, which has a uniform depth and width all the way up the building as Figure 8.16 reveals.

The exhaust stacks connect into a roof plenum from where the stale air is exhausted to ambient. Figure 8.18 shows the plan arrangement at roof level, and Figure 8.19 a part of the busy roofscape of several types of exhaust.

The top floor rooms across the building are detached from the main ventilation system in order to avoid exhaust air in the stacks from lower floors flowing back into level four, an important finding from the Coventry scheme. The rooftop plenum of the library occupies the whole of the pitched roof volume and is exhausted by eight louvred termination devices. These are divided into four

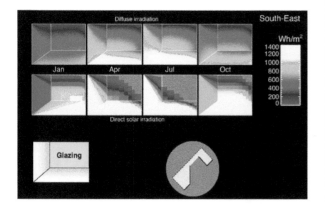

Figure 8.15a
Simulated solar exposure of a southeasterly façade, the deep reveals develop the customary mid 20th century brise soleil geometry for east and west façades which does indeed deliver protection through the summer. From Mardaljevic and Lomas (2006).

Figure 8.15b
Detail of shrouded openings at 10.00 a.m. on 21 March.

Figure 8.16
The precast concrete superstructure of the Judson building, immense thermal mass provided by a competitive subcontractor at a considerable cost in carbon emissions (author).

Figure 8.17
Double façade to the DADA wing under construction (author).

Figure 8.18
The completed DADA building façade from across the new Fenland landscape (author).

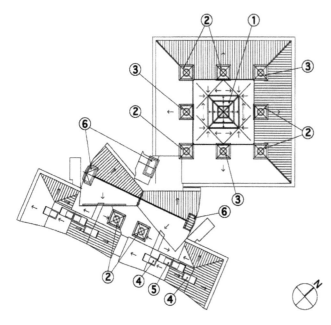

Figure 8.19
Roof plan.
1. Glazed roof to central lightwell to library.
2. Exhaust terminations from roof plenum with four radical compartments.
3. Exhaust termination from Level 4.
4. Single sided exhaust termination.
5. Double aspect termination set on ridge.
6. Termination to double faced classrooms with four clear orientations.

Figure 8.20
View across the busy roof (author).

separate compartments which enable the louvres in the compartments facing downwind to be opened while those facing into the wind to be closed. Five of the terminations exhaust the rooftop plenum, whilst three directly exhaust on Level 4. Similar terminations exhaust the atria in the DADA wing.

A computational fluid dynamics (CFD) model by Cook *et al.* attempted to predict airflow patterns and temperatures across half of the deep plan library element of the building. It certainly picks up the heat sources from occupants and computers, much the most intense in the top floor design studio. The generous ceiling heights are actually essential to permit stratification so that 'vitiated' air is drained out above the breathing zone. Clearly there would be a problem if they were lower. The exercise is discussed in detail in Lomas *et al.* (2006).

The building environmental control strategy is summarized in Figures 8.21. 8.22 and 8.23. The challenge is to predict effective controls responses through the seasonal, and daily cycles to keep the interior within the ASHRAE 55 comfort envelope, seamlessly switching modes and avoiding rapid oscillations, the phenomenon known as 'hunting'. There are three fundamental but dynamic modes of operation for the building: Mechanical Heating and Humidification (MHH), employed primarily in winter; Mechanical Cooling and De-humidification (MCD), the summer mode in effect; and the dominant mid-season mode, Passive Heating and Ventilating (PHV). Seasonal operating strategies will have to be flexible given the volatility of the climate. Changes in mode of operation will occur once or more within a single day. The building is always in passive mode at night. In winter, low ambient temperature and/or humidities will trigger the MHH mode. Occupants are encouraged to be dressed for the occasion and a minimum temperature of 20°C is set, at the bottom of the ASHRAE scale. Temperatures will rise under occupancy until they are above acceptable levels and will need to be vented away. Figure 8.22 explains how the passive mode is triggered, by rising external temperatures but also CO_2 levels. As temperatures rise towards the ASHRAE 55 maximum, mechanical cooling, MCD, is required (Figure 8.23) until temperatures subside and so the cycle continues.

Great expectations and the controls to manage them
The questions posed at the beginning of the chapter remain: how aware are the occupants of the shift in mode, and how forgiving are they in either natural or mechanical mode? Could a culture develop in which collectively the occupants resist the onset of a mechanical mode by raising set points to contribute to carbon reduction targets? This could occur because they are aware that ultimately they have control and are allowed some measure of personal thermal adaptation. The author is optimistic. Hybrid buildings can afford this

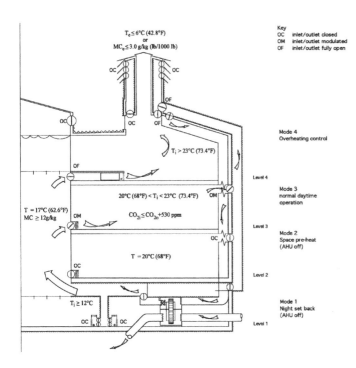

Figure 8.21
Winter operating mode. The MHH mode is triggered by a drop in ambient temperature below 6.0°C or an external humidity level below 3.0g/kg. Humidity is maintained by the recovery of moisture content by the enthalpy wheel shown in the diagram, CO_2 levels set at external + 530ppm. MHH operates in four modes: night-time HW radiators maintain temperatures above 12°C; before the working day commences heating is built to 20°C using the wet system; as occupancy commences the central AHU will supply air to maintain the lightwell at 17°C, machines and people raise the temperature, the exhaust is captured in the roof plenum and recirculated through the AHU; temperatures continue to rise on busy days to 23°C so that the heating is switched off and ventilation rates increased.

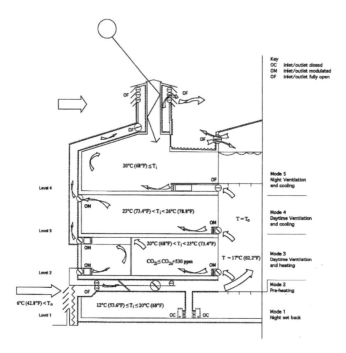

Mode 5
Night Ventilation
and cooling

Mode 4
Daytime Ventilation
and cooling

Mode 3
Daytime Ventilation
and heating

Mode 2
Pre-heating

Mode 1
Night set back

Level 4 — $20°C\ (68°F) \leq T_i$

Level 3 — $23°C\ (73.4°F) < T_i < 26°C\ (78.8°F)$, $T = T_o$

Level 2 — $20°C\ (68°F) < T_i < 23°C\ (73.4°F)$, $CO_2 \leq CO_{2o} + 530\ ppm$, $T \approx 17°C\ (62.2°F)$

Level 1 — $6°C\ (42.8°F) < T_o$, $12°C\ (53.6°F) \leq T_i \leq 20°C\ (68°F)$

Figure 8.22

Mid-season operating mode. As external temperatures rise above 6.0°C the mechanical systems shut down and air moves through the building by natural buoancy driven by internal heat gains, PHV mode, to achieve 20–27°C internally and humidities at or below 12g/kg. There are five PHV modes, the night setback and pre-start modes as the Winter mode with simple perimeter heating by radiators as necessary; on cooler days when external temperatures are below 17°C incoming air will be heated by coils in the plenum and radiators at point of entry, not by the AHU, driven naturally; on warmer days air will flow through unheated; as temperatures rise toward 27°C ventilation rates will rise as dampers fully open until summertime mode is required.

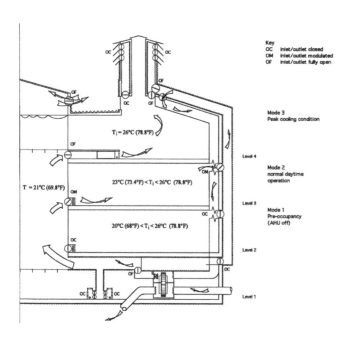

Mode 3
Peak cooling condition

Mode 2
normal daytime
operation

Mode 1
Pre-occupancy
(AHU off)

Level 4 — $T_i = 26°C\ (78.8°F)$

Level 3 — $23°C\ (73.4°F) < T_i < 26°C\ (78.8°F)$, $T = 21°C\ (69.8°F)$

Level 2 — $20°C\ (68°F) < T_i < 26°C\ (78.8°F)$

Level 1

Figure 8.23

Summer operating mode. MCD mode commences as temperatures rise towards 27°C and/or humidities rise towards 12g/kg. External dampers to the supply plenum close and the recirculation link from the attic plenum to the AHU opens. The CFD model suggests that air supply at 6 ac/hr should maintain spaces at 21°C. The sunshading to the lightwell will be in place. The building returns to passive mode as the temperature internally falls below 23°C. MCD has 3 modes: pre-start in passive mode maintaining temperatures between 20°C and 26°C; daytime mode with the various dampers and windows controlling airflows to maintain the comfort envelope; as temperatures rise towards the maximum all dampers will open fully to encourage vigorous flushing of all spaces affected.

opportunity by providing their occupants with overall control of their environment within a programme of induction and readily available information about the internal conditions. The complex arrangements at Judson for supplying fresh air independently to each space pay real dividends. In mechanical mode, the opening of a window will cease the supply of artificially conditioned air. Should one be more lenient in this to allow more personalized environments?

The commissioning exercise was very revealing (Lomas *et al.*, 2009). The designers joined the exercise in measuring airtightness and checking airflow patterns using theatrical smoke generators (Figure 8.24). The vertical envelope tested well for airtightness. However, the proprietary dampers emerged as being not fit for purpose. The blades were out of adjustment, seals were missing and linkages between banks of blades were woefully inadequate, preventing a good seal. Similar leakages emerged through the exhaust dampers to the top floor. Dampers to an exhaust stack were stuck open. The business-as-usual products available in the United States will simply not deliver an energy efficient building, quite the reverse. Partitioning of the roof plenum was missing in part and air from the top floor short-circuited back into the floor. This was almost certainly the result of a general lack of understanding by otherwise high-performing subcontractors. As Lomas *et al.* (2009) observe and the SSEES experience confirms, the industry on both sides of the Atlantic, increasingly focused on delivering discrete packages of work in strict sequence, is not organized to ensure the seamless implementation of a relatively complex design intent in the way other industries have been able to for a long time. Aircraft manufacture is a particular example. This is an important barrier to the delivery of the recovery.

Mechanical ventilation to the WCs disrupted passive airflows, clearly a generic problem for passive buildings. Some building regulation regimes allow passive ventilation of WCs. There is no reason why that should not be very effective. A less efficient option would be to pressurize lobbies between mechanically vented service spaces and the main body of the building. Ventilation stack termination louvres were found to be closing leeward and opening windward, the reverse of what was intended. Perhaps control responses are too slow in turbulent conditions. A line of lightwell window actuators had failed. However, these non-functioning elements were as nothing compared to the SSEES building defects

Figure 8.24
The air in the lightwell may be considerably warmer than ambient in winter, and cooler than ambient in summer, introducing buoyancy forces. The various plena must be tightly sealed. This was not entirely understood and air leakage was problematic during commissioning (Lomas *et al.*, 2009).

described in Chapter 4 and certainly all of the engineers and subcontractors involved were very interested in a productive outcome, and slightly bemused by the poor nature of standard products they had clearly specified many times, products developed in the hubris of 'abundance'.

Performance, cost and payback
Are there real energy benefits from building in this less familiar way? An attempt to quantify the likely performance of the Judson building against a credible 'business-as-usual' building in the same situation used the dynamic thermal model ESP-r (Energy Systems Design Research Unit ESRU, 1998). It is dynamic in that it predicts thermal behaviour through time, in this case using the Midway climate dataset, the Chicago TMY2 as described. Working hours were taken as 8.00 a.m. to 6.00 p.m. every day and weekday internal heat gains at 34W/m^2, reduced at weekends. The control code-compliant building was sealed with lightweight hung ceilings and a central air-handling machine. The Judson building was modelled with the control strategy described in Figures 8.21–23. Figure 8.25 shows the fan power requirement in the Judson building can be much reduced, 1.1 kW/m^2/sec compared to 2.4 kW/m^2/sec in the standard building. Summer cooling loads in the Judson building are less than half those of the standard building, and occur only during the three months of summer rather than seven months. Heat energy is saved because of higher insulation but, in the spring and autumn, heating loads are higher because direct solar gains are blocked by the geometry of the window reveals. Costs for delivering energy to the standard building are higher in every month, not least because fans are running continuously, whereas in the Judson building they are silent for 52% of the year.

What does this improved energy performance actually cost to deliver? Table 8.1 summarizes the costs. The standard building is 'no frills' whereas Judson invested, with restraint, in design ideas. In fact some $175/m^2 could be said to be cost directly attributed to the hybrid strategy, some 9.4% of additional cost. The bespoke controls alone were $109/m^2. Standard controls packages are remarkably cheap in the USA. Inevitably the payback period is not insignificant.

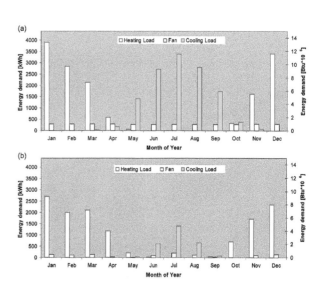

Figure 8.25
Monthly heating, cooling and fan energy loads for a US Standard building operated to a 24°C set point built to Illinois codes above, and for the Judson University building below. There is a marked reduction in summer cooling loads.

Table 8.1 Comparative constructional costs for the Judson College building and a US Standard building.

Element	Description	Judson College[1]		US Standard Building[2]		Judson additional cost and percentage increase
		Cost $/m²	Percent of total	Cost $/m²	Percent of total	
1 Site work	Excavation, site utilities, landscape and retaining structures	$144[3]	6.8%	$102	5.5%	$42 (40%)
2 Sub-structure	Site concrete	$21[4]	1.0%	$16	0.9%	$5 (31%)
3 Concrete	Above ground	$107	5.1%	$81	4.4%	$26 (33%)
4 Structure	i) pre cast concrete	$237		$161	8.7%	
	ii) structural steel including fire proofing	$118 → $355[5]	16.7%	$161	8.7%	$33 (10%)
5 Roofing	Metal roof build-up and coverings	$172[6]	8.1%	$161	8.7%	$11 (7%)
6 Internal Partitions	Framing and drywall	$140[7]	6.6%	$118	6.4%	$22 (18%)
7 Glass and Glazing	Glass, glazing, lightwells, skylights and windows	$137[8]	6.5%	$140	7.5%	−$3 (−19%)
8 Moveable Shading	Shade at top of lightwell	$3	0.2%	Not typical	0.0%	$3
9 External Walls	Masonry and cladding, (for Judson incorporating deep window reveals and return)	$162[9]	7.7%	$129	7.0%	$33 (25%)
10 HVAC	Heating, Ventilating, air-conditioning, controls	$410[10]	19.4%	$301	16.2%	$109 (36%)
11 Other HVAC	Plumbing, alarms, phones, data, elevations, sprinklers	$297	14.0%	$323	17.4%	−$26 (−8%)
12 Finishes	Painting	$145	6.9%	$140	7.5%	$5 (4%)
13 All other construction	Doors, general trades, wc cubicles, flooring, fire shutters	$22	1.0%	$21	1.1%	$1 (1%)
Totals		$2115(11)	100%	$1854	100%	+$261 (14%)

Notes:

1. Judson College building costs (unpublished) are supplied by Shales McNutt Construction (2005).

2. US Standard building costs are assembled from contemporary Dodge Reports (2005).

3. Site works: Additional cost includes the forming of detention ponds, the Fenland landscape, to provide a sustainable drainage scheme; US $319 000 for the import of additional fill/soil to remedy poor site conditions. Excavated fill is used to form Fenland. Site utilities and retaining structures are assumed to be equivalent for both schemes.

4. Substructure: Some additional complexity relates to plan geometry, the orientation of the Division of Art, Design and Architecture (DADA) wing and air intakes; the structural solution of the precast load-bearing wall elements increases the design strength of the perimeter retaining walls; the weight of steel reinforcement rises.

5. Structure: Precast panels are more costly; there is an increased number and pattern of openings in precast panels relating to airflow routes; there is increased sub-contractor engineering input and increased labour content in manufacture. Overall, structural steel content is somewhat lower, whilst the precast concrete element is higher than for a US Standard building. There is a 10% premium rate on the more complex external wall construction to accommodate the advanced environmental strategy.

6. Economies of scale deliver a fairly complex roof form at Judson College for almost the standard rate.

7. Internal partitions: Additional cost relates entirely to increased ceiling heights, which are beneficial to a passive airflow strategy, accommodating temperature stratification.

8. Glass and glazing costs are almost equivalent; the window shapes and expected performance sizes are standard to US suppliers.

9. External walls: The additional cost includes for the light structure and preformed ductwork forming the double façade; all the returns and deep reveals and copper shingle external covering. However, a high degree of repetition enables serial prefabrication off-site in 10ft and 12ft, three-sided elements, lifted into place. This cost is offset, in part, against the full internal ductwork of the Standard building with conventional HVAC.

10. HVAC: US prices for non-standard controls are high, US $500 000 as compared with US $150 000 for the US Standard building (the highest quotation received was of the order of US $1 million). However, there are 1500 sensor points in the Judson College building; the US Standard building envisages only 500; there are more moving elements, actuated dampers and windows, and a requirement for a full three-month commissioning period.

11. In summary of the additional US $261/m², a 14% increase in the overall cost of a US Standard building, additional costs specifically associated with the hybrid environmental strategy emerge as US $175/m², equivalent to a 9.38% premium rate on a US Standard Building costs, made up as follows: US $33/m² premium on structure; US $33/m² on external wall construction; and US $109/m² for controls. In the case of the Judson College building, approximately US $86/m² of the over-cost derives from specific site conditions and client requirements.

Following the standard formula for a 60 year building life expectancy:

$$\frac{\text{Difference in cost including equipment replacement costs}}{\text{Energy saving}} = \text{Payback period}$$

The estimated equipment costs over 60 years, assuming other maintenance costs are equivalent, arise for the US Standard building from the replacement of the chillers every 20 years at $64 500 per exercise and the air handling units every 15 years at $132 000. For the Judson building it was assumed that the chillers are replaced every 35 years at $64 500 and the air handling units every 30 years at $132 000 because of their greatly reduced running times.

Using the difference in construction cost ($175 × 7810 m² = $1 366 600) gives:

$$\frac{\$1,366,600 - ((3 \times 64,500 + 4 \times 132,000) - (64,500 + 2 \times 132,000))}{\$15,800} = 62 \text{ year payback}$$

The pay back is in excess of the life of the building. If energy costs were 25% higher in Illinois, then the cost saving would become $19 800 per annum and the payback period 49 years. If energy costs rose almost immediately by 50%, then the payback period would reduce to 41 years.

The net present value (NPV) analysis of capital and running costs for the Judson building, compared to the base case, is equally revealing. The methodology is derived from CIBSE Guide B18, Owning and Operating Costs (CIBSE 1986). Investments are generally assumed to be viable if the NPV is zero at the appropriate discount rate.

The analysis, described in full in Short *et al.* (2007), is set in the more commercial context of 25 year and 50 year periods and uses both a 5% discount rate (as advised by the College Vice-President, Business Affairs) and a 3.5% discount rate as recommended for UK public sector investment. The discount rates are real rates and all costs are considered in real terms, ignoring inflationary rises. Capital Cost at Year 0 is not discounted.

NPVs are always negative and the running costs multiplier (K) to achieve a NPV of zero is between 5.3 and 7.6, although the US Federal Government's advice is that fuel costs will remain stable in the long term. The non-energy benefits of the building are, however, substantial and around the minimum that the college anticipates achieving through additional student enrolment and the associated additional fee income, staff retention, enhanced reputation and profile.

The college would not have proceeded on energy-saving grounds alone. As Ellingham and Fawcett (2006) explain, uncertainty, irreversibility and the use of options, explicitly or implicitly, in decision-making, makes managers more cautious about energy-saving measures than other capital items. They describe this phenomenon as the 'energy paradox' and refer to the intense frustration it causes energy conservation advocates. In the Judson context the required additional benefits to achieve a zero NPV are relatively containable on an annual basis relative to, say, its annual turnover of between $17.5 million (2003/04) and $18.7 million (2004/05). And, this client has a strong ethical position, but must maintain a healthy economy enabling growth and development to survive.

How will this building strategy perform in a changing climate? The Köppen map updated to 2076–2100 reminds us that the IPCC modelling projects a change in climate. The Chicago climate is now predicted to shift to find itself at the meeting of the Cfa 'Warm Temperate, fully humid, hot summers' and the Dwd 'Snow, winter dry and extremely continental' regions towards the end of the century (Rubel and Kottek, 2010, Figure 2.9). Extreme events will happen more frequently throughout the year and so buildings will need to be robust.

But, there is the possibility that the greater Chicago area, notwithstanding any urban heat island effects, will experience milder and longer mid-season periods, extending into the beginning and end of winter. The opportunity to free-run the Judson building may be extended. This is potentially an extremely important development globally across a very large area within the continental climate zone, permitting very substantial carbon reductions in conditioning building environments, an opportunity to be seized. We now turn to Beijing which lies at the same latitude as Chicago with a similarly intense Continental climate.

An application in Beijing

The first hybrid configuration for the Judson University building, a scheme in which coolth was injected both naturally and mechanically into a large open volume through point sources, 'fresh-air fountains', has now been built as an exhibition headquarters in Beijing, Future House. Beijing, at 40° N, is close to the latitude of Chicago, at exactly 42° N.[4] Although classic 'latitudinal' climate theory has long been discredited, there are marked similarities in the climate of the two city regions, if not necessarily in the respective populations' demeanour. The moral philosopher David Hume (1711–76), thoroughly debunked the proposition of a causal connection between national character, climate and air (Hume, 1994, p. 83): 'The CHINESE have the greatest uniformity of character imaginable; though the air and climate, in different parts of those vast dominions, admit of very considerable variations.'

Beijing is located in the 'Cold Zone' as designated by the Chinese 'Thermal Design Code' (CNS 1993) depicted in Figure 8.26.

The city is subject to the Siberian air masses that move southward across the Mongolian Plateau, winters are cold and dry. But summers are hot, driven by warm and humid monsoon winds from the southeast. IPCC model projections to 2100 (Figure 8.27) mapped onto the Köppen-Geiger climate zone classification, predict, as in North America, a significant expansion of the more temperate zone so that Bsk 'arid steppe, cold' may 'push' the Dwa 'snow, desert, hot summer' zone northeastwards. Beijing will no longer be transitional, in the way that the climate enjoyed by Chicago may also morph. Perhaps 'Latitudinal Belt Theory' will be less redundant than the anti-determinists have argued.

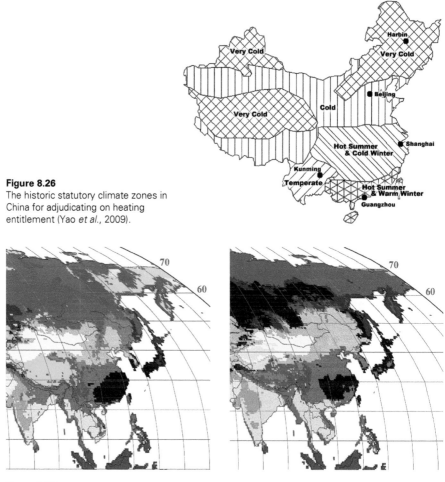

Figure 8.26
The historic statutory climate zones in China for adjudicating on heating entitlement (Yao *et al.*, 2009).

Figure 8.27
China's observed climate regions as of 1901–21 (left) and as predicted by the relevant IPCC team's modelling for 2076–2100 (right) based on the original Köppen-Geiger map (Rubbel and Kottek, 2010).

China's building sector energy use comprised 30% of the country's total energy use in 2013.[5] It was 31% in 2005. As a proportion, it is lower than in a Western economy, but this economy is in a very different developmental stage, growing as successive ten year plans deliver 20 billion m^2 of new building apiece, 2 billion m^2 per year, equivalent as Professor Baizhan Li has calculated, to a New York City replicated annually. China is currently operating at the level of energy consumption enjoyed by the United States in 1980. Is it on the same trajectory? The building sector certainly seems to be, only faster. Nowhere is it more important to achieve the 'recovery'. Hybrid schemes such as that devised for Judson University may just satisfy the Chinese population's legitimate aspirations to greater comfort and changing perceptions of 'value' in the built environment. Both phenomena are imports from the West but, as Judson demonstrates, hybrid buildings have a significantly lower energy and carbon penalty because of their ability to free-run through the periods of more benign external conditions.

Hybridity transferred to Beijing: Future House

The course of the Future House project could be read as a cipher for the early 21st century building dilemma in China. The Expo Centre at the heart of the Future House Exposition, as built, is a three storey plus basement, north–south orientated building of approximately 4000m^2 gross internal floor area. It is technically a retrofit of an existing reinforced concrete frame. Originally designed by Europeans, the intent was to clad the frame in a highly glazed envelope, including a south-facing glass façade to be entirely constructed in photovoltaic panels (which, of course, generate significant heat at their internal face). The Dean of Building Science at Chongqing University observed that the original design would overheat even on mild days and require intense mechanical cooling to permit inhabitation, somewhat more so than required to condition a conventional building to Western standards of comfort, an unwelcome irony at an exposition of sustainable housing technologies. Shove *et al.* (2008) comment ironically on 'the impressive convergence of indoor climates around the world and between one building type and another', which they credit to 'powerful providers' of office floor space with international estates, 'despite the diversity of climates and potential adaptive responses'.

Glass building envelopes are becoming an entrenched 'social practice' in the immense Chinese building economy. Adherence to this imported building type with, as has been briefly surveyed in Chapter 2, its bizarre and highly complex origins in late 18th and early 19th century horticultural buildings, early 20th century Expressionist phantasy, obsessive compulsive behaviours, and the collision of emerging technologies will certainly impede China's construction industry from achieving the mandatory carbon reduction target of 50% of the 1983 figure by 2020 (CNS, 2001). The State is very aware of the dilemma. The Ministry of Urban–Rural Housing Development (MOHURD) is anxious to point to energy efficient building Demonstration Projects. Future House is such a Demonstration Project, in Beiquija on the Fifth Ring Road of Beijing.[6] It lies within the Future House 'Exhibition Village', intended to be a collection of innovative but replicable low energy homes donated by various nations to exhibit their national expertise in low energy design and technology.

The author's design for the retrofit of the Future House building is informed, not unexpectedly, by a close reading of the Chinese National Weather Data, specifically the set within it assembled for building designers, the China Standard Weather Data for Building Thermal Environment Design (CSWD 2005). This generates a typical meteorological year (TMY) based on hourly data collected between 1971 and 2003. Table 8.2 summarizes key external design data from CSWD 2005. There is a strong comparison with Chicago.

Figure 8.28 depicts mean monthly Beijing temperatures with 80% and 90% acceptable 'Operative Temperature' ranges using the Standard and Alternative Methods superimposed on the unforgiving ASHRAE 55 Standard Method chart.

Table 8.2 Key external design data for Beijing (CSWD 2005)

Winter		Summer	
Design temperature for ventilation	−7.6° C	Design temperature for ventilation	29.9° C
Design humidity for ventilation	N/A	Design humidity for ventilation	58%
Design temperature for space conditioning	−9.8° C	Design dry-bulb temperature for air-conditioning	33.6° C
Design humidity for space conditioning	37%	Design wet-bulb air-conditioning	26.3° C
Mean air velocity	2.7 m/s	Daily mean temperature	29.1° C
Extreme temperature	−18.3° C	Extreme temperature	41.9° C

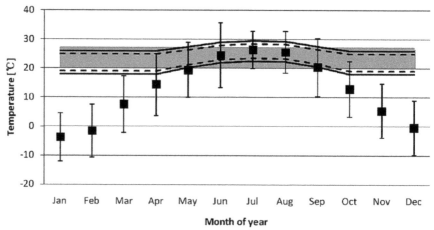

Figure 8.28
Mean monthly Beijing temperatures with 80% and 90% acceptable 'Operative Temperature' ranges, Standard and 'Alternative', ie Adaptive, Method.

The pattern is not unlike that for Chicago. Summer nights are hotter, in July and August remaining resolutely at, or above, 20°C, whilst winters are perhaps less intensely cold and less volatile. In terms of transferring our hybrid proposition to Beijing, the mid-season temperatures are strikingly similar. As in Chicago, periods in April to October fall within the comfort band.

Figure 8.29 shows the frequencies of ambient temperatures in excess of 25°C, and 28°C for the months from April to October. From the figure we can see that in July, the hottest month in the 'typical' year, there are 90 hours between 25–28°C and 187 hours over 28°C, which account for 29% and 60% of working hours respectively.

The crude 28°C cap would pose an undeliverable challenge to a wholly passive scheme. Clearly, cooling is required, but, as in Chicago, not for every hour of the extended summer period. Table 8.3 shows the CSWD data for Beijing input into the ASHRAE adaptive comfort equation for establishing comfort temperatures.[7] For Beijing, comfort temperatures vary through the summer between 22°C and 26°C, a surprising but promising range.

As in Chicago, a relatively crude strategic design exercise scoped the prospects for a hybrid approach, a simple dynamic thermal model, 'Lighting, Thermal and Ventilation' (LTV). The results were expressed as a 'Natural Ventilation Cooling Potential' (NVCP) index, the ratio of the number of hours within the comfort zone over the total occupied hours (Yao *et al.*, 2009). The analysis demonstrated that night-time passive ventilation across sufficient

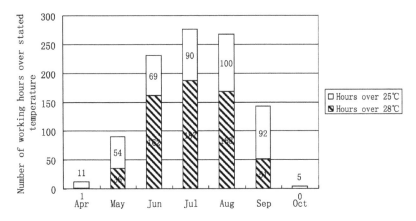

Figure 8.29
Monthly occurrence of ambient temperatures between 25°C and 28°C, and in excess of 28°C in the Beijing CSWD during working hours, 08.00 a.m. to 6.00 p.m.

Table 8.3 The acceptable comfort temperatures in Beijing °C

	April	May	June	July	August	September	October
Monthly mean temperature °C	14.4	19.4	24.5	26.4	25.6	20.4	12.9
Comfort Temperature °C	22.3	23.8	25.4	26.0	25.7	24.1	21.8

exposed thermal mass could extend the passive operating time and, in mid-summer, significantly reduce the load on the air-conditioning system. The dynamic simulation showed that the NVCP through the Beijing summer is 63% to meet the 80% acceptable thermal comfort upper limit, established as 28.8°C, and 81% to meet the 90% acceptable thermal comfort upper limit of 29.8°C.

Figure 8.30, the north–south cross-section through the Future House building illustrates the environmental design concept.

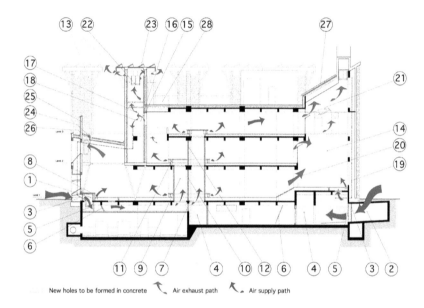

New holes to be formed in concrete   Air exhaust path   Air supply path

Figure 8.30
North–south cross-section: existing concrete frame shown in solid black.
1. Low level air intake to south.
2. Air intake to north.
3. Louvre blades.
4. Fresh air collection chamber to feed plenum.
5. Dampers.
6. Heating elements.
7. Damper connecting north and south sections of plenum to balance pressures.
8. Floor register supplying air.
9. Fresh-air fountain rising to Level 2.
10 and 12. Internal divisions within fresh-air fountains.
11. Air outlet to Level 2.
13. Ventilation exhaust stack.
14. Full height atrium to north.
15. Dampers connecting Levels 2 and 3 to south exhaust stacks.
16. Exhaust stack.
17. Dampers.
18. Open grille access platform.
19. Return air.
20. Level 1 exhaust directly into atrium.
21. Levels 2 and 3 compartmented, opening vents exhaust into atrium.
22. Fans at stack terminations engage when natural ventilation flows stall to prevent downdraughting.
23. Photovoltaic panels installed on stack tops.
24. Stacks shade north lights.
25. Brick masonry screens shade roof glazing.
26. Louvre admits exhaust air from Level 1 to stack.
27. Photovoltaic panels on south-facing atrium roof.
28. Safety guarding at parapet.

The existing frame of two-way coffered slabs, shaded black, acquires a plenum below the upper ground floor and various low level air intakes, numbered 1 and 2 on the drawing, to provide, alternately, a natural or mechanically-driven supply of fresh air as at Judson University. Diminishing shafts climb up out of the plenum through the floors to supply air at specific points, described as the 'fresh-air fountains', numbered 4 and 9–12 on the section. An air-handling unit in the basement will temper incoming air diverted from its direct naturally driven route. Air is exhausted at high level from each floor through stacks built into a double façade at the perimeter and stacks propped on steel legs mid-plan, numbered 13, 15, 16 and 26. The atrium formed to the north also exhausts the lower two levels, 14, 20 and 21. Figure 8.31 shows the section taken at 90° across the deep plan. During construction, the inboard stacks were omitted by the developer, an anxious moment for the design team, and the analysis re-run to establish the implications of this mid-flight value-engineering, the outcomes reported a little later in the chapter.

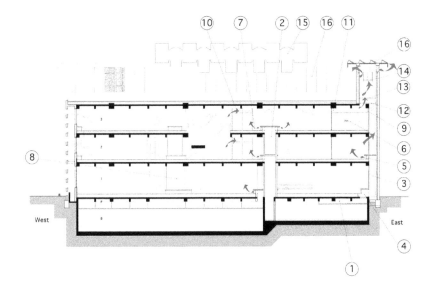

Figure 8.31
East–west cross-section locating plenum, fresh-air fountains and double façade elements.
1. Air supply plenum to north.
2. 'Fresh-air fountain' supplies Levels 1, 2 and 3.
3. Fresh-air supply to Levels 1 and 2.
4. Connection chamber between east double wall and plenum.
5. Air inlet to Level 2.
6. Enclosed office with bypass duct.
7. 'Fresh-air fountain' supply to Level 3.
8. 'Fresh-air fountain' in elevation beyond.
9. Bypass exhaust duct connection to stack.
10. Exhaust Level 3 to atrium.
11. Dampers to bypass duct.
12. Dampers to stack.
13. Open grille access platform.
14. Fan to prevent stalling airflow.
15. Exhaust terminations to north wall beyond.
16. Photovoltaic panels.

The ventilation infrastructure for all ventilation modes is the same as in Chicago, essential in this hybrid approach to maintain costs comparable with the fully artificial equivalent. This is the architectural challenge. It is a dominant design driver. The infrastructure is oversized for mechanically driven flows so that fan energy is reduced, there is less resistance in the 'ducts', grilles and other custom and practice resistances, an unforeseen additional benefit.

There is a 9% cost premium for the Judson University hybrid building, much of it in the bespoke controls but it is essential to achieve parity to enable the widespread market penetration required to begin to achieve the carbon reduction targets. The strategy for the Beijing Future House building has been developed from the first scheme for the Chicago building as discussed. It is materially different to the as-built Chicago building, which employs a central supply atrium and a double façade to distribute ventilation air. The Future House building is a genuinely deep plan building, exploiting concentrated sources of coolth which we refer to as 'fresh-air fountains' and exhausts distributed around the perimeter, and, ideally, inboard, centre-in/edge and centre-out. Readers may note a striking resemblance to Dr Folsom's scheme for the General Wards at the Johns Hopkins Hospital in Baltimore of the mid 1870s described and modelled in Chapter 6. It is a generic robust diagram for a single large space.

Ventilation design strategy

Importantly, the Future House scheme envisages some discernable variability in the temperatures across the deep floorplates, such that occupants in the exhibition spaces would gravitate towards the 'fountains' of cooled air as they wish and become similarly warmed in the winter, a traditional strategy usually based on radiant heat sources and potentially the source of some 'thermal delight'. This represents the recovery of pre-modern concepts of 'good' environments swept away by the industry's assumption that uniformity is required and demonstrates 'sophisticated' mastery and control by new technologies.

Briefly, the design unfolds as follows. The plenum plan (Figure 8.32) is drawn below the Level 1 concrete floor slab and appears as a suspended ceiling and bulkhead within the basement.

Fresh air enters into the plenum from the perimeter of the building. There are two inlets at the south where air enters the plenum from above, through openings already formed in the Level 1 concrete slab (1). To the north, fresh air is introduced into the plenum through the existing vehicle entrance (2). Air intake louvres, with bird mesh are located at the perimeter (3). In order to get sufficient volumes of air from the north and one of the southern inlets it is necessary to create a full height air corridor (4) rather than a bulkhead. Insulated mechanical dampers (5) control the volume of incoming air, actuated by the Building Management System. The air is tempered when entering the plenum by heater batteries (6), installed at an angle to encourage the flow of air into the building rather than reverse on warming. The plenum plan shows the location of new holes required in the existing concrete slab (7) and the reuse of existing holes (8). It is in effect an adaptive refurbishment. The critical connection points between plena compartments are highlighted on the plan (9 and 10), linked to enable air supply to each of the 'fresh-air fountains' regardless of wind direction. Mechanical dampers, fire-rated due to the security requirements of the

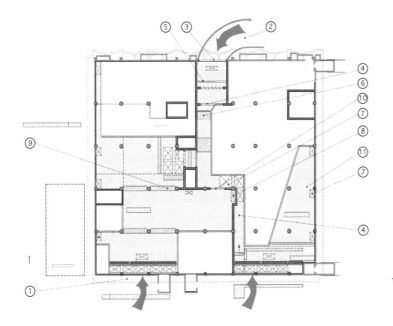

Figure 8.32
Plan of plenum.
1. Air intake to south.
2. Air intake to north.
3. Insect mesh screen.
4. Connection to air supply plenum.
5. Dampers.
6. Heater batteries inclined at 45°.
7. Indicates openings in the slab above.
8. Existing opening utilized.
9. Connections formed through structure.
10. Damper connects north and south plena compartments.
11. Second damper connects plena.

basement, are placed in the existing concrete dividing wall (10) at high level to enable this. Fire dampers may also be required in the other new openings required to the east of the basement (11). The whole of the plenum floor is well insulated, both to keep the plenum cool in the mid-season and peak summer months and warm in the winter months. The importance of achieving air-tightness cannot be overestimated.

Level 1 (Figure 8.33) comprises a separate ventilation compartment and incorporates the atrium to the north (1) and the double height entrance space (2).

The building is split into two ventilation compartments to reduce the potential for overheating of the top level by exhaust air bleeding back in from a shared exhaust, a lesson learned from the near failure of the Lanchester Library design. Fresh air is delivered into Level 1 from the perimeter (3) at the north and south via low level inlets, and from the centre (4) via two fresh-air fountains, rising through the building, diminishing in girth with height. Dampers and small heating elements are integrated within these fresh air inlets to 'fine tune' the incoming air. Ottoman seating is formed around the base of the fresh air inlets, permitting both conversation and silent detachment.

Stale air from Level 1 is exhausted via brick ventilation stacks, the double height space at the entrance (1) and a rooftop exhaust to the atrium (2). In order to balance the exhaust, and to prevent reversing flows down the front stacks, the ventilation stacks on the north and south are prescribed at the same height (Figure 8.30), an important learning from the SSEES building in which they were not. Fixed tables, which can be used as exhibition display space, are provided along the south elevation directly above the air intake into the plenum (6). Fixed seating (7) is also provided at the base of the ventilation stacks, which enable the fresh air from the raised floor void to enter into the base of the stack, maintaining the critical cross-sectional area. A steel transfer structure (8) supports the brick ventilation stacks and carries the load of the brickwork onto the concrete columns directly below in the basement for stacks 5–7 and back to the south retaining wall (9) for stacks 15–18. The mid-plan, free-standing stacks were judged structurally too demanding, too expensive and were therefore omitted.

Levels 2 (see Figure 8.34) and 3 are linked vertically by the voids at the front of the plan at Level 2 (1) and by the large staircase void (2) at Level 2.

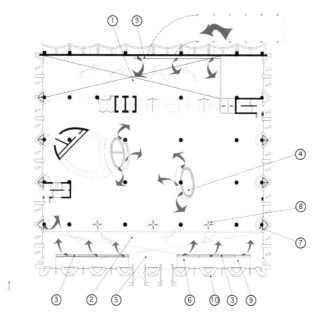

Figure 8.33
Level 1 Plan.
1. North atrium.
2. Void to south side.
3. Intakes at south side.
4. Fresh-air fountain.
5. Entrance area.
6. Bulkhead to distribute fresh air.
7. Distribution bulkhead to east side.
8. Structural supports to south stacks.
9. Bulkhead to distribute fresh air.
10. Air intakes to plenum, south side.

Figure 8.34
Level 2 Plan.
1. Rooflight above.
2. Rooflight above stairwell.
3. Enclosure to atrium.
4. Glazing to south in formed wall depth.
5. Opening vents exhaust level 2 into atrium.
6. Lobby to stair.
7. Intakes at perimeter walls.
8. Fresh air fountain.
9. Exhaust stack.
10. Outline of PV roof glazing above.
11. Exhaust through stair toplight.
12. Steel bracing to stacks.

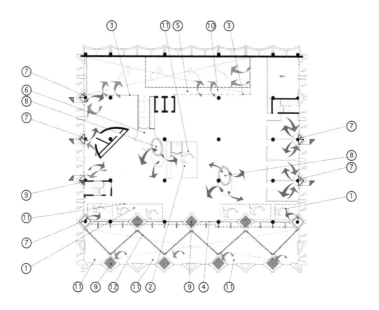

They are separated from Level 1 by a glazed screen along the north of the Level 3 plan (3) and a brickwork wall with large window openings at the south (4) at Level 2. Levels 2 and 3 can however exhaust stale air into the atrium space via Building Management System controlled openable windows (5) in this northern glazed screen (Levels 2 and 3). In order to fully separate the two compartments (Level 1 and Levels 2/3) it is necessary to install a glazed lobby at the top of the staircase from Level 1 to Level 2 (5). Fresh air is delivered to Level 2 in much the same way as it is to Level 1. Air is introduced from the perimeter and from the centre by the fresh-air fountains (8 on Level 2 plan). It is exhausted at the perimeter via the perimeter stacks and through the rear atrium. Stale air from Level 2 is also exhausted to Level 3 through the voids at the front of the plan and through the staircase void. The exhaust for Level 1 can also been seen at this level – automated dampers are placed at high level within one side of the brick ventilation stacks. There are rooflights above which are shown dotted on plan. Diagonal steel bracing (12 on Level 2) is inserted to support the brickwork ventilation stacks at the front of the plan. The perimeter offices are treated separately from the rest of the floor. Air is both introduced and exhausted via the perimeter ventilation stacks.

Tailoring the façade to the latitude

The architecture of the west and south elevations is, in part, determined from the clients' preference for brick as a material and the stack driven exhaust of vitiated air but perhaps the most dominant driver is the exclusion of solar radiation during the overheating season. The geometries of the window openings are configured to admit and exclude sun to a precise schedule through the year. The shading is integral with the wall construction, not an applied lightweight sunshade, but a more fluid and responsive development of the Chicago strategy (see Figure 8.35).

Figure 8.35
Future House Expo Centre, east elevation late morning on 21 June.

The energy required for mechanical cooling in 'peak summer' is significantly reduced and the period in which the building can stay in its most efficient 'mid-season' mode is prolonged. Figure 8.36 shows how the deep inset window geometry at the mid-season becomes increasingly defensive as the seasons unfold towards mid-summer.

As the sunlight is less intense during the winter months, direct sunlight is allowed to penetrate into the building, a useful source of passive heating, but glare may still need to be controlled with internal blinds. The coloured shading represents the direct sunlight at a given time of day, for example orange denotes 12:00pm. Figure 8.37 shows the built outcome, crisp window surrounds pre-formed very precisely in glass reinforced concrete.

There is clearly tremendous potential to derive many more sunshade geometries tuned to their very particular context, climate by climate, latitude by latitude, culture by culture. Graphic exploration is most productive, enabling the designer to 'feel' the interaction of the moving sun with a fluid three-dimensional form, proprietary shadow projection software merely reports after the fact, injecting a misplaced sense of achievement for a banal outcome. Banality in this context, the eponymous horizontal eyebrow, tends to over-restrict useful winter daylight. Figure 8.38 takes a typical bay of the south elevation and shows the direct sunlight penetration permitted at the mid-season onto the glass behind.

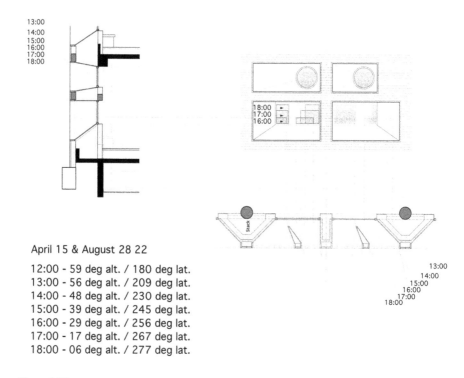

April 15 & August 28 22

12:00 - 59 deg alt. / 180 deg lat.
13:00 - 56 deg alt. / 209 deg lat.
14:00 - 48 deg alt. / 230 deg lat.
15:00 - 39 deg alt. / 245 deg lat.
16:00 - 29 deg alt. / 256 deg lat.
17:00 - 17 deg alt. / 267 deg lat.
18:00 - 06 deg alt. / 277 deg lat.

Figure 8.36
Analysis of solar exclusion to west-facing openings.

Figure 8.37
Detail of glass reinforced concrete reveals to
westerly elevation.

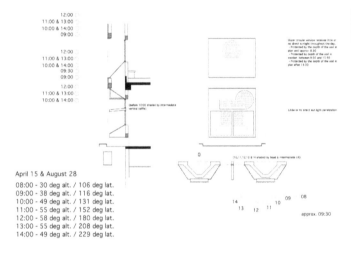

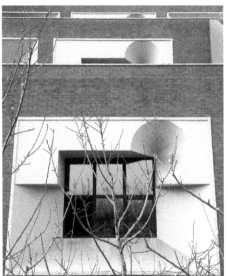

Figure 8.38
South elevation: performance of solar geometry in April/August.

There is a correction, a scooping out of the upper right hand corner of the reveal to allow the entry of early morning low angle sun to enliven the interior. The building is quite highly glazed as the south elevation (Figure 8.39) demonstrates.

Seasonal Strategies
Three main seasonal strategies for operating the environmental design are available to moderate the Beijing climate:

A. **'Winter' (mechanical heating and humidification).** Periodic humidification in addition to heating is required due to the relatively very low humidity of

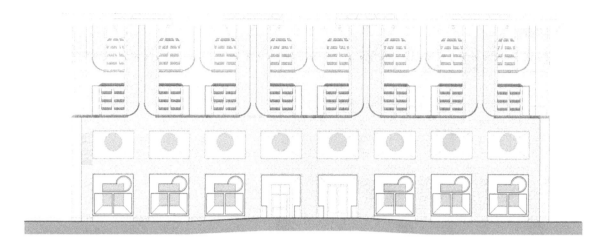

Figure 8.39
General arrangement of the south elevation.

the air at the low external ambient temperatures. The Air Handling Unit (AHU) conditions the external ambient air and supplies it to the plenum below Level 1 ducted to the 'fresh-air fountains' (Figure 8.32) and the lower sections of the perimeter stacks to ensure uniform distribution. In this mode the dampers at the top of the ventilation stacks are closed and the air is returned to the AHU in the basement through return air ducts much as in the Chicago building.

The change of operating mode between 'Winter' (A) and 'Mid-season' (B) will occur when the energy expended in running the mechanical air handling system is greater than the energy required to provide the necessary space and air heating for the building through naturally induced airflows to maintain comfort conditions. There is no heat recovery in 'mid-season'. The exhaust will be naturally stack-driven, developing relatively very low pressures and distributed across the 12 stacks in the design. Effective heat recovery requires a single fan-driven point of exhaust and supply, in the manner of the Passiv-Haus strategy, a completely different prospect for low energy artificial environments in cold climates with cool summers.

B. **'Mid-season' (passive ventilation and cooling).** The building is in 'natural' mode, with little or no additional heating required and no mechanical cooling. By minimizing heat gains through the design of the building envelope and by running an effective 'night purging' strategy, this 'natural' or 'free-running' mode can be prolonged into the summer. The aim is to keep the building in this 'natural' mode for as much of the year as possible. A number of below ground tunnels and chambers incorporated into the original frame are re-used as ground cooling air supply paths.

Night purging exploits the lower temperatures during the unoccupied night to flush out the heat built up in the building from the previous day (from computers, lighting, heating, people and solar gain), and to cool down the exposed concrete structure and most particularly the air supply plenum in readiness for the next day. The pre-cooling of the plenum enables the fresh air to be passively cooled upon delivery into the occupied interior of the building, an effect accelerated by the ground cooling potential of the supply tunnels, in effect a reprisal of the Malta strategy in the preceding chapter.

The change of operating mode between 'Mid-season' and 'Peak Summer' will occur when the internal temperature rises, or is predicted to be rising, towards and above a set comfort level, the predicted adaptive comfort temperatures for the month and season. This can be set on a daily basis – the building need only be in 'Peak Summer' mode when the conditions are extreme, switching back to passive mode whenever possible.

C. **'Peak Summer' (mechanical cooling and de-humidification).** When the external ambient temperature or the relative humidity become too high for internal comfort conditions to be maintained, to the criteria of the adaptive model, the building will cycle into a mechanical air-handling mode. The external air is conditioned by the AHU as described above in the winter mode. An interesting issue, as in Chicago, is whether the occupants will sense the change in mode and their expectations of a comfortable environment change as a consequence. Occupant survey work by the Professor of Architecture at Judson is under way in the Chicago building. We hope to collect similar data in the completed Future House building.

Comparative testing of alternative and business-as-usual designs
Demonstration projects become meaningful through comparison with custom and practice. Five environmental design strategies were modelled using the Energy Plus programme (Energy Plus, 2011), which has the capacity to predict the buoyancy-driven natural ventilation rates in the hybrid strategies, identify the time periods through which the natural ventilation system could operate and inform the detail of the control strategy (Short *et al.*, 2014). The proposed design with 12 stacks was compared with: a theoretical design to Historical Performance Standards (1980s) in China; a design to then Current Performance Standards in China 2011; the original all glass proposition AGT (Advanced Glazing Technology). In addition the design was remodelled with four of its central stacks omitted, as has actually occurred, to gauge the damage to performance through the reduction in stack capacity. Figure 8.40 depicts the models for historically compliant and hybrid schemes.

Internal occupancy gains, 8:00 a.m. to 6:00 p.m. during weekdays, were assumed to be 10W/m², internal artificial lighting gains of 11W/m² and equipment gains about 20 W/m², the offices unoccupied over the weekends. The standard and fully glazed schemes were assumed to be fully air-conditioned to the Chinese design standard (NTMDCC, 2009) requiring set points for heating of 18°C and for cooling of 26°C.

From Table 8.4 we can see that the fully air-conditioned systems, Cases 1–3, consume significantly more energy than the proposed hybrid scheme, up to twice as much. Cases 1 and 2 indicate that the updated insulation regulation significantly improves energy efficiency by some 22%.

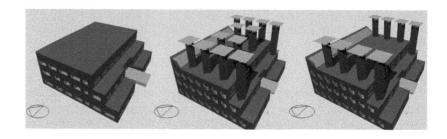

Standard model   Hybrid system with 12 Stacks   Hybrid with 8 Stacks (as built)

Figure 8.40
Massing diagrams of three of the design variants tested using Energy Plus.

Table 8.4 Comparisons of the predicted energy consumptions for the five options modelled. The scheme as designed halves the energy intensity of the compliant 1980s scheme and improves markedly on the current compliant scheme. Surprisingly the omission of a third of the stacks has minimal impact

Case No		Total energy (kWh)	Future House energy saving Total (%)	Total energy per square metre (kWh/m²)	80% of acceptable during transition season	90% of acceptable during transition season
Case 1	Standard model 1	262281	–	62		
Case 2	Standard model 2	205817	22%	49		
Case 3	AGT	188576	28%	45		
Case 4	Hybrid system 12 stacks	126974	52%	31	81%	70.5%
Case 5	Hybrid system 8 stacks	129566	51%	32	78.9%	69.1%

The highly glazed option posits 'Advanced Glazing Technology' in conjunction with super insulation but the cost of configuring such a façade is high and the technology unavailable from within China. Surprisingly, the model predicts the loss of energy saving capacity in the value engineering induced reduction in the number of stacks to be minimal. However, comfort is reduced. The thermal comfort satisfaction level of the hybrid systems during the transition seasons is not, of course 100% simply because indoor air temperatures periodically exceed the upper limit of thermal comfort.

The dynamic thermal model yields encouraging predictions of comparative performance but will air flow passively through the volume in a distributed way, at sufficient velocity, in the directions hoped for? The Chongqing University building scientists constructed a computational fluid dynamics model (CFD) to investigate air movement, not in the 12 stack scheme as designed but the eight stack version built, and compared it with airflows in the Standard compliant and all glass schemes (Essah *et al.*, 2014).

Flows in the latter two arrangements were, unsurprisingly, too sluggish to generate results without fan power driving them.

Figure 8.41 attempts to extract the flow patterns predicted within the ground floor of the model in three dimensions along north–south cross-sections moving east to west.

In winter, velocities remain robust, 0.5 to 2.5m/sec. within 10m of the stacks but falter within the central 10m of the plan even though driving pressures are at their highest theoretically, the omitted stacks and the 'mat-building' characteristic of the original design as explored at Coventry are certainly missed. In summer, pressures are much reduced, internal and external temperatures tending to merge. Here the missing stacks may elevate velocities mid-plan but low velocity fans to lift velocities are clearly required. The environment within the hybrid building in its natural modes is far from constant.

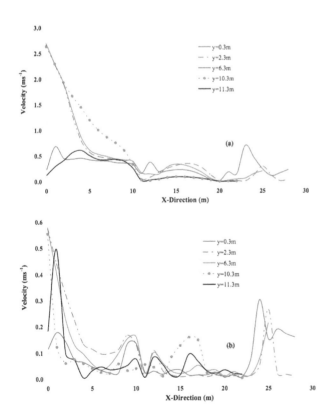

Figure 8.41

CFD model predictions of airflow velocities within the hybrid eight-stack scheme along the x-axis taken at set distances along the y-axis for a) winter and b) summer. Taken from an as yet unpublished paper: Essah, E.A., Yao, R. and Short, C.A., 'Exploring a hybrid ventilation strategy in the continental climate of Beijing using CFD simulations'.

Many fans working continuously are required to achieve the uniform environments of contemporary air-conditioned buildings. To what effect? Even as Saxon Snell explained in his verbal evidence to the 1907 Select Committee on the House of Commons, such unvarying atmospheres are enervating and unpleasant.

In summary, this fierce Continental climate type envelopes a highly significant portion of the occupied land masses and includes some of the most energy intensive built environments in North America and, perhaps more importantly, those on the brink of becoming highly energy intensive in China. China need not replicate Western building practice to satisfy its aspirant population's expectations of comfort. Much closer examination of regional and city climate profiles reveals a much more complex dynamic situation than the transnational comfort standards suggest. Closely tailored solutions can eek out the seasons during which relatively little energy need be expended in climate responsive designs to the extent of halving current good practice building as usual demand. The Judson building demonstrates this; we hope the Beijing building will achieve similar performance. It should do.

Demand reduction is key. Renewable energy gadgets should only be applied to buildings engineered to reduce energy demand, both new and existing. The re-engineered designs must achieve cost parity, the additional design intelligence must be available, the construction industry must step up to the plate and understand the design intent at all levels down to the colleague caulking up the cracks. And there is everything to play for. As the climate changes, it seems likely that more temperate conditions will bleed northwards, the arid steppe Bsk climate will displace the more intense Dwa 'snow, desert, hot summer' type as the century progresses, opening potentially immense potential to build and adapt into flexible hybrid mode with its attendant reductions in carbon and energy consumption. Its occupants will be empowered and will learn how to manipulate their environments to increase their pleasure in experiencing them whilst greatly reducing their carbon impact.

Notes

1. For the Act affecting the meatpacking industry. See www.encyclopedia.chicagohistory. org/pages/804.html (accessed 27 July 2013).
2. See State Climatologist's Office for Illinois. SDee www.isws.illinois.edu/atmos/ statecli/roses/wind_climatology.htm (accessed 23 July 2013).
3. The competition was won by a team led by the author, supported by the Institute of Energy and Sustainable Development (IESD) at De Montfort University, Leicester, UK. Capital cost has been subsidized by a $7.5 million grant from the 2004 Federal Energy and Water Appropriations Bill, two grants totalling $200 000 from the Illinois Clean Energy Community Foundation, and a Kresge Foundation matching grant for $600 000 with an additional $150 000 predicated upon the achievement of a LEED Silver rating or above. The building achieved Gold.
4. Caution is required when making these comparisons. Boia (2005) presents a reminder of the 'climate belt theory' which proposed five parallel and constant belts or zones from pole to pole: frigid–temperate–torrid–temperate–frigid, and then ascribing profound physical, social and psychological differences between the incumbent populations of the three types of belt. Strabo proposed that latitude was the determinant of climate and character (Jones, 1949). Although the idea is sometimes credited to Parmenides (*On Nature*) it is not apparent in John Burnet's translation of 1892.

5. Professor Baizhan Li, opening address at 'The Launch Ceremony for International Research and the National Base for the 111 Program of Low Carbon and Green Buildings', 10 April 2013, Chongqing University. Professor Li is quoting from 'Urgent action on energy saving and green building development', *China Economic Weekly,* 2005 (9), 11 (in Chinese).

6. Future House Expo Centre was ratified in 2003 as a 'National Residence Science and Technology Comprehensive Demonstration Project'. It was simultaneously confirmed by the Ministry of Science and Technology as an 'International Green Architectural Science and Technology Demonstration Project' and a 'China Countryside Architectural Science and Technology Innovation Project' in 2006.

7. $T_{com} = 0.31T_{aout} + 17.8$ (where, T_{com} is comfort temperature and T_{aout} is the mean monthly outdoor temperature).

9

Existing buildings: adapting hospitals

... generation succeeded generation with unwearied enthusiasm, and the cathedral front was at last lost in the tapestry of its traceries, like a rock among the thickets and herbage of spring.

(John Ruskin, *The Stones of Venice*, vol. 2, 1893, p. 208)

INFANT NURSERIES. NEO-PAVLOVIAN CONDITIONING ROOMS ... The Director opened a door. They were in a large bare room, very bright and very sunny; for the whole of the southern wall was a window. Half a dozen nurses, trousered and jacketed in the regulation white viscose uniform, their hair aseptically hidden under white caps, were engaged in setting out bowls of roses in a long row across the floor.

(Aldous Huxley, *Brave New World*, 2007, p. 15)

Aldous Huxley's sinister image of the fully glazed 'modern' hospital in 1932 actually came to pass. Bradford Royal Infirmary's 1960s Maternity Hospital is very lightweight, liberally glazed and very hot in mild conditions. Peak temperatures of 36°C were recorded by Design and Delivery of Robust Hospital Environments in a Changing Climate (DeDeRHECC)[1] researchers in mid June 2010, on a day when external temperatures only peaked at 21.5°C, after a night in which temperatures fell as low as 11°C. The temperature trace (Figure 9.1) shows how rapidly internal temperatures rose.

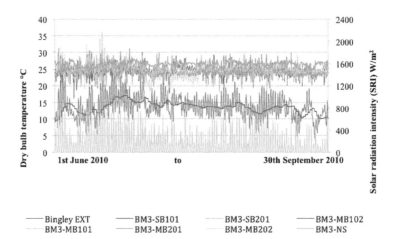

Figure 9.1
Measured temperatures in the Bradford Royal Infirmary Maternity Hospital through the summer of 2010 taken by researchers from the EPSRC funded DeDeRHECC project team.

The authors of the national UK Climate Change Risk Assessment CCRA are unusually candid in assessing the construction world's enthusiasm for adaptation to current and future climate change (Capon and Oakley, 2012, Section 8.4 Summary):

Expert stakeholders consulted during the early stages of the CCRA project expressed the view that the adaptive capacity of the Built Environment sector is low ... For many stakeholders, the issue of climate change lies too far into the future to trigger investment in the current economic climate. Thus the driver of capital cost has a major impact on the degree of adaptation that is currently achieved.

This chapter will demonstrate that viable and economical refurbishment schemes are available for the systematic adaptation of much of the existing hospital building stock to a changing climate, in 'everyday construction' with no sophisticated technologies or gadgets. No doubt, many more options are yet to be invented. Hospitals are a very useful barometer of each modern era generation's building practices, particularly when provided by the state. They differ from non-clinical public building types only in their progenitors' enhanced anxieties about cleanliness and comfort. Many of the ideas and principles presented in this chapter can be applied with modifications to many other non-domestic buildings.

Simulations for future environments in the Bradford building as the century unfolds, predict 163 hours a year above 28°C in the 2050s, 866 hours above the Adaptive Comfort Standard 'Category 1' limit in a typical warming year, in the North of England (Table 9.1). In an atypical hot year in the 2050s, these increase markedly to 620 and 938 hours, equivalent to expectations of the climate of southern Spain, the north coasts of the Black and Caspian seas or the Mississippi floodplain.

The conundrum of the Retained NHS Estate is compounded by the prospect of a changing climate, one which might exhibit more Continental characteristics. Figure 9.2 overlays the NHS Acute Hospital Estate onto predictive climate change maps generated by Graf and Yu in Cambridge University's Centre for Atmospheric Science for this purpose.[2] The versions reproduced here assume the medium IPCC emissions scenario RCP4.5 and predict the geographical reach, through the century, of the exceedance of various significant temperature thresholds: minimum night temperatures above 25°C which interrupt sleep and, in Figure 9.3, daily maxima of 28°C, the blanket threshold derived from the official guidance of a decade ago (HTM 07-02, 2006). The latter criterion, originally a recommendation, came to be institutionalized in Private Finance Initiative style agreements as a contractual requirement, attracting penalties if exceeded. The unanticipated 'revenge effect' of this requirement is the likelihood that NHS buildings will be sealed and refrigerated. The chapter concludes with predictions of the dramatic reversal in achieving national carbon reduction targets if the entire NHS Estate was sealed and refrigerated.

Professor Graf's team concluded that the Norwegian model NorESM1-M for simulated summer daily maximum temperatures gives the best fit to UK historic trends. The variation between the outputs of the five models considered is startling.[3] Perhaps more should be made of these variations in the presentation

Table 9.1 Abridged from simulations of the Bradford Royal Infirmary Maternity Hospital building by R. Renganathan and K.J. Lomas for the EPSRC funded project, 'Design and Delivery of Robust Hospitals in a Changing Climate' (DeDeRHECC)

Years	Max. temp (°C)	Min. temp (°C)	Mean night time temp. (°C)	HTM03: Total hours over 28°C	CIBSE: Night time hours over 26°C	BSEN15251: Total hours above Cat I Upper limit	BSEN15251: Total hours above Cat II Upper limit	Max. temp (°C)	Min. temp (°C)	Mean night time temp. (°C)	HTM03: Total hours over 28°C	CIBSE: Night time hours over 26°C	BSEN15251: Total hours above Cat I Upper limit	BSEN15251: Total hours above Cat II Upper limit
			TRY, no fan								DSY, no fan			
2005	28.6	21.3	24.2	10	10	1041	399	30.1	21.3	24.3	115	60	1198	497
2030s	30.6	21.6	24.3	93	59	859	395	33.0	21.2	24.7	383	231	711	341
2050s	31.0	21.2	24.3	163	87	866	351	34.8	21.8	25.2	620	388	938	445
2080s	31.7	20.7	24.5	232	152	695	285	36.5	20.9	26.0	1035	609	1132	645

Figure 9.2
Probability of daily maximum temperature exceeding 25° C under RCP4.5 projection, downscaled to 5km resolution. England's NHS General Acute Hospitals are shown by blue dots located by NHS grid references (Graf, Short, Xiaoyong Yu [2014] University of Cambridge, unpublished).

Figure 9.3
Same as Figure 9.2 but for daily maximum temperature exceeding 28° C. (Graf, Short, Xiaoyong Yu [2014] University of Cambridge, unpublished).

of climate predictions to the public. The challenge is to predict the phasing of anthropogenic contributions with natural variability. The high resolution bias of the 11-year running mean of daily maximum temperature in 1960–2005 is added to the low resolution RCP4.5 daily maximum temperature for each model and then the high resolution probability of exceeding 25°C and 28°C under RCP4.5 is calculated. Alignment of phasing is critical to give a credible prediction.

The probability of hot summer days with daily maximum temperatures exceeding 25°C increases from 2020–39 to 2080–99. The most dramatic increase is to be expected in the Greater London area, inland East Anglia where a second heat island appears to develop over the dry Fen landscape, and the southeast of England. NorESM1-M shows the lowest increase in probability of daily maximum temperatures exceeding 25°C while the HadGEM2-CC model shows the highest. The probability of any summer day exceeding 25°C is 24–28% across the East Midlands within 2040–59, growing to 28–32% during the two decades of 2080–99. The probability of hot summer days exceeding 28°C over the Greater London area and East Anglia varies startlingly from 2–4%, or two to three days under the NorESM1-M model to 24–28%, or 22–26 days under the HadGEM2-CC model. In 2080–99, it varies from 12–16% or 11–15 days (NorESM1-M) to 34–40% or 31–37 days (HadGEM2-CC).

In terms of the capacity of our adaptation options to sustain low energy resilience, the Norwegian model predictions depict environmental conditions that can probably be absorbed without wholesale recourse to permanent air-conditioning. Hybrid schemes may be resilient in the more intensely affected areas, the heat islands. An annual expectation of a month or more of continuous high temperatures provided by the more aggressive models presents a different prospect for the NHS and the inhabitants of non-domestic building types generally. Hospital buildings generally belong to highly recognizable general building types but subject to heightened sensitivities about internal conditions, in theory if not in practice.

It is entirely reasonable to be concerned about the effect of higher summer temperatures on building occupants. During the 2003 heatwave in France, 14 729 excess deaths were recorded, a 55% rise in expected mortality during that 20 day period, whilst in England 2091 excess deaths occurred (Fouillet *et al.*, 2006). The fit and healthy were also affected (Boyson *et al.*, 2014; Johnson *et al.*, 2005). Public Health England reports that heat-related mortality occurs when outdoor temperatures exceed 17–20°C. Higher internal temperatures over one day have an immediate impact on health but single hot days are not registered within UK heatwave monitoring. There is an annually updated Heatwave Plan for England (Public Health England, 2014). Hospitals are particularly susceptible. The Plan reminds us that the main causes of death in a heatwave result from respiratory and cardiovascular diseases. The vulnerable are those who cannot take action in the face of high temperatures: small children; the bed-bound; the isolated; patients with mental illnesses, not least dementia; those aged 75 and over; residents in care homes, with chronic and severe illness; and those taking certain medications. The Heatwave Plan is severely practical, promoting simple actions within known and very limited resources, in what are clearly assumed to be buildings largely inadequate to the task from, perhaps, bitter experience. The NHS confronts a conundrum: passive inadequacy on the one hand or a surge in carbon emissions from cooling systems on the other. Is this really the stark choice? It will be shown below this is not the case. It could be relatively quick and economical to make the retained NHS Estate resilient whilst significantly reducing carbon emissions.

How can one possibly comment on a distributed estate of this size? Although vast, the NHS estate comprises a recognizable number of recurrent 'type' building forms/plans of different eras. The research project Design and Delivery of Robust Hospitals in a Changing Climate (DeDeRHECC) consistently recorded internal temperatures of 24–28°C and above within 125 hospital spaces across the NHS Acute Hospital Estate. It selected 'exemplar' case studies from a typology of hospital buildings for closer analysis: Low Rise Courtyard buildings of one to two storeys (St Albans West Herts NHS Hospitals Trust); Low Rise Nucleus hospitals as a specific courtyard type (Glenfield Hospital Leicester) (Giridharan *et al.*, 2013); Medium Rise Courtyard buildings of three storeys (The Rosie Maternity Hospital at Addenbrooke's Cambridge) (Short *et al.*, 2015); Deep Plan (St Albans West Herts NHS Hospitals Trust); 'Matchbox on Muffin', towers on podia (The Ward Tower, Addenbrooke's Hospital) (Short *et al.*, 2012); Slabs; Towers; Pavilions (Northwick Park Hospital Maternity Wing) (Short *et al.*, 2010); and pre-1948 hospitals, some 22% of the Estate in 2012, largely comprising linear, narrow section, three to four storey Nightingale type ward buildings (Bradford Royal Infirmary) (Lomas *et al.*, 2012; Short *et al.*, 2014). Thirty-one per cent of buildings were built between 1948 and 1985. High internal temperatures were recorded throughout the Estate except in the pre-1939 stock. Resilience to overheating resides primarily in the 110 or more remaining pre-1939 ward buildings built faithfully to the Florence Nightingale specifications and dimensions as summarized in Chapter 6, and recent buildings enjoying fully air-conditioning, but only for as long as the capacity of the mechanical cooling is not outstripped. Thereafter, having little resilience, they will be increasingly vulnerable.

University of Cambridge researchers studied 248 NHS hospitals (an additional 13 were discounted because of ambiguities in their actual extent), identifying and categorizing each building by type, and then measuring the gross floor areas off scaled aerial views using software developed for physiologists to enable quantification of complex bone shapes. Table 9.2 summarizes the measure, some 12.4 million m^2.[4] The Department of Health Estates and Facilities Policy Division confirmed that this figure was broadly supported by their own ERIC database.[5]

Case Study buildings, representative of the types, were selected outside of the urban heat islands, as there are additional complexities for those within UHIs. The selected buildings were monitored for two to three years, models constructed and calibrated to reflect their observed performance and then modelled against the predictive climate data available for the UK Engineering and Physical Sciences Research Council's (EPSRC) Adaptation and Resilience to Climate Change (ARCC) set of projects. The outcomes were 'diagnosed' by researchers supported by an expert panel.[6] Adaptation strategies were evolved in response to the vulnerabilities exposed. An intelligible and viable diagnostic method is a priority. Understanding the environmental performance of a large building through the diurnal cycle across all seasons is complicated and sometimes counterintuitive, especially in a volatile Temperate climate. Figures 9.4 and 9.5 attempt a graphic summary of key factors in determining environmental failure in both summer and winter for a specific case, many will be universal.

Table 9.2 The NHS Hospital Estate in England, measured floor areas by type
(University of Cambridge)

Building Classification	No of hospitals with buildings in these classifications	m²
Courtyard Low Rise (1-2 Storeys)	114	1,760,371
Courtyard Medium Rise (>3 Storeys)	117	3,012,568
Deep Plan	124	2,395,319
Matchbox Muffin	3	39,911
Nucleus	45	900,283
Pre-1939	110	1,963,244
Slab	70	1,952,891
Tower	25	298,487
Pavilion	19	81,720
TOTAL (all hospitals combined)	248	12,404,794

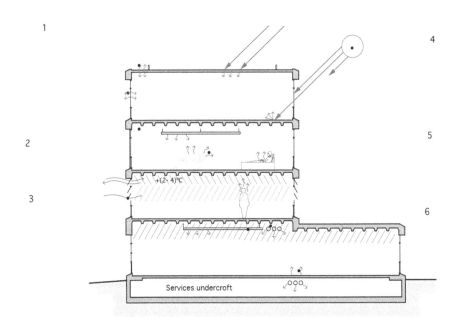

Figure 9.4
Section through maternity wing (as built) showing environmental characteristics during summertime operation (author).
1. Fabric heat losses/gains.
2. Lightweight surfaces.
3. Uncontrolled infiltration.
4. Direct solar gain through large areas of unshaded glazing.
5 and 6. Internal heat gains from patients, staff, equipment and lighting.

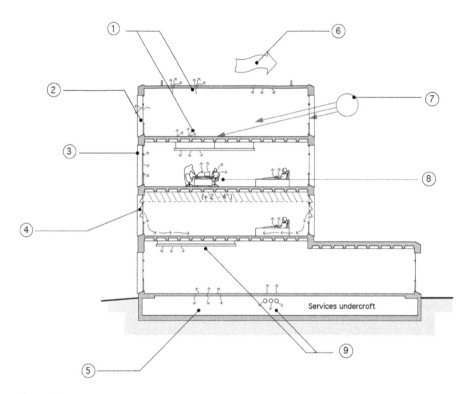

Figure 9.5
Section through maternity wing (as built) showing environmental characteristics during wintertime operation (author).
1. Poorly insulated roof slab radiates heat to night sky.
2. Heat loss through poorly insulated solid envelope, glazing and non-airtight construction.
3. Dry resultant temperature (operative temperature) is lower than air temperature due to cold surfaces.
4. Ill-fitting single glazing causes uncontrollable infiltration and cold downdraughts for patients.
5. Conductive heat losses through uninsulated ground floor slab.
6. Wind effects increase heat loss from building.
7. Glare from low-level sun results in use of blinds which reduces daylight resulting in greater use of artificial lighting.
8 and 9. Internal heat gains from patients, staff, equipment and lighting.

DeDeRHECC revealed that adaptation measures almost invariably deliver mitigation benefits and apparently 'better', more locally controllable environments. This was a surprising finding. Industry-standard, sealed 'Passiv Haus' type schemes, super-insulated with efficient heat exchange performed well in 2010 against 2005 climate data but increasingly less well as the cooling requirement becomes more urgent through the century. Moreover, the refurbishment schemes for all types are relatively inexpensive, £900–1700/m². Why are we so interested in passive ventilation solutions? Lomas and Ji (2009) report:

> … simulations suggest that refurbishment of an overheating building using ANV (Advanced Naturally Ventilation) rather than SNV (Simple Natural Ventilation, opening windows) will result in much greater resilience to future climate change (and to increase in internal heat gains). The strategy also offers a route to up-grading to a hybrid strategy via the introduction of cooled air through the perimeter shafts: the SNV strategy cannot offer this.

Detailed reports on adaptation re-designs and the predicted performance through the century are available in the published papers referenced to each type, but in summary: the Pavilion-type is represented by Northwick Park Hospital Maternity Wing in outer West London (Short *et al.*, 2010). Designed in the late 1960s, medium rise and medium depth in its H-plan, concrete framed, largely lightweight in cladding, extensively single glazed, it presents a bizarre combination of high thermal mass and ultra-lightweight enclosure (Figures 9.6 and 9.7).[7]

Subject to prodigious solar gains in summer and mid season and, despite the glazed envelope, achieving a markedly non-uniform natural light distribution across the wards, it achieves a 30% Daylight Factor (DF) at the glazed perimeter but comfortably over 2% DF at the very rear of the deepest wards, the traditional

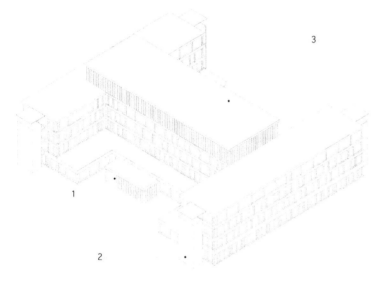

Figure 9.6
Isometric view of Northwick Park Maternity wing.

Figure 9.7
Northwick Park prior to refurbishment.

design light level for school classrooms. As a consequence, all spaces appeared dark and gloomy against the peripheral glare, resulting in a permanently artificially lit interior throughout, driving up internal heat gains.

Spaces were modelled against existing climate data for 2005, results being expressed against the criterion set in HTM03-01, no more than 50 hours a year above 28° C. The building just about satisfied this criterion in 2005 (numbers shown in green indicate compliance) but the majority of spaces fail by 2020 (numbers in shown in red indicate exceedance) and almost all by 2050 by some considerable margin against Test Resultant Year data, the 'normal years' data (Table 9.3).

Table 9.3 The Pavilion-type: Northwick Park Maternity Wing. Modelled thermal performance of the original building design, extensive overheating by 2020

	Hours over 28°C			
Room	**2005**	**2020**	**2050**	**2080**
GF Consulting Room 2a	11	48	186	344
GF Consulting Room 2b	14	54	193	348
GF Consulting Room 3a	15	48	155	262
GF Consulting Room 3b	14	73	211	344
GF Corridor 1a	0	0	19	120
GF Corridor 1b	0	0	19	120
GF Toilets 1a	0	0	23	129
GF Toilets 1b	2	3	33	178
GF Treatment Room 1a	32	69	212	562
GF Treatment Room 1b	32	72	224	584
GF Ultrasound Room 1a	54	107	311	812
GF Ultrasound Room 1b	54	107	311	812
GF Consulting Room 1a	48	188	350	588
GF Consulting Room 1b	13	79	247	420
FF Four Bed Ward 1a	104	291	502	927
FF Four Bed Ward 1b	39	150	313	710
FF Four Bed Ward 2a	93	278	482	890
FF Four Bed Ward 2b	38	141	300	689
FF Four Bed Ward 3a	37	108	235	470
FF Four Bed Ward 3b	30	80	181	381
SF Delivery Room 1a	52	149	310	689
SF Delivery Room 1b	24	71	207	525
SF Four Bed Ward 1a	89	270	471	877
SF Four Bed Ward 1b	38	142	300	685
TF Four Bed Ward 1a	38	120	247	477
TF Four Bed Ward 1b	38	115	231	480
TF Four Bed Ward 2a	49	151	255	472
TF Four Bed Ward 2b	32	94	202	400
FF High Dependency 1a	0	0	0	0
FF High Dependency 1b	0	0	0	0

Researchers attempted to diagnose this deteriorating performance. Their conclusions are those summarized in Figures 9.4 and 9.5. An adaptive re-design/re-engineering strategy was brokered in response, as depicted in Figure 9.8 (Short *et al.*, 2010).

It unwraps and exploits the natural resilience of the structure, reducing radiant temperatures within by stripping the building back to its frame, incrementally as necessary, bay by structural bay, and adding back a deep façade. This is not a fully glazed double façade, which would undoubtedly exacerbate current difficulties without energetic flushing through of the entrapped air throughout the summer. Instead, it is a façade with an air supply and exhaust infrastructure doubling as a deep reveal to shade still significant areas of glazing, a variant of the Continental climate strategies described in Chapter 8.

Figure 9.9 describes the interventions required within a four-bed ward in the extension of exhaust ducts from the exhaust stacks on the perimeter, at ceiling level.

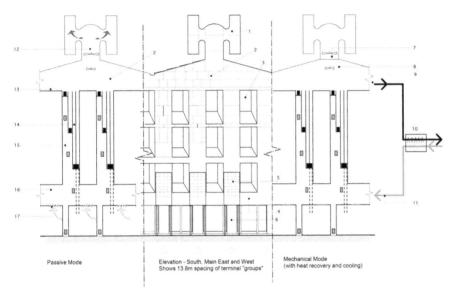

Passive Mode

Elevation - South, Main East and West
Shows 13.8m spacing of terminal "groups"

Mechanical Mode
(with heat recovery and cooling)

Figure 9.8
Re-design/re-engineering proposals, replacement of the existing elevations, reducing glazed area, incorporating air supply and exhaust in an edge-in/edge-out configuration in passive mode. Alternative mechanical control mode shown for UK locations requiring hybrid strategy, London and Southeast, inner East Anglia.
1. H-pot terminal.
2. Exhaust plenum.
3. Windows shaded by being recessed in thick wall.
4. Air inlet 'ring' duct and mechanical supply.
5. Passive air inlets, with acoustic attenuators where significant ambient noise.
6. Air supply 'drop' to ground floor.
7. Control dampers (closed).
8. Mechanical exhaust fan recirculation to a central AHU.
9. Vitiated air gathered in return-air shaft and exhausted through heat exchanger.
10. AHU with heat recovery.
11. Fresh air supplied into system mechanically.
12. Air is exhausted through fan-assisted terminals in soffit.
13. Return shaft.
14. Dedicated exhaust shaft.
15. Supply shaft.
16. Primary supply shaft.
17. Air enters through control dampers in soffit.

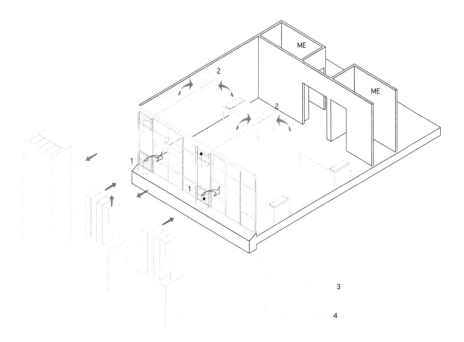

Figure 9.9
The edge-in/edge-out kit of parts applied to existing four-bed ward.
1. Modular low level air inlets.
2. Air exhaust bulkheads could be extended to provide separate exhaust openings closer to outer and inner beds.
3. Air exhaust to external stack.
4. Air supply direct or from shaft, ME Mechanical Extract (say 3 air changes/hr and 10 ac/hr boost for dirty extract), MS Balanced mechanical supply with heat recovery from dirty extract.

This is a hybrid strategy in which air is moving naturally, driven by the stack effect, enhanced by the wind resisting H-pot terminations developed by the team for other building types as earlier chapters report, or cycling into mechanical mode envisaging the use of existing mechanical ventilation within the substantial rooftop plant room. Figure 9.10 takes a lateral section through this deep façade, and Figure 9.11 nominally clothes it. No doubt other more extraordinary architectural possibilities await discovery.

Future building thermal performance and energy consumption were modelled for the existing building, recorded as Case 1, with a building energy consumption of 45 GJ/100m³, and three variants:

Case 2: Advanced Natural Ventilation (ANV), stack driven within the façade, no fans and no cooling in summer, some mechanical ventilation in winter, achieving 35 GJ/100m³ relatively consistently towards 2080. Table 9.4 records predicted performances across a selection of spaces.

Case 3: Fan-assisted Natural Ventilation, fans boost the performance of the ANV scheme and, as a consequence, it is predicted to deliver some 33.5 GJ/100m³.

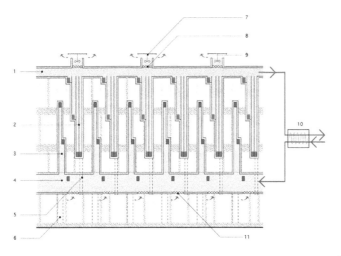

Figure 9.10

Generic adaptation strategy adapted to particular architectural configuration of Northwick Park, section.
1. Mechanical air extract for heat recovery linked back to centralized plant room.
2. Dedicated exhaust path from each space.
3. Direct air inlet to each floor from common air supply shaft.
4. Mechanical air supply to fresh air shaft for heat recovery and cooling, linked back to centralized plant.
5. Exhaust from ground floor is taken internally to the first floor to avoid cross-overs.
6. Supply shaft to ground floor is internal.
7. Stack terminal design to be robust against variable wind direction and rain.
8. Passive exhaust damper closed when in mechanical mode.
9. Low energy fan to minimize airflow reversal in exhaust stack.
10. AHU with heat recovery.
11. Passive supply air damper closed when in mechanical mode.

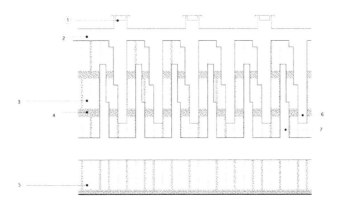

Figure 9.11

Generic pavilion type adaptation strategy adapted to the particular architectural configuration of Northwick Park. One of many architectural arrangements which could flow from the underlying logic of the environmental design strategy, elevation.
1. Fan-assisted ventilation stacks.
2. New clad external ventilation system.
3. New high performance curtain walling system in plane of existing.
4. Existing concrete frame.
5. Ground floor remains relatively untouched. New high performance curtain walling system.
6. Ventilation exhaust air shaft.
7. Ventilation supply air shaft.

Table 9.4 The Pavilion-type: Northwick Park Maternity Wing. Simulation results for the re-engineered building with predominantly Advanced Natural Ventilation (Case 2), the internal environment secured until the middle of the century

| | Hours over 28°C | | | |
Room	2005	2020	2050	2080
GF Consulting Room 2a	7	157	676	1392
GF Consulting Room 2b	6	161	740	1493
GF Consulting Room 3a	22	90	236	406
GF Consulting Room 3b	18	153	314	463
GF Corridor 1a	0	0	13	77
GF Corridor 1b	0	0	13	77
GF Toilets 1a	0	0	10	103
GF Toilets 1b	0	0	12	108
GF Treatment Room 1a	31	64	213	637
GF Treatment Room 1b	31	64	214	640
GF Ultrasound Room 1a	47	104	323	852
GF Ultrasound Room 1b	47	104	323	852
GF Consulting Room 1a	8	22	68	220
GF Consulting Room 1b	7	16	46	184
FF Four Bed Ward 1a	15	23	118	418
FF Four Bed Ward 1b	9	21	99	386
FF Four Bed Ward 2a	15	28	135	451
FF Four Bed Ward 2b	10	24	113	422
FF Four Bed Ward 3a	18	30	81	290
FF Four Bed Ward 3b	12	22	68	262
SF Delivery Room 1a	19	32	128	455
SF Delivery Room 1b	11	24	110	436
SF Four Bed Ward 1a	18	27	100	350
SF Four Bed Ward 1b	13	25	91	337
TF Four Bed Ward 1a	19	37	110	340
TF Four Bed Ward 1b	19	37	113	354
TF Four Bed Ward 2a	17	37	106	346
TF Four Bed Ward 2b	18	35	98	329
FF High Dependency 1a	0	0	0	0
FF High Dependency 1b	0	0	0	0

Case 4: Shared Mechanical Ventilation and Cooling with heat recovery reducing predicted consumption to 34 GJ/100m^3 consistently towards 2080 as the use of gas reduces proportionally to that of electricity. Case 4 is the standard industry response derived from 'Passiv Haus' thinking. Unsurprisingly it protects the majority of spaces until the 2090s. Perhaps surprisingly, Case 3 performs a little more efficiently than Case 4, by some 3%, probably within the model error margin. Figure 9.12 compares modelled energy consumption for the three refurbishment possibilities as against the existing performance.

Figure 9.12

The Pavilion-type: energy demand comparison between the existing building and three adaptive refurbishment options, Cases 2 and 3 ANV, sealed and mechanically ventilated Case 4

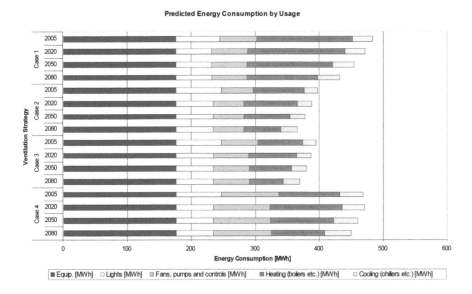

Table 9.5 summarizes predicted costs for the Case 4 option calculated by Davis Langdon AECOM. A construction cost of £1633.46/m² is indicated for Case 3, the fan-assisted ANV refurbishment. Costs are as of the Third Quarter 2008, an inauspicious moment in terms of the global economy, building cost inflation being thereafter suppressed and so these estimates retain perhaps a longer life than in the pre-crash era.

The 'Tower on a Podium, Matchbox on Muffin' type is represented by the ten-storey Addenbrooke's Hospital Ward Tower in the east of England, designed and built between 1967 and 1972 (Short *et al.*, 2012). Some 28 towers, 70 slabs and three full matchbox on muffin combinations were identified across the NHS Estate in England alone. All of the Ward Tower floors have the same overall geometry, a long central corridor to which, on the south side, multi-bed wards (10.2m deep from corridor to window wall) occupy the wider end parts of the building. Private rooms and offices face north. The occupied levels have a structural floor-to-ceiling height of 3.66m with, as designed, a 0.90m void above the suspended ceiling (Figure 9.13).

The windows run as a continuous ribbon incorporating opaque panels at all levels on both north and south façades, yielding a substantial glazing ratio of 57%. Precast concrete panels form the spandrels. Originally, large centre-pivot windows offered generous opening area as required but were replaced in the late 1990s with new low maintenance frames restricted to 100mm opening (due to health and safety requirements), radically disrupting the natural ventilation capability. External sun shading was planned but omitted during construction. The priority was to achieve adequate winter heating: a lesson from the brutally cold winter of 1962–63. Air is mechanically supplied from a central plant room

Table 9.5 The Pavilion-type, comparative cost estimates as of July 2012 (Davis Langdon AECOM)

		Cost for one quadrant	Cost for four quadrants	
				£m²
Base construction cost		1,747,110	6,988,438	816.79
Preliminaries	20%	349,422	1,397,688	163.36
		2,096,531	8,386,126	980.15
Overheads and profit	6.5%	136,275	545,098	63.71
		2,232,806	8,931,224	1,043.86
Location factor	11%	245,609	982,435	114.82
Design contingency	5%	123,921	495,683	57.93
Construction cost at current pricing		**2,602,335**	**10,409,341**	1,216.61
Design fees	15%	390,350	1,561,401	182.49
Non-works costs	1.5%	39,035	156,140	18.25
Equipment cost	0.0%	–	–	–
Planning contingency	0.0%	–	–	–
Sub-total		**3,031,721**	**12,126,883**	1,417.35
VAT (not on design fees)	17.5%	462,240	1,848,959	216.10
		3,493,961	**13,975,842**	1,633.46
Optimism bias	0.0%	–	–	1,633.46
Scheme cost at current pricing		**3,493,961**	**13,975,842**	1,633.46

into the central corridors. There is no organized exhaust except in bathrooms and utility rooms. Patent radiant ceilings, still in operation, helped to heat wards to 65°F (18.3°C) when the external temperature was 30°F (−1°C).

Monitoring reveals the building to be surprisingly resilient. Level 8 ward temperatures varied between 21.4°C and 28.5°C but for 45% of the hours from 1 July to 15 August 2010, internal temperatures exceeded 25°C. For 38 night-time hours (taken as 9.00 p.m. to 6.00 a.m.) temperatures exceeded 26°C, some 8% of the total period, well above the 24°C established as the threshold for unimpeded sleep. The nurses' stations mid-plan were consistently warmer. Temperatures appear to have been arrested by the very high air leakage through the construction. The windows were observed to be open throughout. Energy consumption is close to the maximum recorded in the NHS, over 100 GJ/100m³.

Five refurbishment options were devised by the team, ranging from the industry. Option 1 is standard 'Passiv Haus' type approach, 'Sealed building, Mechanical Ventilation with Heating and Cooling' (SMVHC) (sealing the building

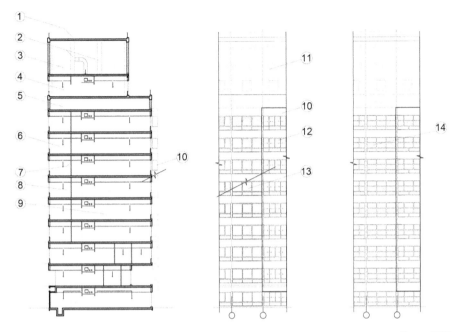

Figure 9.13
The Ward Tower, Addenbrooke's Hospital, as originally built (showing also omitted shading), and as modified by 2011.
1. Uninsulated roof deck.
2. Flat concrete piers in place of cross walls.
3. Plant room: air handling units delivering warmed air throughout the Tower.
4. Upper floor designed as isolation ward for infectious patients.
5. Services distribution plenum, half floor.
6. Typical ward floor: small rooms to N, four to six-bed wards to S.
7. Wards designed with perforated metal ceiling panels connected to LPHW pipework to provide radiant heating.
8. Warmed air ducted to wards.
9. Bathrooms mechanically ventilated.
10. Fully glazed projecting bays to six-bed wards. Originally enclosed and highly cross-ventilated. Possibly to accommodate smokers.
11. Facing brick in concrete frame.
12. Original horizontal centre pivot opening windows, 1.1m square, operated by occupants.
13. 1990s refurbishment substituted double-glazed windows in aluminium frames, lower casements only openable, top-hung.
14. Refurbishment in 2004 substituted double glazed windows in aluminium frames, lower casements only openable, top-hung, restricted to 100mm.

within a heavily insulated jacket with very efficient heat recovery). Its annual predicted energy demands and emissions were 59 GJ/100m^3, within the Department of Health (DH) target, and 137 kgCO$_2$/m^2.

Option 2 is a hybrid: 'Sealed Mechanically Ventilated Environment, Radiant Ceilings active in winter (Heating) and summer (Cold water for cooling) and heat recovery' (SMVRHC), exploiting the radiant action of the ceilings both in heating and cooling mode 46 GJ/100m^3 and 102 kgCO$_2$/m^2 respectively.

Option 3 is another hybrid, with 'Natural Ventilation and concurrent Mechanical Ventilation supply, heat recovery, opening windows and Perimeter Heating' (NVMVPH) delivering 4 ACH (air changes per hour) with all windows openable, achieving 40 GJ/100m^3 and 111 kgCO$_2$/m^2 respectively.

Option 4 is a more innovative passive scheme, 'Natural Cross Ventilation, Perimeter Heating' (CVPH) dispensing with the mechanical ventilation system altogether and enabling cross-ventilation by threading crossover ducts in alternating directions across the width of the floor plate to achieve just $20\,GJ/100m^3$ and $44\,kgCO_2/m^2$ respectively, the omission of fans being the key to such low energy demand.

Option 5, also an innovative passive scheme, is 'Natural Stack Ventilation with Perimeter Heating' (SVPH). It adds external exhaust stacks to develop more reliable airflows as required in banks of four storeys, removes the envelope of every fifth floor (at mid-height) to provide a free air environment in which the stacks to the lower four floors, each dedicated to one space only, can terminate, its performance being similar to Option 4 (Figures 9.14 and 9.15).[8]

All five schemes dramatically improve the performance of the tower but these predictions do not include energy use for matters unconnected with space conditioning (small power, medical equipment, restaurants, etc.), which can be 44% of the total. Given this, it is likely that only Options 2 to 5 could plausibly meet the NHS energy target and CO_2 emissions benchmark of 55 to $65\,GJ/100m^3$. Importantly this excludes Option 1 with a predicted energy demand in excess of $55\,GJ/100m^3$ which has a mechanical supply of 6 ACH as stated in HTM03-01, accumulating fan power and mechanical cooling energy penalties through the coming century. Whilst Options 2 and 3 approach 40 $GJ/100m^3$, Options 4 and 5, the naturally ventilated options, achieve 20–30 $GJ/100m^3$. Option 5 can achieve much greater ventilation rates and presents a nascent diagram for making a relatively very low energy high-rise building. It introduces a void at the sixth floor so that the lower five floors exhaust into free

Figure 9.14
Option 5: SVPH, natural stack ventilation, perimeter heating.
1. H-pot stack termination, robust to wind turbulence.
2. Insulated perimeter stacks external to existing elevation.
3. BMS controlled damper at entrance to stack, manual override.
4. Level 6 envelope removed to create free space connecting north and south sides.
5. Lower bank of five floors connected to stacks.
6. All glazing openings under user control.
7. Perimeter heating connected to actuated vents.
8. Terminations connect to form full cross H-pot arrays, dividers guard against cross-flows and risk of airborne cross-infection.
9. Each stack comprises four discrete cells isolated from point of entry to cross pot.
10. Actuated points of entry into stacks.
11. Void through tower.
12. Lower ranges of stacks terminate in plane of void to enable 360° access to stack termination.
13. Glazing reduced in size and shaded within thickness of superimposed 'double façade'.
14. Highlighted floor modelled, slightly a typical as above void. U-value for floor $0.07\,W/m^2\,K$.
Notes: 1. Glazing reduced in size and shaded within thickness of superimposed 'double façade'; 2. Highlighted floor modelled, slightly atypical as above void. 3. U value for floor $0.07\,W/m^2\,K$.

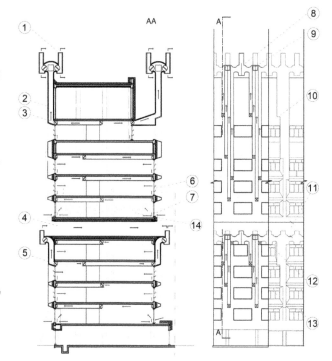

Figure 9.15
Still from animation portraying Option 5. For the
Addenbrooke's Ward Tower from the film 'Robust
Hospitals in a Changing Climate'.

air regardless of wind direction. Bundles of five floors (four inhabited floors, one floor open for natural ventilation) enable external stacks to gather across the elevation whilst retaining adequate glazing, even at the uppermost floor. An interesting theoretical study is commencing in the context of Chinese mega-cities in the hot summer/cold winter zone.[9]

DeDeRHECC methodology was applied to an example of the strongly recurring 'Courtyard Medium Rise type', the late 1970s concrete framed three-storey double courtyard, brick-clad Rosie Maternity Hospital on the Addenbrooke's campus (Figue 9.16). Short *et al.* (2015) report significant overheating in mild conditions across the 26 spaces monitored through the summers of 2010 and 2011, concealed, as in the Ward Tower, between 24°C and the customary 28°C threshold with a peak of 30.7°C.

None of the existing spaces met the adaptive comfort standard Category I (British Standards Institute, 2007). The Intensive Care Units are mechanically cooled. Perhaps surprisingly, for clinical reasons the heating set point throughout the building is 24°C, so that it is heated at night during heatwaves. Summer overheating even in mild 'normal' conditions in the Rosie hospital is a major concern. Figure 9.16 summarizes the principle summer environmental pressures. The existing building enjoys a relatively low energy and low carbon performance against the DH guidance benchmarks but at the cost of comfort. It is unable to shed heat so that internal temperatures reach 28°C in relatively mild external conditions. Night temperatures are consistently uncomfortable, 23.5–26°C, suggesting that simple measures against the recommended threshold may be inadequate. For example, one bedroom records 1992 hours above 25°C, 888 of which occur at night. In the 2030 DSY the existing building

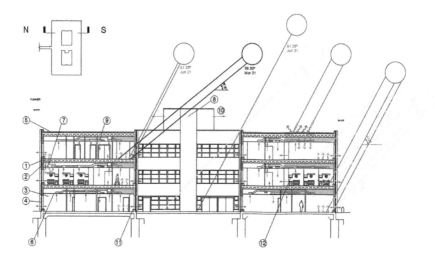

Figure 9.16
The Rosie Maternity Hospital, Addenbrooke's, Cambridge. Existing building, summer condition.
1. Reinforced concrete frame: waffle slab 150mm with 300mm downstands, from inner leaf to external cavity wall above glazing spanning full structural bay.
2. External leaf of 100mm brick, 50mm polyurethane insulation in cavity, inner leaf 140mm concrete blockwork. Occasionally steel stud and two-layer plasterboard.
3. Non-structural internal partitions vary: plasterboard on 100mm softwood studwork or 140mm concrete blockwork plastered both sides.
4. External windows double glazed in hardwood frames, mullions and transoms (75 × 100mm), opening lights restricted to 100mm since late 1990s.
5. Roof: 150mm RC slab, waffle 300mm deep, 50mm roof deck of pre-screeded slabs with reinforced edges, asphalt, 50mm extruded polystyrene 'Roofmate' held down by 50mm of washed gravel.
6. Lightweight suspended gypsum ceiling tile grid.
7. HW pipes suspended within ceiling void, original insulation, some missing, water circulated at 55°C or 60°C continuously to avoid bacterial growth.
8. Service riser to east and to west sides. Each connects to a plant room on ground floor (Level 1). 3 AHU units in east, fresh air supply drawn down riser ducts from roof level.
9. Supply air, 2 AHUs heating, 1 AHU cooling per plant room. Unconditioned air supplied when external temperature below 22°C. Conditioned air delivered at 21°C through variable speed fan.
10. Exhaust air vented mechanically. No heat recovery.
11. Perimeter heating: thermostatic radiator valves recently installed (2011) supplied by HW at minimum 65°C. Target temperatures 24–25°C in all rooms/wards.
12. Internal gains: TV 2.2 w/m²; bed lamp 2.8 w/m²; general ceiling mounted lighting 3.3 w/m² as calculated for multi-bed ward.

has a night-time mean temperature in excess of 25° C. The mechanical ventilation rate is too low. In most of the bedrooms it does not even reach a minimum value of 10 l/s/p (litres per second per person).

The building was modelled to predict likely internal temperatures to the 2030s using UKCP09 data. The results indicate increasing peak internal temperatures. Four adaptive intervention schemes were subsequently developed, commencing as per other type case studies with an 'enlightened' industry standard 'Passiv Haus'-type.

Option 1: SMVHC (Sealed Mechanical Ventilation Heating and Cooling).
This provides super-insulation, sealed glazing, good airtight construction and 6 ACH with 60% heat recovery, predicting that DSY peak temperatures

will oscillate around 28°C, so that additional cooling capacity will be required, and will be increasingly necessary (Figure 9.17). Does the building have the capacity to accommodate it? Predicted energy demand is high and associated CO_2 emissions very high, 130kgCO_2/m², almost two-thirds resulting from its electrical demand which will, of course, rise as cooling capacity is increased, the 'worm in the bud' of this approach, it has no other means of defence.

Option 2: NCVPH (Natural Cross-Ventilation retaining Perimeter Heating). Figure 9.18 shows a lower technology-based scheme promoting natural cross-ventilation by providing greater opening glazing area, opening to 45° with window guards, removing some cellularization of the plan adjacent to courtyard façades to promote cross-ventilation with transfer grilles, exposing some thermal mass at concrete soffits, providing sun-shading and additional insulation. Peak DSY temperatures hover between 33.5–34.9°C so that additional cooling will be required, but night ventilation cooling, an obvious response, is wholly excluded by current practice. A peak of 31°C is predicted in TRY summer conditions. More research is required into the clinically safe night conditions in maternity wards. Could cooler temperatures be contemplated? Incubators in ICUs provide close environmental control for vulnerable infants. However, Option 2 has the lowest energy penalty and markedly lower CO_2 emissions than Option 1.

Option 3: ANCSVPH (Advanced Natural Cooling Summer Ventilation). Figure 9.19 shows an enhanced natural ventilation scheme glazing over the two courtyards to provide supply air preheating 'winter gardens', unheated, with liberal opening area to dissipate solar gains, and much of the intervention in Option 2 to promote cross-ventilation of the floors. Option 3 offers similar

Figure 9.17
Medium Rise Courtyard type: Option 1 SMVHC Sealed Mechanical Ventilation Heating and Cooling.
1. Ventilation system delivers 6 ach⁻¹ as required by HM 03-01.
2. Return air ducts in corridors and wc/bathrooms.
3. Windows are sealed shut.
4. South and west facing glazing shielded by interstitial blinds. Air-tightness improved by continuous mastic seal.
5. Perimeter heating not utilized in this option.
6. Existing HW and steam supply pipes insulated to high standard.
7. Additional 100mm insulation added to 50m Roofmate extruded polystyrene slabs.
8. Walls receive 100mm mineral fibre insulation and proprietary render treatment. Lower opaque glazed panels to window frames receive additional 100mm insulation.

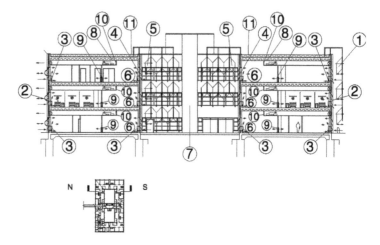

Figure 9.18
Medium Rise Courtyard type: Option 2 NCVH Natural Cross-Ventilation Perimeter Heating.
1. External sunshading applied to south, southeast, southwest elevations. Wing geometry of shade designed to maximize view to upper sky, in translucent coloured material.
2. Existing timber window units, currently four upper panels open, restricted to 100mm. All eight glazed panels made openable to 45° with window guards to fulfil NHS safety regulations. Lower opaque elements to open to admit air across perimeter heating units.
3. Perimeter heating below glazing.
4. Courtyard glazing units: all glazed panels opening to 45° protected by window guards.
5. Sunshading as (1) applied to southwest and southeast elevations.
6. Cellular rooms to centre of each elevation to courtyards removed to open circulation areas directly to courtyard fresh air supply, become patient day areas.
7. Service risers maintained.
8. Suspended lightweight ceiling removed to expose concrete waffle slab.
9. Transfer grilles enable fresh air from courtyards to cross-sections.
10. Transfer ducts within suspended ceiling exhaust opened courtyard patient areas.
11. Additional 100mm extruded polystyrene added to roof.

DSY peak conditions but TRY peaks in line with current guidance. Predicted energy demand and CO_2 emissions are only marginally higher than Option 2.

Option 4: NVPDCPH (Natural Ventilation incorporating Passive Downdraught Cooling and Perimeter Heating). Figure 9.20 proposes the enclosure of the courtyards but in a more active way, developing the low energy cooling strategy, Passive Downdraught Cooling, of the UCL School of Slavonic and East European Studies (SSEES) Building and the atrium project in Central London described in Chapter 4. Cooled water batteries at high level openings within the rooflights induce a downward flow of pre-cooled air contained by a lightweight, acoustically absorbent, fabric shroud. The cooled air is then drawn across surrounding occupied spaces. The diagram suggests ground-sourced cooling could supplement the action of the PDC by utilizing thermal storage, shown as readily available water tanks, crossing the seasons so that heat gained from summer hot spells is dissipated in winter and winter coolth utilized in summer as in the London scheme. Banks of passive solar water heaters, mounted on the roof of each PDC rooflight, supplement warming of winter supply tanks. Recovered heat from all sources is gathered in winter to supplement the supply to the

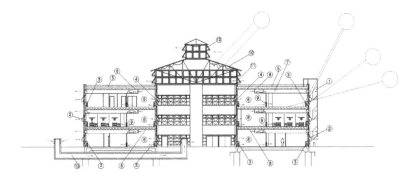

Figure 9.19

Medium Rise Courtyard type: Option 3 ANCSVPH: Advanced Natural Cooling Summer Ventilation.
1. External sunshading applied to South, South-East, South-West elevations. Geometry of shade designed to maximize view to upper sky, in translucent coloured material.
2. Existing window units, currently 4 upper panels open, restricted to 100mm. All 8 glazed panels made openable to 45° with window guards to fulfill NHS safety regulations. Lower opaque elements to open to admit air across perimeter heating units.
3. Perimeter heating below glazing.
4. Courtyard glazing units: all glazed panels opening to 45° protected by window guards.
5. Additional 100mm extruded polystyrene insulation added to roof.
6. Cellular rooms to centre of each elevation to courtyards removed to open circulation areas directly to courtyard fresh air supply, become patient day areas.
7. Transfer ducts exhaust inboard spaces (6) to exterior.
8. Suspended lightweight ceiling removed to expose concrete waffle slab.
9. Transfer grills admit supply air from enclosed courtyard to wards.
10. Lightweight steel framed double glazed roof across internal courtyards with actuated fabric awnings to exclude direct sunlight in overheating season.
11. Low level actuated vents cross vent 'atrium', perform as smoke vents.
12. High level lantern vents summer heat gains from upper part of atrium.
13. Low level supply to atrium formed in spun concrete pipework.
(Note: might one seal the perimeter of the building with heat recovery in winter, but open all perimeter glazing in summer?)

perimeter heating system. Air is permitted to circulate within the twin summer and winter storage chambers to a predetermined regime. Option 4 offers lower TRY and DSY peaks than Options 2 and 3.

However, improved performance is achieved at a cost. Although the existing building, unaltered, has the worst thermal performance, it has the lowest energy consumption, and the lowest CO_2 emissions. Figure 9.21 reveals Option 1 has the highest energy consumption and annual CO_2 emissions of approximately 130 $kgCO_2/m^2$. Option 2, the most free-running, achieves 70 kg CO_2/m^2, Option 3 at approximately 78 $kgCO_2/m^2$ and Option 4 which delivers cooling reaches 90 $kgCO_2/m^2$. The benchmark set for hospitals by CIBSE in 2010 was 142.4 $kgCO_2/m^2$ so that the most straightforward adaptive Option can halve that best practice performance, but at what cost?

Full elemental cost analyses were conducted to yield relative 'value for money' guidance to NHS Trusts (Table 9.6). Option 1 is less than £1000/m² whilst Option 2, yielding the lowest carbon penalty, but at a cost in comfort by the 2030s, is estimated at £1152/m². It will need cooling, whereas Options 3 and 4 include for low energy cooling, both at less than £1800m², well within NHS refurbishment norms. Option 4 is the most resilient but the most ambitious technically.

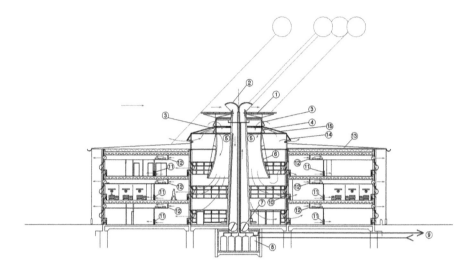

Figure 9.20
Medium Rise Courtyard type: Option 4 NVPDCPH: Natural Ventilation incorporating Passive Downdraught Cooling and Perimeter Heating.
1. Solar water heating panels supply 'winter' tanks below courtyard.
2. Wind-catchers flush thermal storage chambers as required.
3. Opening glazed vents to exterior supply cooling batteries (5).
4. Internal opening glazed vents allow re-circulation.
5. PDC cooling batteries.
6. Cone of acoustic absorbent material on light frame to direct PDC airflow.
7. Vent connects cooled water tanks chamber to external environment as required.
8. Cooled and warmed water tank enclosures formed below courtyard.
9. Flow and return to geothermal array via horizontal pipe to field or borehole.
10. Air admitted through opening vents in each bay of elevation.
11. Transfer ducts through outer corridor partitions.
12. Exhaust ducts connect courtyard-facing spaces to outer envelope.
13. Lightweight insulation on roof.
14. High level vents to flush upper part of courtyard when PDC not engaged.
15. Glazed roof with retractable shading.

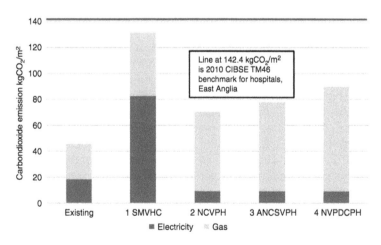

Figure 9.21
Predicted CO$_2$ emissions for existing and refurbishment options for year 2010, Cambridge (Bedford weather file). The CO$_2$ values are the average of the three wards studied.

Table 9.6 Medium Rise Courtyard type: Rosie Maternity Wing, comparative costs of four adaptation options

Option	Area in m²	Cost £/m²	Total £ sterling
1. SMVHC	8536	953.38	8,138,016
2. NCVPH	8536	1152.40	9,836,855
3. ANCSVPH	9324	1568.85	14,627,985
4. NVPDCPH	9324	1776.59	16,564,897

'Nucleus-type hospitals' comprise connected attenuated cruciform plan templates, each a Latin cross, of 1000m², forming residual square courtyards between, mat buildings in effect (Mumford, 2001). Nucleus was a hospital planning system accommodating virtually any set of hospital activities, at a pinch, devised in response to the 1973–74 recession by slimming down the Department of Health Hospitals Building Division's 1969 Harness system, of which only two of 70 planned were realized. Capital expenditure was restricted to £6 million in the first instance to provide the 'Nucleus' (Francis *et al.*, 1999). The same authors go on to report that 130 Nucleus schemes were built across the UK. The DeDeRHECC research team selected the 1984 Glenfield Hospital in Leicestershire for closer study as Giridharan *et al.* (2013) report. The wards have a hybrid ventilation strategy with a low rate of mechanical ventilation and are passively cooled using Simple Natural Ventilation, manually opening windows at will, but their opening was subsequently restricted to 100mm (due to health and safety concerns initiated in the late 1990s). Between June and September 2010, a historically mild summer, the maximum indoor temperatures in the case study spaces occurred as external temperatures exceeded 20°C, varying between 27.3°C and 29.3°C in late June. The nurses' station was found to be the hottest area. A minimum indoor temperature of 18.2°C was recorded at 1:00 a.m. in mid July when external night-time temperatures dropped to 14.7°C suggesting that heating was fired up at this point, the set point being 18°C. However, the majority of the monitored spaces performed within the thermal comfort threshold defined by HTM03-01, 50 hours above 28°C annually. The simulation results demonstrate that light-touch, low carbon interventions could produce comfortable conditions in Nucleus bedrooms into the 2050s in the UK Midlands, simple sunshading of vulnerable glazing, ceiling fans, although there is an anxiety about increasing the risk of airborne infection and the consequences of relying on automated window opening gear linked to the Building Management System. Giridharan *et al.* (2013) explain window opening coupled with the opening of internal doors is dominant in suppressing peak temperatures. Flows are well below 6 ACH. Clearly there is an adaptation scheme available which organizes and controls airflows through and across the wards without infringing privacy and enabling further subdivision of open wards. The Pavilion and Tower schemes suggest how that might be achieved. However, it is essential to avoid filling the courtyards with new building, tempting as this might be to Directors of Estates because, of course, this will rapidly make a very deep plan hospital, evaporate any natural resilience to overheating thereby introducing severe overheating problems as the century progresses, certainly south of the Liverpool latitude. The Nucleus template was designed to enable natural day-lighting and cross-ventilation.

Masonry 'Nightingale' Wards

A parallel DeDeRHECC investigation revealed that the masonry-built 'Nightingale' ward type has considerable resilience with the potential to contribute to the solution of the NHS conundrum but only if reconfigured to deliver to NHS Modernization policy goals (Lomas et al., 2012; Short et al., 2014).[10] In the early 2000s the UK NHS was directed to abandon the traditional collective healthcare model of an open shared ward for 24–30 patients and adopt the single room model hitherto reserved for the very unwell and the privately insured.[11] Pre-1939 wards built to Florence Nightingale's original very detailed specifications were declared to be no longer fit for purpose by the British government.[12]

The DeDeRHECC research team measured and modelled the four-storey Nightingale ward buildings at Bradford Royal Infirmary over two full years. They retain the original Nightingale glazing configuration offering prodigious free areas for cross-ventilation through quadruple banked hopper windows alternately top- and bottom-hung. Across all eight measuring points in one representative ward the temperature only varied from 20.1°C to 27.4°C with a mean of 23.7°C. The mean night-time temperature was 23.2°C and the maximum diurnal swing recorded was just 5.2°C. The temperatures in a second ward were similar. Overall the temperatures in all the spaces were well controlled and well within the wide range recommended for wards by HTM03-01 of 18–28°C. The predicted building energy demand in 2010 was an extraordinarily low 14 GJ/100m^3 with over 90% of this being for space heating, significantly below the NHS target of 55–65 GJ/100m^3 for refurbished buildings and well below the target of 35–55 GJ/100m^3 for new buildings. In other words, the building type delivers the NHS carbon reduction target. Building environment CO_2 emissions plus that sustaining the clinical function approximate 53 kgCO_2/m^2, just below the best DH target.

Three incremental refurbishment options were devised for the Nightingale wards. Option 1 adds 100mm of insulation to the walls and 300mm to the roof, opens up the triple light windows and protects occupants with an architectural external steel grillage and provides a sunshade at each opening. Fresh air supply in winter is provided by the re-opening of a trickle vent behind a perimeter heating element. Space heating demand dropped from about 13 GJ/100m^3 to an extremely low 5 GJ/100m^3, CO_2 emissions to about 15 kgCO_2/m^2. Option 2 adds ceiling fans operable by the patients to this strategy (Figure 9.22). Option 3, designed for well into the century in Bradford, sooner in the south, introduces 100mm diameter high-level air inlets above each bed space, between each window, with a damper, and a simple convective heating device fixed to the internal face to enable supply air to be preheated and/or recirculation within the space.

Primary heating and cooling is delivered through the installation of radiant panels. The addition of radiant cooling eliminates entirely the risk of overheating. The further adaptation work reported here takes the second option as the base treatment of the envelope.

Simulations of 'normal' TRY and hot DSY years were undertaken for 2010, 2030, 2050 and 2080. The results clearly indicate that neither the existing nor

Figure 9.22
Nightingale Ward:
Bradford Royal Infirmary
adaptation scheme
Option 2.

the refurbished building will overheat in typical years, as judged by the HTM03 and BSEN15251 criteria. However, in the 2050s warmer night-time temperatures may be experienced (although these might be ameliorated easily with a refined window opening regimen if the windows are openable to a useful degree in sufficient numbers). In the extreme temperature years (i.e. the DSYs) however, HTM03 shows overheating will occur in the existing building and in refurbishment Option 1 as early as the 2030s whilst Options 2 and 3 are resilient throughout.

The highly resilient Option 2 was then taken as the basis for exploring reconfiguration possibilities with staff at the Bradford Royal Infirmary. The aim was to couple this resilient envelope with an internal layout delivering greater privacy and dignity. Evidence from ward staff suggests a value in open wards for the care of geriatric patients. Figure 9.23 shows:

(a) The Nightingale layout as originally built, a little narrower than prescribed in 1858.
(b) 'Partitions': The plan subdivided into one bed cubicles, the full 16 foot in height, with the addition of external bathroom towers.
(c) The 'Pullman': The arrangement of a compartment railway carriage, the incorporation of an internal corridor and subdivision into six two-bed rooms served by split bathroom towers.
(d) 'Zig-zag': Preserving the full open volume but arranging the beds at 45° either side of a wardrobe-high central partition set out to a zig-zag plan, offering visual if not acoustic privacy with five external bathroom towers.
(e) 'External corridor': The recovery of more usable floorspace by adding an external corridor to each floor, enabling ward rooms of three to five beds.

The options were modelled to assess airborne infection implications at the University of Leeds. They perform well. The British Government Cabinet Office is enthusiastic about the zig-zag arrangement, almost a kind of aeroplane 'business class' rethought for NHS patients (Figure 9.24).

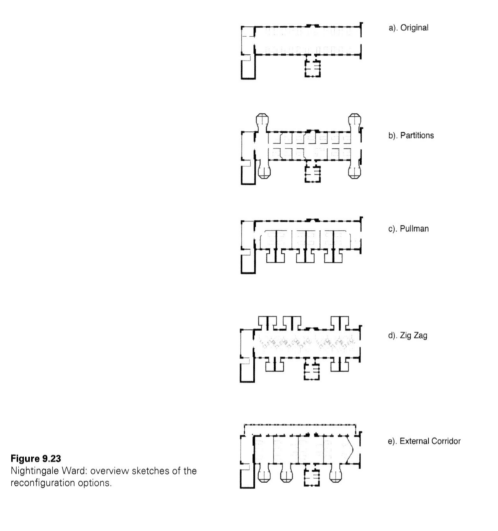

a). Original

b). Partitions

c). Pullman

d). Zig Zag

e). External Corridor

Figure 9.23
Nightingale Ward: overview sketches of the
reconfiguration options.

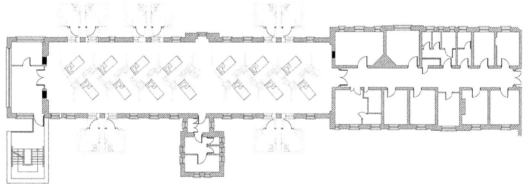

0 1 2 3 4 5 m

Figure 9.24
Nightingale Ward: adaptive reconfiguration: central spine, zig-zag option, 'Business Class for the NHS'.

DeDeRHECC monitoring of lightweight modular buildings reveals low resilience and a significant overheating risk despite the attractions of rapid installation and flexibility. Measurements in the relocatable modular building at the Bradford Royal Infirmary reveal high internal temperatures in mild conditions.

Adaptation: a view of the whole NHS Estate

Table 9.7 reveals that if all the adaptation schemes were implemented across the building types some 337.735 GJ, or 93,816 million kWh, or 93.8 MWh of energy could be saved annually, equivalent to 456.99 million $kgCO_2e$, some 11.23% of all NHS emissions associated with its 28 million m^2 estate, estimated as 4.07 $MtCO_2e$ in 2010.

To remain on course to achieve the Carbon Reduction Target, the NHS has to reduce emissions associated with its buildings by 0.1628 $MtCO_2e$ annually (as described previously in Chapter 6). These schemes, which might be considered rather radical by the UK construction world, will deliver 2.8 times that saving annually, but only saving about 1% of total NHS CO_2e emissions. From this sobering statistic one can start to appreciate the scale and invasiveness of the intervention required. First, the NHS needs very good information at a fine enough level of resolution to be able to tailor these 'cures', and secondly, the skills and insight to use it. It has neither. It also needs adequate resources: £17.5 billion for the ANV schemes, which is £1.6 billion less than for 'Passiv Haus' type interventions. This is a vast sum, but the DH division responsible reports that NHS Trusts receive some £4–5 billion per annum for backlog maintenance. It would then seem to be quite possible to enact this adaptation programme over a five-year period absorbing the health and safety upgrades intended from this funding. Moreover, some £4.5 billion of the total would deliver a comprehensive reconfiguration of all extant Nightingale Wards from which may flow other benefits core to the mission of the NHS, particularly with respect to caring for older people.

Ironically, the 2013–14 Department of Health's Energy Efficiency Fund Scheme administered by the author and his Cambridge colleagues would not have funded any of the adaptation options described here.[13] The Treasury requires a return on investment of 2.4 times the original expenditure realized within four to five years of implementation. This is potentially a huge barrier to achieving the adaptation of the public non-domestic building stock, which is unfortunate given that the NHS Retained Estate would seem to be a particularly promising place, perhaps the most promising place, to implement an effective public sector adaptation scheme.

Table 9.7 Initial overview of current and future performance of the NHS England acute hospital estate (University of Cambridge Departments of Architecture and Engineering)

Hospital building type	Gross floor area as measured[1]	Number of buildings identified	Resilience: from recent observed data	Current likely performance in GJ/100m³/pa and carbon emissions in kgCO$_2$/m²/pa	Potential performance post-adaptation: adaptation schemes designed to limit peak temperatures to 28°C to 2050	
					Industry standard: [sealed, mechanically ventilated with efficient heat recovery]	Passive/hybrid ANV [promoting stack and cross-ventilation, solar shading, natural light, exposed thermal mass]
Pre-1939	1,963,244	110	Resilient: Nightingale wards peak temp. of 29°C for 4 hours only predicted for 2010, peak night time temp. 3hrs. only above 26°C (delivered by heating system). 2050 prediction 66hrs. above 28°C in TRY and 364hrs. in DSY	14 GJ/100m³/pa 30 kgCO$_2$/m²/pa	7.5 GJ/100m³/pa 15 kgCO$_2$/m²/pa	6.5 GJ/100m³/pa 15 kgCO$_2$/m²/pa
Courtyard Low rise (1–2 storeys)	1,760,371	114		*As nucleus*	*As below*	*As below*
Nucleus	900,283	45	Peak temp. of 29.2°C June 2010 in 6-bed ward, external temp 24.3°C. Predicted TRY peak of 32.1°C in 2050s, DSY peak 34.2°C. Opening windows in diurnal temp range of 4°C+ very effective if doors kept open to enable cross-ventilation.	36.8 GJ/100m³/pa 58.29 kgCO$_2$/m²/pa	NHS best target 55 GJ/100m³/pa 143 kgCO$_2$/m²/pa?	40 GJ/100m³/pa 63 kgCO$_2$/m²/pa (Note: interventions reduce 2050s TRY peak to 28.9°C and DSY peak to 31.9°C)
Courtyard Medium rise (>/= 3 storeys)	3,012,568	117	Rosie multi-bed ward S. facing 48 hrs above 28°C in 2010, TRY 195hrs in 2030, DSY 706hrs in 2030	30 GJ/100m³/pa 45 kgCO$_2$/m²/pa	65 GJ/100m³pa 130 kgCO$_2$/m²/pa	25 GJ/100m³/pa 30 kgCO$_2$/m²/pa

Evidence for current predicted performance	Costs £/m² industry standard	Costs £/m² Passive/hybrid	Savings in energy/carbon	
	Total construction costs across building type for industry standard adaptation excl VAT/fees	Total construction costs across building type for passive/hybrid adaptation excl VAT/fees	Industry standard	Passive /hybrid
Lomas, K.J., Giridharan, R., Short, C.A., Fair, A.J. (2012) Resilience of 'Nightingale' hospital wards in a changing climate, *Building Serv. Eng. Res. Technol.* 33,1, pp. 81–103 C. Alan Short, Catherine J. Noakes, Carl A. Gilkeson (2014) Functional recovery of a resilient hospital type, *Building Research & Information*, DOI: 10.1080/09613218.2014.926605	£2763/m² £5424.44 million	£2281/m² £4478.16 million	12.76 million GJ 29.45 million kgCO$_2$	14.72 million GJ 29.45 million kgCO$_2$
Very approximately equivalent to nucleus, probably includes some nucleus hospitals	£1400/m² £2464.52 million	£1200/m² £ 2112.45 million	−32.04 million GJ −149.10 million kgCO$_2$	−5.63 million GJ −8.27 million kgCO$_2$
Based on study of Glenfield Hospital, Leicester, light touch interventions, reduced internal gains at nurses' stations, solar shading and user controlled fans, but no costings available. Giridharana, K.J. Lomas, C.A. Short, A.J. Fair (2013) Performance of hospital spaces in summer: A case study of a 'Nucleus'-type hospital in the UK Midlands, *Energy and Buildings* 66 (2013) 315–328	*Assume* £1400/m² £1260.4 million	*Assume* £1200/m² £1080.34 million	−16.39 million GJ −£76.25 million	−2.88 million GJ −£4.23 million
Short, C.A., Lomas, K.J., Renganathan, G. (2015) A medium-rise1970s maternity hospital in the east of England: resilience and adaptation to climate change, *Building Serv. Eng. Res. Technol.*, special issue 'Indoor Temperature and Air Quality', DOI: 10.1177/0143624414567544	£953.38/m² £2872.13 million	£1152.40/m² (£1776.59/m² for Passive Downdraught Cooling option) £3471.68 million (£5352.1 million)	−105.44 million GJ -256.07 million kgCO$_2$	15.06 million GJ 45.189 million kgCO$_2$

Table 9.7 *continued*

Hospital building type	Gross floor area as measured[1]	Number of buildings identified	Resilience: from recent observed data	Current likely performance in GJ/100m³/pa and carbon emissions in kgCO₂/m²/pa	Potential performance post-adaptation: adaptation schemes designed to limit peak temperatures to 28°C to 2050	
					Industry standard: [sealed, mechanically ventilated with efficient heat recovery]	Passive/hybrid ANV [promoting stack and cross-ventilation, solar shading, natural light, exposed thermal mass]
Pavilion	81,720	19	4-bed ward in 2005 might experience 39hrs above 28°C, but 150hrs in 2020, 313hrs in 2050, a SW facing ward 104hrs, 291hrs and 502hrs respectively.	45 GJ/100m³/pa 55.52 kgCO₂/m²/pa	34 GJ/100m³/pa 49.05 kgCO₂/m²/pa	32.5 GJ/100m³/pa 47.47 kgCO₂/m²/pa
Matchbox on Muffin	39,911	3	*As slab below*	*As slab below*	*As slab below*	*As slab below*
Slab	1,952,891	70	Addenbrooke's Ward Tower currently rarely exceeds 28°C but suffers high night time temperatures of 24°C. In TRY 163hrs above by 2050s, in DSY 383hrs above by 2030s peaking at 33°C	102 GJ/100m³/pa 180 kgCO₂/m²/pa	56 GJ/100m³/pa 140 kgCO₂/m²/pa	25 GJ/100m³/pa 50 kgCO₂/m²/pa
Tower	298,487	25	*As for slab*	*As above*	*As above*	*As above*
Deep Plan (incl. relatively recent PFI hospitals)	2,395,319	124	No direct data, extrapolated from low rise courtyard	85 GJ/100m³/pa assumed see ERIC data for known deep plan hospitals	55 GJ/100m³/pa DH target	50 GJ/100m³/pa lower DH target
Total above	**12,404,794**	**627**				

Evidence for current predicted performance	Costs £/m² industry standard	Costs £/m² Passive/hybrid	Savings in energy/carbon	
	Total construction costs across building type for industry standard adaptation excl VAT/fees	Total construction costs across building type for passive/hybrid adaptation excl VAT/fees	Industry standard	Passive /hybrid
Based on Northwick Park Hospital Maternity Block West London Short, C. Alan, Cook, Malcolm, Cropper, Paul C. and Al-Maiyah, Sura (2010) Low energy refurbishment strategies for health buildings, *Journal of Building Performance Simulation*, 3: 3, 197–216, doi: 10.1080/194014909033 18218	£5154.42/m² for hybrid scheme with cooling £421.22 million	£5154.42/m² for hybrid scheme with cooling £421.22 million	0.899 million GJ 0.529 million kgCO₂	1.022 million GJ 0.658 million kgCO₂
In effect slabs and towers on podium, podia included as 'deep plan'	£1056.88/m² £ 42.18 million	£1286.95/m² £51.36 million	1.836 million GJ 1.596 million kgCO₂	3.073 million GJ 5.188 million kgCO₂
Based on the Ward Tower at Addenbrookes Cambridge Short, C.A., Lomas, K.J., Renganathan, G., Fair, A. (2012) Building resilience to overheating into 1960's UK hospital buildings within the constraint of the national carbon reduction target: adaptive strategies. *Building and Environment*, doi: 10.1016/j.buildenv.2012.02.031	£1056.88/m² £2064 million	£1286.95/m² £2513 million	89.83 million GJ 78.12 million kgCO₂	150.37 million GJ 253.87 million kgCO₂
Towers considered equivalent to slabs at this stage but racetrack corridor plans around a solid core more challenging to cross-ventilate	£1056.88/m² £315.47 million	£1286.95/m² £384.14 million	13.73 million GJ 11.94 million kgCO₂	22.98 million GJ 38.8 million kgCO₂
Short, C.A., Al-Maiyah, S. (2009) Design Strategy for low energy ventilation and cooling of hospitals *Building Research and Information*, 37(3), pp.1–29. doi: 10.1080/09613210902885156	£1600/m² assumed £3832.51 million	£1250/m² assumed £2994.15 million	71.86 million GJ 71.86 million kgCO₂	83.84 million GJ 83.836 million kgCO₂
	£18.697 billion	£17.507 billion	**–1.771 million GJ** **–137.43 million kgCO₂**	**337.735 million GJ[2]** **456.99[3] million kgCO₂**

Notes to Table 9.7

1 Acute hospital buildings identified by Prof. C.A. Short on each site to typology identified by the EPSRC
DeDeRHECC project as depicted in Google Earth and Bing views and floor areas estimated directly using
Image-J software by researchers at Cambridge University Department of Engineering: Florian Hirmer, Anne
Grau, Gerard Casey and Stephanie Hirmer.

2 337.735 million GJ equivalent to 93,816 million MWhrs

3 NHS SDU 'Carbon footprint update NHS England 2012' records 25M Tonnes CO_2e of which 4.07MTonnes
relates to built environment 17% of carbon footprint. 0.457MTonnes saved CO_2e is equivalent to 11.23%
of the total for the NHS built environment and 1.83% of the NHS total carbon footprint (excluding 2.66
million m^2 of Nucleus and low rise courtyard acute hospital buildings for which data requires more
validation).

Notes

1. The Adaptation and Resilience to Climate Change (ARCC) project, Design and Delivery of Robust Hospital Environments in a Changing Climate (DeDeRHECC) 2009–13, funded by EPSRC, NHS Trusts and the Department of Health. Co-investigators were Professor Alan Short, University of Cambridge; Professor John Clarkson, University of Cambridge; Professor K. Lomas, Loughborough University; Dr Claudia Eckert, Open University; Dr Catherine Noakes, University of Leeds. The film of the project 'Robust Hospitals in a Changing Climate', winner of the tv/e Global Sustainability Film Award 2013, is available at: www.sms.cam.ac.uk/media/1446036 (accessed 12 October 2016).

2. Professor Graf writes, 'For the calculation of the historic and future frequency of daily temperature maxima at the sites of UK NHS hospitals we used historic data that are available from 1960–2005 in the UKCP09 data set, see www.metoffice.gov.uk/climatechange/science/monitoring/ukcp09/download/daily/gridded_daily.html (accessed 12 October 2016) at spatial resolution of 5km across the UK and historic (driven by observed greenhouse gas trends and natural drivers as solar activity and volcanic eruptions) and future (driven by greenhouse gas emissions according to the RCP4.5 emission scenario) coupled ocean–atmosphere climate model data at 1 degree resolution available for several climate models from the CMIP5 model inter-comparison project. See http://pcmdi9.llnl.gov/ (accessed 12 October 2016).' Of the four models tested, the Norwegian model NorESM1-M yielded a temperature bias of −2°C and when bias corrected, a correlation of 0.96 against historic trends.

3. Forthcoming paper in preparation, Xiaoyong Yu, S., Graf, H. and Short, C.A., 'The future threat of hot summer days to UK NHS Acute Hospitals'.

4. The exercise was led by the author, Professor Peter Guthrie and Stephanie Hirmer of the Centre for Sustainable Development at the University of Cambridge Department of Engineering, the methodology devised by Florian Hirmer, with Anne Grau and Gerard Casey who conducted the measure. A full paper is in preparation.

5. Major mental health institutions are excluded from this exercise but should be analysed in a similar way as a matter of urgency.

6. Comprising Dr Simos Yannas of the Architectural Association, Professor Kevin Lomas of Loughborough University, and Dr Nick Baker of the University of Cambridge.

7. Northwick Park is of course affected by the London urban heat island, the exercise pre-dating the succeeding DeDeRHECC offensive.

8. Animations of the options are available in the project film, 'Robust Hospitals in a Changing Climate' at: www.sms.cam.ac.uk/media/1446036 (accessed 12 October 2016).

9. The adaptation potential within the high-rise tower type is a key objective of the UK-China EPSRC-NSFC Low Carbon Cities project for which the author is Principle Investigator, 'Low carbon environmentally-responsive heating and cooling of cities (LoHCool)', 2015–2018.

10. Department of Health (2010). *Privacy and dignity.* See http://webarchive. nationalarchives.gov.uk/fl/www.dh.gov.uk/en/Managingyourorganisation/Workforce/ Leadership/Healthcareenvironment/DH_4116444 (accessed 21 July 2011).
11. Labour Party (2001). *Renewing public services: NHS reform.* See www.labourparty. org.uk/manifestos/2001/2001-labour-manifesto.shtml (accessed 25 November 2013).
12. Department of Health (2002). £40 million to eliminate outdated Nightingale wards. Press release. See www.prnewswire.co.uk/news-releases/40-million-to-eliminate-outdated-nightingale-wards-154713785.html (accessed 18 November 2013).
13. Department of Health (2012). Gateway information 1838: 'Improving energy efficiency in the NHS, applications for capital funding 2012–13'. Foreword by Dr Dan Poulter, Under Secretary of State for Health, issued 17 December 2012 for circulation within the NHS only.

10

Delivering the 'recovery'

The morbid state of men of design injures themselves only ... but the modern English fact-hunter, despising design, wants to destroy everything that does not agree with his own notions of truth, and becomes the most dangerous and despicable of iconoclasts, excited by egotism.

(John Ruskin, 1893, *The Stones of Venice*, vol. 2, p. 18)

Correct attributions generally appear spontaneously and 'prima vista' ... such unthinking recognition may be regarded as unscientific whilst the 'method' that Giovanni Morelli imagined, or asserted, that he had found may be admired as scientific. I am convinced that Morelli for all his method would have achieved nothing had he not been a talented connoisseur, more, I am convinced that he never used his method but that he shrouded the results of intuitive connoisseurship in a mantle of false erudition to make them appear unassailable to the naïve mind.

(M. Friedlander, 1956, p. v)

Ruskin captured all too vividly the prejudice of aesthetes against the threatened contamination of the creative 'will to form' by science. Perhaps a bad dinner encounter or two at Oxford sealed this belief, still widely held by many architects. Friedlander was contemptuous of any possible assistance from science in intellectual judgement in 1916. A century before, Goethe had applied his theory of the 'Ur-phenomenon' to the perception of coloured light, ridiculing Newton's depiction of the spectrum as 'cucumber salad'. Goethe's biographer Friedenthal (1993, p. 403) records Hegel's corroboratory view that a prism was, 'the triangular glass cudgel which the satanic angel carries in his hand to strike the physicists', and that Goethe had threatened, 'to nail the physicists to the table by their donkeys' ears'. Might this outrage at Newton's dehumanized account of the spectrum extend to the diktat of the algorithm claimed to predict 'comfort' in the minds of the 'anti-factual' and all the measuring devices which contributed to it?

What immense opportunities this traditional prejudice excludes.[1] This book has taken a quite contrary interdisciplinary approach in its attempt to build an argument through the preceding chapters:

- A key question was 'How did we get here?' Why are our buildings made as they are today? Answering this meant exhuming the 'pre- and early modern' fears that miasmatic atmospheres bearing fomites could bleed into buildings and corrode their occupants' health. Fears which were prefigured by Burton's 'aerial devils', Alberti's 'infectious winds' and Arbuthnot's involuntary 'Ingesta', 18th century speculations about phlogiston and carbonic acids, and degeneracy caused by exposure to bad air. These were arguments which extended well into the 1890s and into the early era of attempts to make artificial environments.

- Classical perceptions of what might constitute a 'good' environment were considered from Hippocrates' medicalized deductions on the importance of seasonal diversity to Herodotus' accounts of difference between regional populations in different climates and Strabo's prescription for the optimal climate for civilizing progress. Universally this was believed to be that of the Mediterranean, most perfectly realized at Baetica and Rome. These ideas had an extraordinary longevity, and after two millennia, the cooling of Strabo's optimal climate for civilization into the misty mountain atmospheres revered by the German Romantics, refrigerated into the high Alpine domain of Nietzsche's *Zarathustra*. As the 20th century unfolded, overtly racist writings promoting Temperate conditions by Köppen, Ratzel, Huntington, Semple, Hellpach and others were popularized in North America and beyond by Mills and others, with enormous, if unrecognized, consequences for the built environment.

- Unsurprisingly, given the fear of bad air, a sustained campaign from the later 18th century was uncovered to make safe, naturally ventilated environments in public buildings. The stakes couldn't have been higher, the potential rewards imagined to be the guarantee of public health, longed for social recognition and a possibility of tremendous financial return. The campaign became increasingly bullish in its scientific content, pooling evidence and understanding of fundamental physics, fluid dynamics, chemistry and human biology. It progressed through the irritating but visionary Dr Reid's simple but prescient diagrams, the very much more sophisticated treatise of General Morin with its vivid practical examples, the internationally researched guidance disseminated by the medical authority and known bully Major J.S. Billings and, of course, the contributions of many others, not least the tragic Dr Pettenkofer.

- The ambitious claims underpinning some key examples of early modern 'Advanced Naturally Ventilated' public buildings in Chapters 3, 4 and 5 were tested. Examples were considered but most particularly in Chapter 6, the search for an exemplary 19th century hospital design by the Johns Hopkins Trustees in Baltimore. Two of the most intriguing propositions put to them in 1875 were digitally reconstructed and then subjected to the latest analytical techniques to predict the spread of pathogens. Two viable, very low energy general ward designs were revealed, both of which comply comfortably with contemporary NHS environmental standards in the UK. Major Billings' scheme, as built, invites a positive reassessment of the sizeable estate of Nightingale ward buildings in Britain, reported on in Chapter 9. They offer material clues as to how to reform hospital design and redesign to achieve the immense carbon savings required of the UK health service but Dr Folsom's scheme offers a radical reconfiguration of the communal ward, a type in which interest is reviving as older people, some with dementia, come to dominate hospital populations in Western societies as they are already tending towards in Japan.

- A start was made to unravel the complex and diverse factors which de-railed this well-evidenced interest in making natural environments in buildings. This was found to be dominated by commercial interests. Walter Bernan propagandized his vision of a completely sealed artificial environment from the middle of the 19th century. Heating and ventilating 'engineers' flourished. Companies with an international reach formed as the technology of boilers, valves and controls advanced. Powerful fan types were evolved, not least the 'Sirocco', which enabled North America to pursue ducted warm air rather than radiant systems, systems awaiting the manufacture of cool dehumidified air, refrigeration being already well established technically and commercially. Ventilation and humidity problems in industrial production prompted ingenious responses in mechanical environmental control from Willis Carrier and his rivals. There is a strong circumstantial but unproven parallel with the thinking of those 'scientizing' ancient beliefs in climate determinism. The air-conditioning industry transferred its technologies from industry to commerce, cinemas, department stores and then to office environments. Air-conditioned environments were not wholly welcome, as Arsenault and other commentators report. The industry was marketed relentlessly, picking at insecurities at all levels of society from the most corporate to the most intimate personal anxieties. One strand was based on providing the means of personal 'deodorization' to address a new, induced anxiety about social exclusion caused by odours emanating from the mouth and the torso. Another strand promised domestic and business success from working and living in a chilled environment and certain failure if condemned to an uncomfortable and malodorous existence in a natural environment, even in the home.
- Discussion in some arenas had become ideologically and/or culturally unacceptable in its assertions of individual or societal determinism in relation to climate. Implicit, but difficult to evidence, was the comforting notion that cool temperate climates could be manufactured anywhere within International Style building envelopes so that no population need be 'disadvantaged' by birth or geographical accident. Strabo had argued that civilization could only prosper in a very particular climate, but now this precious environment could be manufactured at will. This was perceived to be a huge social good but, we now know, at an immense cost in energy and carbon and ironically, perhaps, in well-being.
- The body of knowledge accumulated over a century about making naturally conditioned public buildings appears to have been lost very rapidly. This occurred in the United States, perhaps by the early 1920s, but the knowledge was retained for longer in Europe, no doubt as recession and debt delayed market penetration by the new technology. Eventually, the market value of air-conditioned commercial floorspace exceeded that of the naturally conditioned equivalent sufficiently to persuade developers to adopt the whole sealed artificial package unequivocally. Guy and Shove (2000, Chapter 7) identified, 'property agent-developer-institutional investor communities' operating in city centres and their design and construction industry dependents as a particularly intransigent and influential network, 'agents … have a critical role in managing this supply driven approach to property valuation'. Guy and Shove (2000, Chapter 7) object to the 'homogenization of office design' encouraged by this small community, revealing, from their perspective, a social practice.

- As a consequence, the architectural means to achieve substantial naturally conditioned buildings is largely, but not wholly, extinguished. But it has never been more possible to do so, as shown in Chapters 3–9, due to the new capabilities to test and evaluate the performance of a design through modelling and simulation. The environmental consequences of that loss have been immense and one can hazard with reasonable confidence estimates for particular sectors such as the vast NHS Estate in Chapter 9.

- The technology of mechanical cooling was united with a thin, vulnerable, energy intensive mid 20th century model for non-domestic architecture, regardless of place or function. They are interdependent. The extreme manifestation of the union is the all-glass tower, the global icon of commercial 'Sturm und Drang', power and influence. It was an unlikely candidate to achieve this architectural prominence, given its bizarre and complex quasi-mystical origins and its fragility. Has this now become the dominant preference of the tiny commissioning elite? It has certainly infected Asia and the Middle East, in highly unpromising harsh Continental climates.

- This raises profound questions and challenges for the small community involved in the creation of buildings, and their 'social practices' (Schatzki, 2002).[2] Prospective financial outcomes dominate with all the behaviours associated with that goal, conservatism, adversity to risk, short-termist responses to fluctuations in supply and demand.

- The evidenced review of recent attempts to achieve natural or near natural environments across the principal climate zones is encouraging. Large buildings can be made to be comfortable with very much less energy.

- What about the existing building stock? In the developed world, the annual replacement rate is low, in parts of Asia it would be financially and environmentally ruinous to replace the large amount of recent building. Although the resilience of the UK's existing building stock to higher temperatures is poor, its capacity for physical configurational, not technological, adaptation is high. This is particularly true for the pre-1939 stock with high thermal mass and the residue of a once effective environmental strategy. These buildings are an unrecognized resource. This may be our most important discovery.[3]

- The close analysis of climate data recurs as a theme throughout the book, attempting to predict future conditions in a chaotic system. The predictions in this book for the UK are for change, but at a slower rate of warming than the published predictions, if the natural variability exhibited by the climate over the last quarter of a century continues to follow the same pattern and emissions continue in the middling scenario. The very important implication for our argument is that the strategies for new-build and adaptive rebuild are more viable. They will have a significantly longer period of acceptable performance.

- What are the prospects for the promising environmental design strategies described here in the predicted future climates, as climate zones shift through the coming century? This is a question of huge import. The IPCC models suggest that the temperate climate zones in the northern hemisphere will grow northeastwards as comparison of the domains of temperate mean daily minima in the northern hemisphere generated for this book indicates, a significantly broader geographical reach (Figures 10.1 and 10.2).

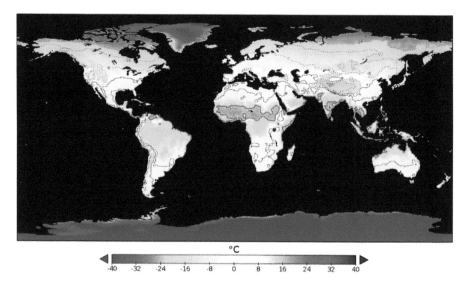

°C

-40 -32 -24 -16 -8 0 8 16 24 32 40

Figure 10.1
Recorded data 1966–2005 April to September mean daily temperatures below 18°C (i.e. white implies mean daily temperatures 14–18°C, pale blue 10–14°C), based on ERA reanalysis. Clearly there should be no need to mechanically cool buildings in 'normal' usage outside the yellow to orange zones where some greater ingenuity is required. Graf, Xiaoyong Yu and Short, University of Cambridge, unpublished.

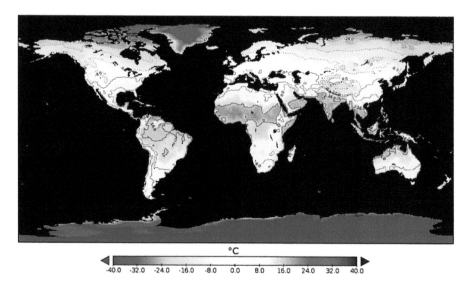

°C

-40.0 -32.0 -24.0 -16.0 -8.0 0.0 8.0 16.0 24.0 32.0 40.0

Figure 10.2
Predicted 2090s April to September mean daily minimum temperatures below 18°C, changes from 1996–2005 are apparent in the northern half of the USA, its eastern seaboard, and the former Soviet Union, South America, India and Southeast Asia. Graf, Xiaoyong Yu and Short, University of Cambridge, unpublished.

- The implication is that the strategies for naturally ventilated and mixed-mode buildings will become more applicable across vast territories enjoying warming conditions as our modelling and that behind the recasting of Köppen's map reveals. As Arrhenius observed, a changing climate will deepen existing challenges but open new opportunities. There is a tremendous opportunity for energy reduction, whilst passive and hybrid strategies may become even more critical in the rapidly warming Mediterranean-type climate zone.

All of which suggests that a full or partial recovery of natural environments is possible in most climates. These new/old environments will be made in what will be materially different buildings, different to today's custom and practice outputs and different to each other, varying by place, weather, climate and the intensity of activity. This recovery is likely to be both economically viable and politically desirable as a potentially important component of the response to the global imperative to cut carbon emissions, in the face of a changing climate in which the demand for cooling in particular is predicted to increase. It is not viable to deliver to that demand mechanically on a number of counts – particularly the significant energy and CO_2 entailed in mechanically moving large volumes of air – but the careful interrogation of and extrapolation beyond established, if largely forgotten, environmental design strategies with recent advances in scientific understanding and contemporary technology could deliver to that demand. The examples presented in this book have demonstrated that it is both viable and preferable to harness the morphology (built form) of buildings, along with the opening/closing of windows and vents, to harness air movement and deliver 'comfort' and healthy indoor environments.

Could the principle obstacle to the recovery be 'taste', even more than perceptions of 'comfort'? Throughout the book, the outputs of the poetic imagination were combined with an understanding of the operation of the natural physics of the world. Without an appreciation of both, one is left in the hands of the air-conditioning industry whose simple aim, not unreasonably, is to capture market share and generate profit, or a gestural architecture with no environmental presence or meaning, no productive 'lacing of human and other phenomena' (Schatzki, 2002, p. xiii).

The social practice interpretation clearly applies to the very small community who commission non-domestic city buildings. Most of us have no control whatsoever over the social and economic regime which makes the environments within which we gather and work. The majority of building designers and their constructor confederates follow custom and practice of necessity and out of a sense of 'oughtness'. They are assembling proprietary products within a predetermined price range, which enables investment formulae to be solved. But architecture is ultimately led by thinking designers, they hold the pencil and thus have some control and responsibility for these decisions. They switch their ambitions in response to rational and, simultaneously, entirely subjective stimuli. The design discipline needs to present an alternative model. Guy and Shove's (2000) isolated community does not invent form.

Driving designers' ideas, their 'artistic will', in a new direction is a different order of challenge as Worringer (1908) argued over a century ago. Worringer insisted that these changes in 'will', which foment new artistic movements, are not arbitrary but caught in a prevailing 'atmosphere' of common interest and preference. It is geographic, relating to cultures rooted in their land, their climate and a myriad other factors not to be suppressed. Might it be that widespread disenchantment will arise with sealed artificial environments, the potential implications for health and well-being, revulsion at the profligacy in energy and carbon and realization that homogenized and insubstantial transparent building forms no longer capture 'the spirit of the age'? This would force the crystallization of what Worringer called a new 'World Conception'. Louis Sullivan couldn't understand why an authentic new American architecture hadn't emerged by 1896. American architects were 'handcuffed in the asylum of a foreign school'. He thought that if his colleagues pursued the actual intended purpose and activities of a new building then this new expression would materialize, so powerful would this driver be. A similar argument has been advanced in this book: what is going to, and/or could happen in a building sensitive to its broader environmental context, invoking all the richness that that context has and could give to society so that, recovered, it

> opens up the airy sunshine of green fields, and gives to us a freedom that the very beauty and sumptuousness of the outworking of the law itself as exhibited in nature.
>
> (Sullivan, 1896)

Notes

1. C.P. Snow's 1959 Rede Lecture, 'The Two Cultures and the Scientific Revolution 5', described a more general divide between 'the literary intellectuals' and the 'natural scientists', a term of the 1830s. He describes the intellectuals as 'natural Luddites'. F.R. Leavis gladly assumed that mantle with malevolent intent. Stefan Collini's detailed introduction to the 1998 edition tracks what he describes as a 'fissure in types of knowledge', from its essentially 19th century origins, not least through the evolving structure of secondary education in Britain. Architectural training has attempted to bridge this gap, at a cost.
2. Schatzki defines a 'practice' as a '"bundle" of activities … an organized nexus of actions' (p. 70) and 'a practice is a set of doings and sayings' (p. 73) which 'hang together', social phenomena. His illustrative examples come from Shaker herbal medicine enterprises and aggressive day trading on the Nasdaq stock market where profit, success and self-esteem goals predominate to apparently little effect.
3. Within three years more should be known about the adaptive potential within the Chinese megacities as the author's new EPSRC-NSFC research project 'Low carbon climate responsive heating and cooling of cities' (LoHCool) 2015–18 reports.

REFERENCES

ASHRAE (2004) *Thermal Conditions for Human Occupancy. Standard No. 55-2004*. Atlanta, GA: American National Standards Institute (ANSI) and ASHRAE.

Ackerman, M.E. (2002) *Cool Comfort: America's Romance with Air-Conditioning*. Washington, D.C. and London: Smithsonian Institution Press.

Alexander, C., Ishikawa, S., Silverstein, M. with Jacobsen, M., Fiksdahl-King, Angel, S. (1977) *A Pattern Language. Towns-Buildings-Construction*. New York: Oxford University Press.

Allen Brookes, H. (1982) *The Le Corbusier Archive, A Series in Garland Architecture Archives*. 32 volumes, New York and London: Garland Publishing, Inc., and Paris: the Fondation Le Corbusier.

Altenmueller, U. and Mindrup, M. (2009) The City Crown by Bruno Taut. *Journal of Architectural Education*, **63**(1), 121–134.

Anderson, Sean (2010) *The Light and the Line: Florestano Di Fausto and the Politics of Mediterraneità*. eScholarship: University of California.

Arbuthnot, J. (1733) *Concerning the Effects of Air on Human Bodies*. London: J. Tonson.

Arnot Robertson, E. (1982) *Ordinary Families*. London: Virago Press Ltd, first published 1933 by Jonathan Cape Ltd in Great Britain.

Arrhenius, S. (1896) *On the Influence of Carbonic Acid in the Air upon the Temperature of the Ground*, reprised for the general reader in: Arrhenius S. (1908) *Worlds in the making: the evolution of the universe*, transl. Borns, H., Harper and Brothers, London, pp. 51–53.

Arrhenius, S. (1908) *Worlds in the Making: the Evolution of the Universe*. Translated by Borns, H. London: Harper and Brothers.

Arsenault, R. (1984) The End of the Long Hot Summer: The Air Conditioner and Southern Culture. *The Journal of Southern History*, **50**(4), 597–628.

Awad, M.F. (1990) Italian Influence on Alexandria's Architecture (1834–1985). *Environmental Design: Journal of the Islamic Environmental Design Research Centre*, 1990, 72–85.

Bachelard, G. (1971) *On Poetic Imagination and Reverie*. Translated and introduced by Colette Gaudin, Dallas, TX: Spring Publications, Inc.

Banham, R. (1969) *The Architecture of the Well-tempered Environment*. London: The Architectural Press.

Barber, J.M., Ferris, F.J., Forster, R., Roberts, B.M. (2000) *Wilson Weatherley Phipson, MICE: 1838–1891, a Victorian Building Services Engineer*. Dublin: CIBSE/ASHRAE Joint National Conference.

Barnard, F.A.P. (1972) The Germ Theory of Disease and its relations to Hygiene, reproduced in Brieger, G.H. (1972), 278–292.

Beguin, F. (1979) Les Machines a guerir, aux origins de l'hopital moderne, Brussels: Pierre Mardaga, 1979, 7–18, translated from *Dits et Ecrits*, © Gallimard 1994 (no. 257, vol. III, pp. 725–742) translated by Richard A. Lynch 2014 ISSN: 1832–5203 *Foucault Studies*, No. 18, pp. 113–127, October 2014, http://rauli.cbs.dk/index.php/foucault-studies/article/viewFile/4654/5087.

Benton, T. (1987) Cite de Refuge, Paris, in *Le Corbusier Architect of the Century*, Arts Council of Great Britain and authors, Great Britain.

Bernan, W. (1845) *On the History and Art of Warming and Ventilating Rooms and Buildings...*, *Vol. 1*. London: George Bell.

Berrios, G.E. (1989) Obsessive-Compulsive Disorder: Its Conceptual History in France During the 19th Century. *Comprehensive Psychiatry*, **30**(4), 283–295.

Billings, J.S. (1883) The Registration of Vital Statistics. *Am. J. Med. Sci.* 85, 33–59.

Billings, J.S. (1884) The Principles of Ventilation and Heating and their Practical Application, *The Sanitary Engineer*.

Billings, J.S. (1890) *Description of the Johns Hopkins Hospital*, 4 volumes. Baltimore, MA: Johns Hopkins Hospital.

Billings, J.S. (1893) *Ventilation and Heating*. London: The Engineering Record. Reprinted in book form from *The Engineering Record*.

Billington, N.S. and Roberts, B.M. (1982) *Building Services Engineering: A Review of its Development*. Oxford: Pergamon Press.

Blunt, A. (1968) *Sicilian Baroque*. London: Weidenfeld and Nicholson.

Blunt, A. (1975) *Neapolitan Baroque and Rococo Architecture*. London: A Zwemmer Ltd.

Boesiger, W. and Girsberger, H. (1967) *Le Corbusier 1910–65*. Translated by Gleckman, W.B. London: Thames and Hudson.

Bohme, G. (2013) The art of the stage set as a paradigm for an aesthetics of atmospheres, *Ambiences*. Available at http://ambiances.revues.org/315 (accessed 27th April 2014).

Boia, L. (2005) *The Weather in the Imagination*. London: Reaktion Books Ltd.

Boissevain, J. (1965) *Saints and Fireworks, Religion and Politics in Rural Malta*. London School of Economics Monograph on Social Anthropology No. 30. London: The Athlone Press, University of London. Reissued 1993 by Progress Press Co Ltd., Valletta, Malta.

Bordass, W. (12 June 2013). Inaugural George Henderson Memorial Lecture: Improving building performance: Sparing no expense to get something on the cheap? London: UCL Energy Institute.

Boyson, C., Taylor, S. and Page, L. (2014) The National Heatwave Plan – A Brief Evaluation of Issues for Frontline Health Staff, PLOS. *Currents Disasters* 13, Edition 1.

Brieger, G.H. (1965) The Original Plans for The Johns Hopkins Hospital and Their Historical Significance. *Bulletin of the History of Medicine* 39, 518.

Brieger, G.H. (ed) (1972) *Medical America in the Nineteenth Century; Readings from the Literature*. Baltimore, MA: The Johns Hopkins University Press.

Brieger, G.H. (ed.) (1976) *Theory and Practice in American Medicine*. New York: Science History Publications.

British Standards Institute (2007) *Indoor Environmental Input Parameters for Design and Assessment of Energy Performance of Buildings Addressing Indoor Air Quality, Thermal Environment Lighting And Acoustics*. British Standard BSEN15251. Brussels: British Standards Institute.

Brook, P. (1972) *The Empty Space*. Harmondsworth: Pelican Books.

Brook, P. and Estienne, M.-H. (2002) *The Man Who: A Theatrical Research (Modern Plays)*. Bloomsbury: Methuen Drama.

Brookman, J. (1991) Polys start building work. *The Higher Education Supplement, News*, 22 February 1991, p. 5.

Building News (1904, January) Analysis of report and evidence. [Report of the Select Committee on Ventilation: appointed by the House of Commons] London: Hickson, Ward & Co.

Burton, R. (1621) Some Anatomies of Melancholy, in *Quantity of Diet a Cause*. London: Penguin Books (2008).

Butterfield, H. (1973) *The Whig Interpretation of History*, Penguin Books Ltd, Harmondsworth, Middlesex, England. First published 1931 by George Bell & Sons.

CIBSE (2002) *CIBSE Guide J: Weather, Solar and Illuminance Data*. London: CIBSE.

CIBSE (2005) *TM36: Climate Change and the Internal Environment, a Guide for Designers*. London: CIBSE.

CIBSE (2006) *CIBSE Guide A: Environmental Design*. London: CIBSE.

Capon, R. and Oakley, G. (2012) *Climate Change Risk Assessment for the Built Environment Sector*. UK 2012 Climate Change Risk Assessment (CCRA). Defra Project Code GA0204.

Chalfant, R.W. and Belfoure, C. (2006) *Niernsee and Neilson, Architects of Baltimore: Two Careers on the Edge of the Future*. Baltimore, MD: Baltimore Architecture Foundation.

Chang, H. (2009) We have never been Whiggish (About Phlogistion), *Centaurus*, Vol. 51: pp. 239–264; doi:10.1111/j.1600-0498.2009.00150.x

Chang, H. and Jackson, C. (eds) (2007) *An Element of Controversy: the Life of Chlorine in Science, Medicine, Technology and War*. BSHS Monographs No. 13, British Society for the History of Science.

Chapman, T. (2000) *Architecture 00*. London: The Royal Institute of British Architects.

Clarke, K. (1964) *The Gothic Revival*. London: Pelican.

Collini, S. (2006) *Absent Minds: Intellectuals in Britain*. Oxford: Oxford University Press.

Conant, J.B. (ed) (1950) *The Overthrow of Phlogiston Theory: The Chemical Revolution of 1775–1789*. Cambridge: Harvard University Press.

Cook, J. and Hinchcliffe, T. (1995) Designing the well-tempered institution of 1873. *Architectural Research Quarterly*, **1**(2), 70–78. (published online 19 August 2008).

Cook, M.J., Lomas, K.J. and Eppel, H. (1999a) Design and operating concept for an innovative naturally ventilated library. CIBSE Conference, October, Harrogate, UK.

Cook, M.J., Lomas, K.J. and Eppel, H. (1999b) Use of Computer Simulation in the Design of a Naturally Ventilated Library. PLEA Conference, September, Brisbane, Australia.

Cooper, G. (1998) *Air-Conditioning America: Engineers and the Controlled Environment, 1900–1960*. Johns Hopkins Studies in the History of Technology. Baltimore, Maryland: Johns Hopkins University Press.

Cope, V.Z. (2012) *John Shaw Billings, Florence Nightingale and the Johns Hopkins Hospital*, (Medical History, News Notes and Queries, medhist00185-0086.pdf, pp. 367–8). For Sir Vincent Zachary Cope see David Hamilton, 'Cope, Sir (Vincent) Zachary (1881–1974)', rev. *Oxford Dictionary of National Biography*, Oxford University Press, 2004. Available at www.oxforddnb.com/view/article/30968 (accessed 28 July 2014).

Corbin, A. (1996) *The Foul and the Fragrant: Odour and the Social Imagination*. London: Papermac, MacMillan Publishers Ltd.

Cox, S. (2010) *Losing our Cool: Uncomfortable Truths about Our Air-Conditioned World (and Finding New Ways to Get Through the Summer)*. New York: The New Press.

Davies, C. (1995) Green Gothic. *Architecture, The Journal of the American Institute of Architects*, July, 88–97.

de Dear R.J. and Brager, G.S. (1998) 'Developing an adaptive model of thermal comfort and preference,' *ASHRAE Trans.*, V.104(1a), pp. 145–167.

De Lucca, D. and Carapecchia, R. (1999) *Carapecchia: Master of Baroque Architecture in Early Eighteenth Century Malta*. Valletta: Midsea Books.

Department of Health, Estate and Facilities Division (2006) *Statistics on energy performance and carbon and CO_2 emissions, NHS England, 1999/00 to 2004/05 (with prediction to 2009/10)*. London: The Stationery Office.

Dexter, D.G. (1904) *Weather Influences. An Empirical Study of the Mental and Physiological Effects of Definite Meteorological Conditions*. New York and London: The MacMillan Company.

Di Fausto, F. (1937) Visione Mediterranea Della mia Architettura. *Libia*, **1**(9), 16–18.

Di Lampedusa, G. (1958) *Il Gattopardo (The Leopard)*. Translated by Colquhoun, A. (1961). Milan: Feltrinelli Editore.

Diamond, Jr., A.M. (2013) *Keeping Our Cool: In Defense of Air Conditioning*. Department of Economics University of Nebraska at Omaha. Available at www.artdiamond.com/DiamondPDFs/InDefenseOfAirConditioning2013-08-14.pdf (accessed 4 November 2016).

Dietrich, N. (1992) Three Early Projects by Mies van der Rohe, *Perspecta* (Journal of the Yale School of Architecture), vol. 27, 76–97.

Dinsmoor, W.B. (1950) *The Architecture of Ancient Greece, an Account of its Historic Development*. New York: Biblo and Tannen.

Dlugolecki, A. *et al.* (2009) *Coping with Climate Change: Risks and Opportunities for Insurers*. London: Chartered Insurance Institute, CII_3112.

Donaldson, B. and Nagengast, B. (1994) *Heat & Cold: Mastering the Great Indoors*. Atlanta: American Society of Heating, Refrigerating and Air Conditioning Engineers.

Douglas, M. (1966) *Purity and Danger. An Analysis of Concepts of Pollution and Taboo*. London: Routledge and Kegan Paul.

Draper, J.W. (1868) *History of the American Civil War*. New York: Harper Brothers.

Dumas, A. (1844) *The Count of Monte Cristo*. Translated by Buss, R. (1996). London: Penguin Books.

Dunn, G., Bleil De Souza, C., Knight, I. and Marsh, A. (2006) Measured building and air conditioning energy performance: an empirical evaluation of the energy performance of air conditioned office buildings in the UK. Presented at the International Conference on Electricity Efficiency in Commercial Buildings (IEECB 2006), Frankfurt.

Durand, J.N.L. (c.1801) *Recueil et Parallèle des Édifices de Tout Genre, Anciens et Moderne: Remarquables par Leur Beauté, par Leur Grandeur ou par Leur Singularité, et Dessinés sur Une Même Échelle*. Paris: Vincent, Freal et cie.

Eastham, A. (2010) Walter Pater's Acoustic Space: 'The School of Giorgione', Dionysian 'Anders-streben', and the Politics of Soundscape. *The Yearbook of English Studies*, **40**(1/2), 196–216.

Ellingham, I. and Fawcett, W. (2006) *New Generation Whole Life Costing, Property and Construction Decision-Making Under Uncertainty*. Oxford: Taylor and Francis.

Essah, E.A., Yao, R. and Short, A. (2014) Exploring a hybrid ventilation strategy in the continental climate of Beijing using cfd simulation. Unpublished report.

Fair, A.J. (2014) A Laboratory of Heating and Ventilation: The Johns Hopkins Hospital as Experimental Architecture, 1870–90. *The Journal of Architecture*, **19**(3), 1–24. doi: 10.1080/13602365.2014.930063.

Falconer, W. (1781) *Remarks on the influence of climate: situation, nature of country, population, nature of food, and way of life, on the disposition and temper, manners and behavior, intellects, laws and customs, form of government, and religion, of mankind*. London: C. Dilly.

Fergusson, J. (2006) A brief history of air-conditioning. *Prospect Magazine*, 24 September 2006.

Fleming, J.R. (1998) *Historical Perspectives on Climate Change*. Oxford: Oxford University Press.

Foucault, M., Beguin, F., Fortier, B., Thalamsy, A. and Barret-Kriegel, B. (1979) *Les Machines a Guerir, aux Origins de l'Hopital Moderne*. Paris: Institut de l'Environnement.

Fouillet, A., Rey, G., Laurent, F., Pavillon, G., Bellec, S., Guihenneuc-Jouyaux, C., *et al.* (2006) Excess Mortality Related to the August 2003 Heat Wave in France. *Int. Arch. Occup. Environ. Health*, **80**(1), 16–24.

Frampton, K. (1985) Towards a critical regionalism: six points for an architecture of resistance. *Postmodern Culture* 16.

Francis, S., Glanville, R., Noble, A. and Scher, P. (1999) *50 Years of Ideas in Healthcare Buildings*. London: The Nuffield Trust.

Fricke, F., Nannariello, J. and Cabrera, D. (2006) A Statistical Approach to Concert Hall Acoustical Design. Proceedings of ACOUSTICS 2006, 20–22 November 2006, Christchurch, New Zealand.

Friedenthal, R. (1993) *Goethe, His Life and Times*. London: Weidenfeld. First published 1963.

Friedlander, M. (1956) *Fiedländer On Early Netherlandish Painting From Van Eyck To Bruegel*. London: Phaidon. First published 1916.

Fussel, H.-M. (2011) *Global maps of climate change impacts on the favourability for human habitation and economic activity*. Munich: Munich Personal RePEc Archive. MPRA Paper No. 29888. Available at http://mpra.ub.uni-muenchen.de/29888/ (posted 5 April 2011, accessed 12 October 2016).

Garrison, F.H. (1915) *John Shaw Billings; A Memoir*. New York and London: G.P. Putnam's Sons, The Knickerbocker Press.

Gibson, J. (1986) *The Ecological Approach to Visual Perception*. New York, Hove: Psychology Press, Taylor and Francis Group.

Gilkeson, C.A., Camargo-Valero, M.A., Pickin, L.E. and Noakes, C.J. (2013) Measurement of Ventilation and Airborne Infection Risk in Large Naturally Ventilated Hospital Wards. *Building and Environment* 65, 35–48.

Gillies, D. (1999) *Radical Diplomat. The Life of Archibald Clark Kerr Lord Inverchapel, 1882–1951*. London: I.B. Tauris.

Giridharan, R., Lomas, K.J., Short, C.A. and Fair, A.J. (2013) Performance of hospital spaces in summer: A case study of a 'Nucleus'-type hospital in the UK Midlands. *Energy and Buildings* 66, 315–328.

Grainger, H. (2011) *The Architecture of Sir Earnest George*. Reading: Spire Books Ltd.

Griffero, T. (2014) *Atmospheres: Aesthetics of Emotional Spaces*. Translated by de Sanctis, S. from Griffero, T. (2010) *Atmosferologia: Estetica degli spazi emozionali*. Farnham UK and Burlington Vermont USA: Ashgate.

Griscom, J.H. (1850) *The Uses and Abuses of Air*, 2nd Edn. New York: J.S. Redfield, Clinton Hall.

Guy, S. and Shove, E. (2000) *A Sociology of Energy, Buildings and the Environment*, London: Routledge.

Haddad, E. (2010) Christian Norberg-Schulz's Phenomenological Project In Architecture. *Architectural Theory Review*, **15**(1), 88–101.

Hales, S. (1727) *Vegetable staticks: or, an account of some statical experiments on the sap in vegetables: being an essay towards a natural history of vegetation. Also, a specimen of an attempt to analyse the air, By a great Variety of Chymio-Statical Experiments; Which were read at several Meetings before the Royal Society. By Steph. Hales, B.D.F.R.S. Rector of Farringdon, Hampshire, and Minister of Teddington, Middlesex (London)*. London: W. & J. Innys and T. Woodward.

Halliday, S. (2001) Death and miasma in Victorian London: an obstinate belief. *British Medical Journal*, **323**(7327), 1469–1471.

Hamilton, D. (2004) Cope, Sir (Vincent) Zachary (1881–1974). *Oxford Dictionary of National Biography*. Oxford: Oxford University Press.

Harbison, R. (2006) Bloomsbury's Baltic Passages. *Architecture Today* 167, April, p. 48.

Harries, K. (1983) Thoughts on a Non-Arbitrary Architecture. *Perspecta* 20, 9–20.

Hayes, E. and Nimis, S. (2013) *Hippocrates' On Airs, Waters, and Places and The Hippocratic Oath: An Intermediate Greek Reader*. Oxford, OH: Faenum Publishing Ltd.

Hellpach, W. (1938) Kultur und Klima, in Klima-Wetter-Mensch (ed) (1938) *Heinz Wolterek*, pp. 428–429.

Hellpach, W. (1977) *Geopsyche: die Menschenseele unter dem Einfluss von Wetter und Klima, Boden und Landschaft*. Stuttgart: Enke. Republication of Hellpach, W. (1911) *Die geopsychischen Erscheinungen: Wetter, Klima und Landschaft in ihrem Einfluß auf das Seelenleben*. Leipzig: Engelmann.

Henderson Floyd, M. (1974) *Architecture after Richardson: Regionalism before Modernism—Longfellow, Alden, and Harlow in Boston and Pittsburgh*. Chicago: University of Chicago Press.

Henry, T. (1776) *Opuscules Physiques et Chimiques (Essays Physical and Chemical)*. London: printed for Joseph Johnson.

Herodotus. *Herodotus: The Histories, A New Translation by Tom Holland*. (2013) London: Penguin Classics.

Hitchcock, H.-R. and Johnson, P. (1932) *The International Style*. New York: The Norton Library, W.W. Norton & Company Inc.

Hildebrand, G. (1974) *Designing for Industry: the Architecture of Albert Kahn*. Cambridge, MA: MIT Press.

Hix, J. (1996) *The Glass House*. London: Phaidon.

Holmdahl, G., Lind, S.I. and Odeen, K. (1950) *Gunnar Asplund Architect 1885–1940*, 2nd Edn. Translated by Grubbstrom (1981). Stockholm: AB Tidskriften Byggmastaren.

Holmes, M.J. and Hitchin, E.R. (1978) An example year for the calculation of energy demands in buildings. *Building Services Engineering* 45, 186–190.

Honiball, C.R. (1907) The Mechanical Ventilation and Warming of St George's Hall, Liverpool. *The Heating and Ventilating Magazine*, Volume 4, October 1907.

Howard, L. (1833) *The Climate of London Deduced from Meteorological Observations made at Different Places in the Neighbourhood of the Metropolis in Two Volumes*. London: W.Phillips.

Hoy, S. (1995) *Chasing Dirt, The American Pursuit of Cleanliness*. New York: Oxford University Press.

Hulme, M. (2011) Reducing the Future to Climate: A Story of Climate Determinism and Reductionism. *Osiris*, **26**(1), 245–266.

Hume, D. (1994) 'Political Essays' in K. Haakonsen (1994) *Of National Characters*, Cambridge Essay 12.

Humphreys, W.J. (1920) *Physics of the Air*. Philadelphia: The Franklin Institute of the State of Pennsylvania.

Huntington, E. (1919) *World-Power and Evolution*. New Haven, CT: Yale University Press.

Huxley, A. (2007) *Brave New World*. London: Vintage Classics. First published 1932.

Huygens, C. (1932) *Hollandsche Maatschappij Der Wetenschappen. Oeuvres Completes De Christiaan Huygens*. Tome Dix-Septiemme La. Haye: M. Nijhoff.

IPCC, Pachauri, R.K. and Meyer, L.A. (eds). (2014) *Climate Change 2014: Synthesis Report. Contribution of Working Groups I, II and III to the Fifth Assessment Report of the Intergovernmental Panel on Climate Change*. Geneva: IPCC.

Ingels, M. (1952) *Willis Haviland Carrier: Father of Air Conditioning*. Garden City, NY: Doubleday and Co. Inc.

Izenour, G.C. (1992) *Roofed Theaters of Classical Antiquity*. New Haven: Yale University Press.

Jarzombek, M. (2000) *Psychologising Modernity*. Cambridge: Cambridge University Press.

Jeffreys, J. (1858) *British Army in India: its preservation by an appropriate clothing, housing, locating, recreative employment, and hopeful encouragement of the troops; with an appendix on India*. London: Longman, Brown, Green, Longmans, & Roberts.

Jencks, C. (1995) *The Architecture of the Jumping Universe, A Polemic: How Complexity Science is Changing Architecture and Culture*. Chichester, West Sussex: Academy Editions.

Jencks, C. (2013) Architecture Becomes Music. *The Architectural Review*, 6 May 2013.

Jensen, T.B. (2009) *P.V. Jensen-Klimt, The Headstrong Masterbuilder*. Denmark: The Royal Danish Academy of Fine Arts School of Architecture Publishers.

Ji, Y. and Lomas, K. (1999) Current and likely future performance of advanced natural ventilation. *Building Simulation 2009*. Glasgow, Scotland: Eleventh International IBPSA Conference July 27–30, 2009.

Johnson, H., Kovats, S., McGregor, G., Stedman, J., Gibbs, M. and Walton, H. (2005) The impact of the 2003 heat wave on daily mortality in England and Wales and the use of rapid weekly mortality estimates. *Euro Surveill*. 2005 **10**(7), 168–171.

Horace Leonard Jones, London, 1949 book 4, 5-4

Jones, W.H.S. (1923) *Hippocrates with an English translation by W.H.S. Jones*. London: William Heinemann and New York: G. Putnam's Sons.

Khalil, M.A.M. (2009) The Italian Architecture in Alexandria, Egypt (the conservation of the Italian residential buildings). Thesis submitted for second level master's degree in architecture restoration (2008–09) University Kore of Enna, Sicily, Italy. Supervised by Professor Teotista Panzeca, Professor Manuela Garofalo and Professor Daniela Villari.

King, F. (1875) 'Letter addressed to the Authors of the Essays', reproduced in *Hospital Plans. Five Essays Relating to the Construction, Organization and Management of Hospitals*, contributed by their authors for the use of the Johns Hopkins Hospital of Baltimore, New York: William Wood and Co., 27 Great Jones Street.

King, A.J. (1966), Hospital Planning: Revised Thoughts on the Origin of the Pavilion Principle in England. *Medical History* 10, 360–373.

Kirwan, R. (1787) *An Estimate of the Temperature of Different Latitudes*. London: J. Davis.

Klautke, E. (2012) History of European Ideas Defining the Volk: Willy Hellpach's Völkerpsychologie between National Socialism and Liberal Democracy, 1934–1954, *History of European Ideas*, doi: 10.1080/03044181.2012.735086.

Köppen, W. (1884) Die Wärmezonen der Erde, nach der Dauer der heissen, gemässigten und kalten Zeit und nach der Wirkung der Wärme auf die organische Welt betrachtet (The thermal zones of the earth according to the duration of hot, moderate and cold periods and to the impact of heat on the organic world). *Meteorol. Z.* **1**, 215–226. Translated and edited by Volken, E. and Bronnimann, S. (2011) *Meteorol. Z.* **20**, 351–360.

Krausse, B., Cook, M. and Lomas, K. (2006) Environmental performance of a naturally ventilated city centre library. Proceedings of the International Conference 'Comfort and Energy Use in Buildings – Getting them Right', 27–30 April, 2006, Windsor, UK.

Krausse, B., Cook, M. and Lomas, K. (2007) Environmental Performance of a Naturally Ventilated City Centre Library. *Energy and Buildings* 39, 792–801. doi: 10.1016/j. enbuild.2007.02.010.

Lamb, H.H. (1959) Our changing climate, past and present. *WEATHER* 14, 299-318, reproduced in (1966) *The Changing Climate: Selected Papers*. London: Methuen.

Latour, B. (1993) *We Have Never Been Modern*. Translated by Porter, C., Cambridge, MA: Harvard University Press.

Le Corbusier. (1927) *Towards A New Architecture,* 13th edition (1972). Translated from *Vers une Architecture* (1923). The Architectural Press, London.

Le Corbusier. (1930) *Precisions, on the State of Architecture and City Planning*. Translated by Edith Schreiber Aujame. Paris: Cres et Ciep.

Lefevre, P. (1995) Ecology and Architecture in Europe. *ARCHICREE,* November.

Leoni, G. (1726) *The Architecture of Leon Battista Alberti, in Ten Books: of painting, in three books: and of statuary, in one book / translated into Italian by Cosimo Bartoli; and now first into English, and divided into three volumes by James Leoni ... to which are added several designs of his own*. London: T. Edlin.

Livingstone, D.N. (2002) Race, Space and Moral Climatology: Notes toward a Genealogy. *J. Hist. Geogr.* 28, 159–180.

Lomas, K.J. (2007) Architectural design of an advanced naturally ventilated building form. *Energy and Buildings* 39, 166–181.

Lomas, K.J. and Cook, M.J. (2005) Sustainable buildings for a warmer world. Proceedings of the World Renewable Energy Conference, Aberdeen, UK, May 2005.

Lomas, K.J., Cook, M.J. and Fiala, D. (2006) Low energy architecture for a severe climate: design and evaluation of a hybrid ventilation strategy. *Energy and Buildings*, **39**(1), 32–44.

Lomas, K.J., Cook, M.J. and Short, C.A. (2009) Commissioning Hybrid Advanced Naturally Ventilated Buildings: A US Case Study. *Building Research & Information*, 37(4), 397–412.

Lomas, K.J., and Ji, Y. (2009) Resilience of Naturally Ventilated Buildings to Climate Change: Advanced Natural Ventilation and Hospital Wards. *Energy and Buildings*, doi: 10.1016/j. enbuild.2009.01.001.

Lomas, K.J., Giridharan, R., Short, C.A. and Fair, A.J. (2012) Resilience of 'Nightingale' hospital wards in a changing climate. *Building Serv. Eng. Res. Technol.* **33**(1), 81–103.

Lovins, A., von Weizsacher, E., Hunter Lovins, L. (1997) *Factor Four, Doubling Wealth, Halving Resource Use: The New Report To The Club Of Rome*. London: Earthscan Publications Ltd.

MacKay, D.J.C. (2008) *Sustainable Energy – without the Hot Air*, Cambridge, UK: UIT.

Manley, G. (1944) Some Recent Contributions to the Study of Climatic Change. *Quarterly Journal of the Royal Meteorological Society* 70, 197–220.

Manley, G. (1958) The Revival of Climatic Determinism. *Geogr. Rev.* 48, 98–105.

Mann, T. (1901) *Buddenbrooks*. Translated by Lowe-Porter, H.T. London: Vintage, Random House (1999).

Mardaljevic, J. and Lomas, K. (2006) *Precision Irradiation Modelling of Self-Shading Facades*. 37th HVAC Congress, Beograd.

Marion, W. and Urban, K. (1995) *User's Manual for TMY2s, Typical Meteorological Years*. Golden CO: National Renewable Energy Laboratory. Available at http://rredc.nrel.gov/ solar/pubs/tmy2/overview.html (accessed 4 November 2016).

Marshall, J. (1878) *On a Circular System of Hospital Wards Including Remarks and Illustrations by P. G. Smith*. London: Smith Elder.

McCoskey, D. (2012) *Race, Antiquity and its Legacy*. Oxford: I.B. Tauris.

McLaren, B.L. (2006) *Architecture and Tourism in Colonial Libya: An Ambivalent Modernism*. Seattle: University of Washington Press.

Meinig, D.W. (ed) (1971) *On Geography: Selected Writings of Preston E. James*. Syracuse, NY: Syracuse University Press.

Menagh, M. (1995) Better Building for the Planet. *Newsweek*, 6 November 1995.

Meurer, P. (2004) *The Strabo Illustratus Atlas*. Haas, P., Le Bail, D. and Weissert, F. (eds). Bedburg-Hau: Antiquariat Gebr. Haas.

Meyer, W.B. (2000) *Americans and Their Weather*. Oxford: Oxford University Press.

Mills, C.A. (1944) *Climate Makes the Man*, London: Victor Gollancz Ltd.

Mills, G. (2008) Luke Howard and the Climate of London. *Weather* 63, 153–157.

Miner, H. (1956) Body ritual among the Nacirema, *American Anthropologist*, 58, 503–507.

Mingotti, N. (2011) On the energetic performance of single and double-glazed facades exposed to solar radiation. Unpublished doctoral research at University of Cambridge, supervised by Short and Woods.

Monk, T. (2001) Cunning plan. *Brick Bulletin*. Windsor: EMAP Ltd for The Brick Development Association.

Monk, T. (2004) *Hospital Builders*. Chichester: John Wiley & Sons Ltd.

Morabia, A. (2007) Epidemiologic interactions, complexity, and the lonesome death of Max von Pettenkofer. *Oxford Journal, American Journal of Epidemiology*, **166**(11), 1233–1238.

Mordaunt Crook, J. (2003) *The Architect's Secret: Victorian Architects and the Image of Gravity*. London: John Murray Ltd.

Mordaunt Crook, J. (2013) *William Burges and the High Victorian Dream*. London: Frances Lincoln Ltd. First published 1981 by John Murray (Publishers) Ltd.

Morin, A. (1863) *Etudes sur la Ventilation*. Paris: L. Hachette et Cie.

Mumford, L. (1938) *The Culture of Cities*. London: Secker & Warburg. (1953 printing).

Mumford, E. (2001) The emergence of mat or field buildings, in Hashim Sarkis; Pablo Allard eds., *Case: Le Corbusier's Venice Hospital and the Mat Building Revival*, Munich; New York: Prestel, pp. 48–65.

Nash, G.D. (1999) *The Federal Landscape: An Economic History of the Twentieth Century West*. Tucson: The University of Arizona Press.

Neumann, D. (1992) Three Early Designs by Mies van der Rohe. *Perspecta* 27, 76–97.

Nicholson-Lord, D. (1995) Medieval building wins green award. *The Independent on Sunday,* 28 May 1995.

Nickson, A. (2011) Cities and Climate Change: Adaptation in London, UK. Case Study prepared for *Cities and Climate Change: Global Report on Human Settlements 2011*. Available from www.unhabitat.org/grhs/2011 (accessed 4 November 2016).

Nielsen, H.N. (ed) (2002) *Even More Studies in the Ancient Greek Polis*. Papers from the Copenhagen Polis Centre, 6. Historia Einzelschriften, 162. Stuttgart: Franz Steiner, 2002, pp. 294. ISBN 3-515-08102-X. EUR 64.00 (pb): Die Deutsche Bibliothek.

Nietzsche, F. *Thus Spake Zarathustra*. Republished by Wordsworth Classics of World Literature, translated by Common, T. (1977).

Nightingale, F. (1858) Sanitary Condition of Hospitals and Hospital Construction. Two papers read before the National Association for the Promotion of Social Science.

Noakes, C.J., Sleigh, P.A., Escombe, A.R. and Beggs, C.B. (2006) Use of CFD analysis in modifying a TB ward in Lima, Peru. *Indoor and Built Environment*, **15**(1), 41–47.

Norris, F. (1903) *The Pit*. New York: Doubleday, Page and Co.

Nuffield Provincial Hospitals Trust (1955) Studies in the Functions and Design of Hospitals, The Report of an Investigation sponsored by the Nuffield Provincial Hospitals Trust and the University of Bristol.

O'Gorman, J.F. (1973) *The Architecture of Frank Furness*. Philadelphia: University of Pennsylvania Press.

Overy, P. (2007) *Light, Air and Openness: Modern Architecture between the Wars*. London: Thames and Hudson.

Pearman, H. (1993) The Battle of the Buildings. *The Sunday Times, Culture Section*, 10 October 1993, 26–27.

Pearman, H. (1999) *The Sunday Times, Culture Section*, 18 July 1999.

Peet, R. (1985) The social origins of environmental determinism. *Annals of the Association of American Geographers*, **75**(3), 309–333.

Pehnt, W. (1973) *Expressionist Architecture*. Translated by Underwood, J.A. and Kustner, E. London: Thames and Hudson.

Pereira, M.L. and Tribess, A. (2005) A review of air distribution patterns in surgery rooms under infection control focus. *Thermal Engineering*, **4**(2), 113–121.

Perez-Gomez, A. (1983) *Architecture and the Crisis of Modern Science*. Cambridge Mass., London: MIT.

Pevsner, N. (1960) *Outline of European Architecture*. 1960 Jubilee edition. Harmondsworth: Penguin Books.

Pevsner, N. (1970) *Outline of European Architecture*. London: Pelican.

Phipson, W.W. (1859) *Remarks on Ventilation with Extracts from Official Reports on the Combination of Ventilation ad Warming System Van Hecke*. London: H.M. Pollett.

Pidwell, S. (2001) Lanchester Library by Short and Associates. *Architecture Today*, 115, February, 38–49.

Polano, S. (1988) *Hendrick Petrus Berlage, Complete Works*. London: Butterworth.

Potter, D.M. (1954) *People of Plenty. Economic abundance and the American character*. Chicago: University of Chicago Press.

Powell, K. and Strongman, C. (2007) *New London Architecture 2*. London: Merrell.

Powers, A. (1995) The New Gothic. *Perspectives,* February, 42–45.

Provincial Hospitals Trust, Nuffield (1955) *Studies in the Functions and Design of Hospitals*. Oxford: Oxford University Press.

Public Health England (2014) *Heatwave Plan for England: Supporting Vulnerable People Before and During a Heatwave – Advice for Health and Social Care Professionals*. PHE publications gateway number: 2014074 2902329, May 2014.

Rapoport, J. (1990) *The Boy who Couldn't Stop Washing: the Experience and Treatment of Obsessive-Compulsive Disorder*. London: Fontana/Collins.

Ratzel, F. (1882) *Anthropo-Geographie oder Grundzüge der Anwendung der Erdkunde auf die Geschichte*. Translated by Ellen Chruchill Semple as Ratzel, F. (2010) *Anthropo-geography: or Broad Application of Geography to History*. Whitefish, Montana: Kessinger Publishing.

Reid, D.B. (1844) *Illustration of the Theory and Practice of Ventilation with remarks on warming, exclusive lighting and the communication of sound*. London, Paternoster Row: Longman, Brown, Green, & Longmans.

Reid, D.B. (1858) *Ventilation of American Dwellings with One Hundred Diagrams, Presenting a Series of Examples in Different Classes of Habitations, to which is added An Introductory Outline of the Progress of Improvement in Ventilation by Elisha Harris MD*. New York: Wiley and Halsted.

Reiter, P. and Wellmon, C. (2015) How the Philologist became a Physician of Modernity: Nietzsches Lectures on German Education. *Representations* 131, Summer 2015.

Richardson, V. (2006) Eccentric Architecture. *Blueprint*, 242, 78–82.

Richmond, P.A. (1976) American Attitudes toward the Germ Theory of Disease (1860–1880), in Brieger (1976).

Ritchie, R. (1862) *A Treatise on Ventilation, Natural and Artificial*. London: Lockwood.

Rivoira, G.T. (1910) *Lombardic Architecture: Its Origins, Development and Derivatives Vol. II*. Translated by Rushforth, G.McN. London: Heinemann.

Ross, W.D. (1924) *Aristotle's Metaphysics*. Oxford: Clarendon Press.

Rubel, F. and Kottek, M. (2010) Observed and projected climate shifts 1901–2100 depicted by world maps of the Koppen-Geiger climate classification. *Meteorologische Zeitschrift*, **19**(2), 135–141.

Ruskin, J. (1893) (Fifth Edition) *The Stones of Venice*. Three volumes. Sunnyside, Orpington and London: George Allen. (Originally published 1851–1854).

Russel, F. (1979) *Art Nouveau Architecture*. London: Academy Editions.

Ruttan, H. (1862) *Ventilation and Warming of Buildings*. New York: G.P. Putnam.

SCSA (1990) *The Athenian Agora: A Guide to the Excavation and Museum*, 4th Edn. Athens: SCSA.

Sabine, W.C. (1922) *Collected Papers on Acoustics*. Cambridge, MA: Harvard University Press.

Sarkis, H., Allard, P. and Hyde, T. (eds) (2001) *Le Corbusier's Venice Hospital and the Mat Building Revival*. Munich: Prestel.

Saxon Snell, H. (1885) Circular Hospital Wards. *The Lancet,* **126**(3239), 590–593.

Schatzki, T.R. (2002) *The Site of the Social, a Philosophical Account of the Constitution of Social Life and Change*. Pennsylvania Park: The Pennsylvania State University Press.

Scheerbart, P. (1914) *Graues Tuch und zehn Prozent Weiss (The Gray Cloth and Ten Percent White: A Ladies Novel)*. Translated by Stuart, J.A. (2001) Cambridge, MA: The MIT Press.

Scheerbart, P. (1971) *Glasarchitektur*. Reproduced by Rogner & Bernard, Munich.

Schoenefeldt, H. (2007) The All-Glass Building – its Predisposing Causes. Unpublished MPhil dissertation, University of Cambridge, supervisor Short, C.A.

Schulze, F. (1985) *Mies van der Rohe: A Critical Biography*. Chicago: University of Chicago Press.

Scott, G. (1914) *The Architecture of Humanism*, London: Constable and Co. Ltd. (1980 edition, London: The Architectural Press).

Semple, E. (1911) *Influences of Geographic Environment on the Basis of Ratzel's System of Anthropo-Geography*. New York: H. Holt and Co., London: Constable and Company Ltd.

Sewell, W.R.D., Kates, R.W. and Phillips, L. (1968) Human response to weather and climate: geographical contributions, *The Geographical Review*, **LVIII**(2), 262–280.

Shapin, S. and Schaffer, S. (2011) *Leviathan and the Air-Pump, Hobbes, Boyle, and the Experimental Life*, Princeton and Oxford: Princeton Univ. Press (first publ. 1985).

Short, C.A. (1991) A well-tempered environment: Peake Short at Leicester. *Architecture Today* 23, 30–35.

Short, C.A. (2004) A typology of design strategies for low energy, efficient, deep plan public buildings in various climates, in Hao, L. (ed) *World Architecture*. Beijing: Tsinghua.

Short, C.A. and Al-Maiyah, S. (2009) Design Strategy for low energy ventilation and cooling of hospitals. *Building Research and Information*, **37**(3), 1–29.

Short, C.A. and Chiddick, D. (1995) Client and architect working together: the Queens Building, De Montfort University, Leicester, in Coleman, R. (ed) *Proceedings of Royal Fine Art Commission Conference, Design Quality in Higher Education Buildings*. London: Thomas Telford, 22–29.

Short, C.A. and Cook, M.J. (2005) Design guidance for naturally ventilated theatres. *Building Services Engineering Research and Technology*, **26**(3), 259–270.

Short, C.A. and Lomas, K. (2007) Exploiting a hybrid environmental design strategy in a US continental climate. *Building Research and Information*, **35**(2), 119–143.

Short, C.A., Barrett, P. and Fair, A. (2011) *Geometry and Atmosphere*. Gower: Ashgate.

Short, C.A., Barrett, P., Dye, A. and Sutrisna, M. (2007) Impacts of value engineering on five capital arts projects, *Building Research and Information*, **35**(3), 287–315.

Short, C.A., Cook, M. and Lomas, K.J. (2009) Delivery and performance of a low-energy ventilation and cooling strategy. *Building Research and Information*, **37**(1), 1–30.

Short, C.A. Cook, M., Cropper, P.C. and Al-Maiyah, S. (2010) Low energy refurbishment strategies for health buildings. *Journal of Building Performance Simulation*, **3**(3), 197–216.

Short, C.A., Goldrick, A., Sharratt, P., Jones, P., Alexander, D. and Jenkins, H. (1998) Design of naturally ventilated theatre spaces, in *Building a New Century, 5th European Conference – Solar energy in Architecture and urban planning*. Bonn: Eurosolar-Verlag.

Short, C.A., Graf, H., Xiaoyong, Yu, Guthrie, P. and Pencheon, D. (2016) NHS Estate: resilience and adaptation to climate change, *unpublished*.

Short, C.A., Lomas, K.J. and Woods, A. (2004) Design strategy for low energy ventilation and cooling within an urban heat island. *Building Research and Information*, **32**(3), 187–206.

Short, C.A., Lomas, K.J., Renganathan, G. and Fair, A. (2012) Building resilience to overheating into 1960s UK hospital buildings within the constraint of the national carbon reduction target: adaptive strategies. *Building and Environment*, pp. 1–23. doi:10.1016/j.buildenv.2012.02.031.

Short, C.A., Lomas, K.J. and Renganathan, G. (2015) A medium-rise 1970s maternity hospital in the east of England: resilience and adaptation to climate change. *Building, Services, Engineering, Research, Technology (BSERT)*, special issue 'Indoor Temperature and Air Quality'. 0(0) 1–28, SAGE, doi: 10.1177/0143624414567544. Available at http://bse.sagepub.com/cgi/reprint/0143624414567544v1.pdf?ijkey=DxAW2Gpz4xsz5XW&keytype=finite (accessed 4 November 2016).

Short, C.A., Noakes, C.J., Gilkeson, C.A. and Fair, A. (2014) Functional recovery of a resilient hospital type. *Building Research & Information*, **42**(6), 657–684.

Short, C.A., Whittle, G. and Owarish, M. (2006) Fire and smoke control in naturally ventilated buildings. *Building Research & Information*, **34**(1), 23–54.

Short, C.A., Yao, R. Runming, Luo, G. and Baizhan, L. (2013) Hybrid environmental design as an alternative to full air-conditioning in the continental climate: a case study in Beijing. *Ecocity & Green Building*.

Shove, E. (2003) *Comfort, Cleanliness and Convenience*. Oxford: Berg.

Shove, E., Chappells, H., Lutzenheiser, L. and Hackett, B. (2008) Comfort in a low carbon society. Editorial, *Building Research and Information*, **36**(4), 307–311.

Shove, E., Pantzar, M. and Watson, M. (2012) *The Dynamics of Social Practice*. London: Sage.

Sinclair, U. (1906) *The Jungle*. New York City: Doubleday, Jabber & Company.

Snow, C.P. (1998) *The Two Cultures*. Cambridge: Cambridge University Press.

Spotts, F. (1994) *Bayreuth: A History of the Wagner Festival*. New Haven: Yale University Press.

Steele, J. (2005) *Ecological Architecture: A Critical History*. London: Thames and Hudson Ltd.

Stehr, N. (1996) The ubiquity of nature: climate and culture. *J. Hist. Behav. Sci. 32*, 151–159.

Street, G.E. (1914) *Some account of Gothic Architecture in Spain*. King, G.G. (ed) Vol. 1. London: J.M. Dent and Sons Ltd.

Sturrock, N. and Lawson-Smith, P. (2006) The grandfather of air-conditioning – the work and influence of David Boswell Reid, physician, chemist, engineer (1805–1863). *Proceedings of the Second International Conference on Construction History at Cambridge 2006*.

Sturtevant, B.F. (1906) *Ventilation & Heating*. London: B F Sturtevant Co.

Sullivan, L.H. (1896) The Tall Office Building Artistically Considered. Reprinted in *Inland Architect and News Record 27* (May 1896), 32–34.

Summerson, J.N. (1945) *Georgian London*. London: Pleiades Books.

Summerson, J. (1957) The case for a theory of modern architecture. *RIBA Journal*, June 1957.

Swenarton, M. and Rickaby, P. (1993) Low energy gothic. *Architecture Today* 41.

Taut, B. (1929) *Modern Architecture*. London: The Studio Ltd, New York: Albert & Charles Boni Inc.

Taylor, J.R.B. (1988) Circular hospital wards: Professor John Marshall's concept and its exploration by the architectural profession in the 1880s. *Medical History* 32: 426–448.

Thielemann, C. (2015) *My Life with Wagner*. London: Weidenfeld and Nicholson.

Thomas, R. (ed) (1996) *Environmental Design*. London: E & FN SPON.

Thomas, R. (2000) *Herodotus in Context*. Cambridge: Cambridge University Press.

Thompson, J.D. and Goldin, G. (1975) *The Hospital: A Social and Cultural History*. New Haven: Yale.

Tozer, Rev. H.F. (1893) *Selections from Strabo, with an Introduction on Strabo's Life and Works*. Oxford: Clarendon Press.

Ulrich, R. (2006) Evidence-based health-care architecture. *The Lancet*, **368**(1), 538–539.

Ulrich, R.S., Zimring, C., Zhu, X., DuBose, J., Seo, H-B., Choi, Y.-S. and Joseph, A. (2008) A review of the research literature on evidence-based healthcare design. *Health Environments Research & Design Journal*, **1**(3).

Veblen, T. (1899) *The Theory of the Leisure Class*. New York: Macmillan.

Venturi, R. (1977) *Complexity and Contradiction in Architecture*. London: The Architectural Press Ltd. First published in 1966 by the Museum of Modern Art, New York.

Vinikas, V. (1992) *Soft Soap, Hard Sell: American Hygiene in an Age of Advertisement*. Ames: Iowa State University Press.

Walpole, H. (1811) *The Castle of Otranto*, 2nd Edn. Edinburgh: Ballantyne.

Wampler, C. (1949) *Dr. Willis H. Carrier, Father of Air Conditioning*. New York: The Newcomen Society of England, American Branch.

Ward-Larson, G. and Shakespeare, R. (1998) *Rendering with Radiance: The Art and Science of Lighting Visualisation*. San Francisco: Morgan Kaufmann.

Watkin, D. (1980) *The Rise of Architectural History*. London: The Architectural Press Ltd.

Watkins, R., Palmer, J., Kolokotroni, M. and Littlefair, P. (2002) The balance of the annual heating and cooling demand within the London urban heat island. *Building Serv. Eng. Res. Technol.* **23**(4), 207–213.

Wettinger, G. (1989) Aspects of Maltese Life, in Mangion G. (ed) *Maltese Baroque. Proceedings of a Seminar on 'The Baroque Route in Malta'* held at the Ministry of Education, Beltissebh, Malta, 3 June 1989. Malta: Ministry of Education, and Strasbourg: The Council for Cultural Cooperation of the Council of Europe.

Willcox, W.F. (1933) *Introduction to the Vital Statistics of the United States 1900–1930.* Ithaca: NY, United States Department of Commerce, United States Bureau of the Census.

Wilson, C. St.J. (1985) Waterhouse's Law Courts Project. *Architects' Journal*, **26**(26), 181.

Woods, A., Chenvidyakarn, T. and Short, A. (2003a) Reversing flow in a naturally ventilated building with multiple stacks. PLEA 2003 The 20th Conference on Passive and Low Energy Architecture, Santiago, Chile, 9–12 November 2003.

Woods, A., Short, A. and Gladstone, C. (2003b) Stack Driven Natural Ventilation with pre-cooled inflow from a central atrium. PLEA 2003 The 20th Conference on Passive and Low Energy Architecture, Santiago, Chile, 9–12 November 2003.

Worringer, W. (1908) *Abstraktion und Einfuhlung, ein Beitrag zur Stilpsychologie (Abstraction and Empathy, a Contribution to the Psychology of Style).* Translated by Bullock, M. (1953). Munich: R. Piper & Co.

Xiaoyong Yu, S. (2014) Progress report: low energy building related temperature metrics under present and future climate. For Graf, H. and Short, C.A. at Cambridge University.

Yanni, C. (2007) *The Architecture of Madness: Insane Asylums in the United States.* Minneapolis, MN: University of Minnesota Press.

Yao, R., Li, B., Steemers, K. and Short, A. (2009). Assessing the natural ventilation cooling potential of office buildings in different climate zones in China. *Renewable Energy*, 34, 2697–2705.

Young, L. (1994) Basic green. *Metropolis*, **14**(2).

Young, J. (2010) *Friedrich Nietzsche: A Philosophical Biography.* Cambridge: Cambridge University Press.

INDEX